Robert Smith
377-70-1703

[redacted]

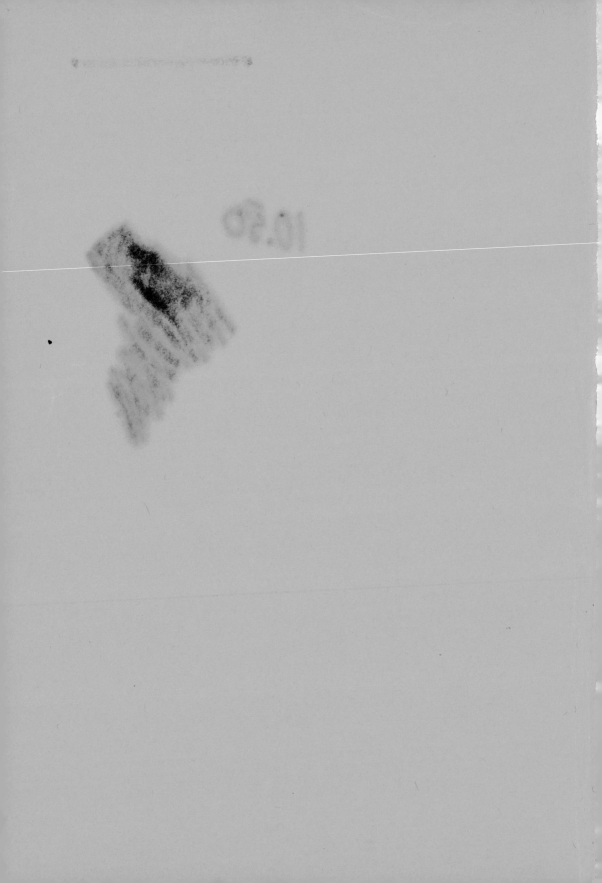

BEGINNING ALGEBRA

SECOND EDITION

BEGINNING ALGEBRA

MUSTAFA A. MUNEM

WILLIAM TSCHIRHART

MACOMB COUNTY COMMUNITY COLLEGE

WORTH PUBLISHERS, INC.

BEGINNING ALGEBRA

SECOND EDITION

PRINTED IN THE UNITED STATES OF AMERICA

SECOND PRINTING JANUARY 1978

LIBRARY OF CONGRESS CATALOG CARD NO. 76-27110

ISBN: 0-87901-063-0

DESIGN BY MALCOLM GREAR DESIGNERS

WORTH PUBLISHERS, INC.

444 PARK AVENUE SOUTH

NEW YORK, NEW YORK 10016

PREFACE TO THE SECOND EDITION

PURPOSE: This textbook was written to provide every possible assistance to students who are first being introduced to the principles of algebra and to students who need to review some of the topics covered in high school algebra. In this second edition some topics have been added, others rearranged, and others have been completely rewritten. The inspiration for these improvements came not only from the authors' continued teaching of the course since publication of the first edition, but also from the many excellent recommendations received from teachers throughout the world who have provided us with the insights gained from their classroom experience with the book.

PREREQUISITES: No previous background in algebra is required. Some familiarity with the basic concepts of arithmetic is assumed.

OBJECTIVES: The basic objectives of the first edition have been maintained. First we build a good foundation in the basic operations of arithmetic. Then we carefully begin to introduce algebraic notation and format in the course of demonstrating the properties of real numbers using illustrative examples from arithmetic. In this way, the student is prepared for the transition from arithmetic concepts to the generalized algebraic form. This sort of careful progression is used throughout the book, in examples, explanations, and problems. By making the transition between topics gradual, so that no difficult hurdles remain, we believe that student interest can be kept from flagging. All the additions, deletions, and refinements in the second edition have been made simply because they seemed to us to contribute to the realization of these basic objectives. Carried over from the first edition are certain features which have won a uniformly favorable reception:

Throughout, students are carefully led to conclusions by the use of meaningful illustrations and are not merely given a set of rules to be memorized;

The extensive review problem sets at the end of each chapter are designed to help students gain confidence in their newfound abilities, or to indicate those topics in which additional study is required;

Several worked-out examples are given to illustrate the solution to each type of problem contained in the problem sets;

Every effort has been made (1) to enable the average student to learn to solve routine problems and develop computational skills with little, if any, assistance from the instructor, (2) to provide the average student with an appreciation of the concepts and logical reasoning involved in this level of mathematics, and (3) to enhance the ability of the average student to apply these concepts and this orderly way of thinking to new situations and new problems.

NEW FEATURES: Color is used to highlight definitions, properties, important statements, and parts of graphs. There are more examples and problems in this edition. Examples have been reworked and reorganized so that every concept is immediately illustrated by many examples. Definitions and properties are carefully stated and in each case they are clarified and justified with specific examples. The problems are related more closely to the examples. The problem sets have been reviewed and revised as experience with the first edition dictated. The grading of the problems, from the very elementary to the more challenging, has been done with greater care. Problem sets are now arranged so that the odd-numbered problems strive for the level of understanding desired by most users while the even-numbered problems probe for deeper understanding of the concepts. This arrangement should simplify the task of making assignments. The answers to odd-numbered problems are given at the back of the book, and the even-numbered answers are provided for the instructor in a supplemental booklet.

CONTENTS: The presentation of topics has been improved in several respects. Much of the material has been rewritten to increase clarity. Because the operations of signed numbers are closely related to the properties of real numbers, Chapters 1 and 2 of the first edition have been combined into Chapter 1 of this edition. Chapter 1 also contains a section on order of operations. Chapter 2 has been rearranged so that it includes all operations of polynomials. Because fractions (Chapter 4) require a knowledge of factoring, Chapter 3 has been reworked to include all types of factoring and special products. Chapter 5 has been reworked and rearranged. The appendixes now contain the metric system and formulas from geometry.

PACE: The pace of a course and the selection of topics depends on the curriculum of the individual school and on the academic calendar. The following suggested pace, based on the authors' and their colleagues' classroom experience, is meant only as a general guide for a three-credit, one-semester course:

Chapter 1:	4 lectures	Chapter 6:	3 lectures
Chapter 2:	5 lectures	Chapter 7:	3 lectures
Chapter 3:	5 lectures	Chapter 8:	4 lectures
Chapter 4:	6 lectures	Chapter 9:	5 lectures
Chapter 5:	5 lectures	Chapter 10:	3 lectures

If students are well prepared, Chapters 1, 2, and 3 can be reviewed briefly in three lectures so that topics in Chapters 9 and 10 can be covered more extensively.

ADDITIONAL AIDS: A *Study Guide* is available for students who need more drill and assistance. The *Study Guide* is written in a semiprogrammed format and conforms to the arrangement of topics in the textbook. It contains fill-in statements and problems, true-or-false statements, and a test for each chapter. All answers are provided in the *Study Guide* to encourage self-testing.

ACKNOWLEDGMENTS: By the time a book reaches its second edition, a great many people have contributed to its development. Of course there would be no second edition if the book's first edition had not been so widely adopted. For this reason, we have asked our publisher to reprint below our acknowledgments to the first edition. In addition, we wish to thank those who were particularly helpful in the writing of this edition: Professors Dolores L. Matthews of Robert Morris College; William G. Cunningham, Walter R. Rogers, and James H. Soles of DeKalb Community College; and Joseph Buggan of the Community College of Allegheny County (Allegheny Campus).

We again would like to express our thanks to our colleagues at Macomb College and to the staff of Worth Publishers for their many contributions.

ACKNOWLEDGMENTS TO THE FIRST EDITION: Discussions with colleagues at many two-year colleges about the problems they face in teaching similar courses have greatly influenced the content and the pace of this text. The guidance of the Committee of Undergraduate Programs in Mathematics (CUPM) was also very helpful.

We wish to express our gratitude to Professors Jerald T. Ball of Chabot College, Curtis S. Jefferson of Cuyahoga Community College, George W. Schultz of St. Petersburg Junior College, James W. Snow of Lane Community College, and Jay Welch of San Jacinto College for their suggestions and criticisms. We are indebted to our colleagues at Macomb County Community College, especially Professor John von Zellen, for his help in reading the manuscript and for his valuable suggestions and criticisms. Finally, we wish to acknowledge Robert C. Andrews of Worth Publishers, Inc., for his excellent cooperation.

Warren, Michigan — Mustafa A. Munem
January 1977 — William Tschirhart

CONTENTS

CHAPTER 1

Fundamental Operations of Real Numbers

1 FUNDAMENTAL OPERATIONS OF REAL NUMBERS

Introduction

The study of arithmetic requires a high degree of mastery of the fundamental operations of numbers—adding, subtracting, multiplying, and dividing. Algebra is simply a generalized form of arithmetic that uses variables, usually letters of the alphabet, to represent numbers. Just as arithmetic has to do with the operations of numbers, in a like manner the study of algebra concerns itself with the operations of algebraic expressions. It is worthwhile, then, to become thoroughly familiar with the operations of real numbers before we begin the study of algebra. The objective of this chapter is to study the operations of real numbers and to introduce the basic language and symbols of sets.

1.1 Symbols and Variables

An understanding of *symbols* is essential in the study of algebra. Symbols are the language of mathematics and are used to express mathematical statements in simple and short form. For example, using symbols the statement "six plus four" is written in the form $6 + 4$. A *numeral* is a symbol with a name used to represent the idea of a number. For example, the number "ten" is denoted by the numeral 10.

Four symbols are used for the basic operations in arithmetic:

1 $+$ is used for addition.
2 $-$ is used for subtraction.
3 $\times$ is used for multiplication.
4 $\div$ or —— is used for division.

Additional symbols are used to convey different ideas. For instance, the symbol $=$ is used to represent the idea of *equality*, whereas the symbol $\neq$ is used for not equal.

EXAMPLES

Write the following statements in mathematical symbols.

1 Seven equals four plus three.

SOLUTION. The statement is written in symbols as

$$7 = 4 + 3$$

2 Twenty-one equals seven times three.

SOLUTION. The statement is written in symbols as

$$21 = 7 \times 3$$

3 Nine is not equal to five.

SOLUTION. The statement is written in symbols as

$$9 \neq 5$$

In Example 2, in the statement $21 = 7 \times 3$, the number 21 is called the *product* of 3 and 7. This product can also be written using parentheses:

$$21 = 3(7) \quad \text{or} \quad 21 = (3)7 \quad \text{or} \quad 21 = (3)(7)$$

Order of Operations

Some numerals include a combination of operational symbols. For example, to find the value of $5 \times 2 + 3$, we have the option of first multiplying 5 and 2, then adding 3; that is,

$$5 \times 2 + 3 = 10 + 3 = 13 \qquad \text{which is correct}$$

or, first adding 2 and 3, then multiplying by 5:

$$5 \times 2 + 3 = 5 \times 5 = 25 \qquad \text{which is not correct}$$

To avoid such ambiguous situations and to assure that everyone who works such a problem always gets the same result, we shall adopt the following *order of operations*:

1 Perform all operations inside parentheses.
2 Perform multiplications or divisions from left to right, and then
3 Perform all additions or subtractions from left to right.
4 In the case of problems involving fractions, perform the operations above and below the fraction bar separately; then simplify, if possible, to obtain the final result.

The following examples will illustrate these operations.

EXAMPLES

Find the value of each expression. Write the answer in the simplest possible form.

1 $5 + 7 \times 4$

SOLUTION. First, we perform multiplication, working from left to right, and then add:

$$\begin{aligned} 5 + 7 \times 4 &= 5 + 28 && \text{Multiplication} \\ &= 33 && \text{Addition} \end{aligned}$$

2 $5(4 + 7) - 8 + 2$

SOLUTION. First, we perform the operation inside the parentheses:

$$\begin{aligned} 5(4 + 7) - 8 + 2 &= 5(11) - 8 + 2 && \text{Working inside parentheses} \\ &= 55 - 8 + 2 && \text{Multiplication} \\ &= 49 && \text{Addition or subtraction} \end{aligned}$$

3 $\dfrac{5(7 - 3) + 4}{3 \times 16 \div 8}$

SOLUTION. In this case, we simplify the top and bottom of the fraction separately. To simplify the top, we have

$$\begin{aligned} 5(7-3)+4 &= 5(4)+4 && \text{(Why?)} \\ &= 20+4 && \text{Multiplication} \\ &= 24 && \text{(Why?)} \end{aligned}$$

The bottom will be simplified as follows:

$$\begin{aligned} 3 \times 16 \div 8 &= 48 \div 8 && \text{(Why?)} \\ &= 6 && \text{Division} \end{aligned}$$

Therefore,

$$\frac{5(7-3)+4}{3 \times 16 \div 8} = \frac{24}{6} = 4$$

Concept of a Variable

Up to this point, we have encountered no difficulty in evaluating and simplifying numerical expressions. However, in dealing with expressions such as $5x + 3$, it is not clear which number the letter x represents or, for that matter, which number the expression $5x + 3$ represents. Because these situations occur frequently in algebra, we shall establish a vocabulary to cover such occurrences. A *variable* is a letter, such as x, y, or z, that represents an unknown number. For example, to translate statements such as "the sum of an unknown number and seven" we could write "the sum of x and 7" or, using symbols, express the statement as "$x + 7$." In this case, we definitely do not know the unknown number, so we use the letter x to stand for it until we can find out the number.

In algebra, we indicate the multiplication of x and y by a raised dot or by using no sign of operation, simply writing x and y next to each other. Thus, the multiplication of x and y is written

$$x \cdot y \quad \text{or} \quad xy$$

Also, we indicate the multiplication of 5 and x by

$$5 \cdot x \quad \text{or} \quad 5x$$

In this book, we shall frequently use the notation illustrated by xy and $5x$ to indicate the preceding products.

A combination of variables, numbers, and symbols for operations is called an *algebraic expression.* Thus, $5 + x$, $x - 2$, $12x + 13$, and $8x + 7x$ are algebraic expressions. Each of the expressions above has different numerical values for different values of x. For example, if x represents the number 5, the numerical value of each expression can be found as follows:

$$\begin{aligned} 5 + x &\quad \text{becomes} \quad 5 + 5 = 10 \\ x - 2 &\quad \text{becomes} \quad 5 - 2 = 3 \\ 12x + 13 &\quad \text{becomes} \quad 12(5) + 13 = 60 + 13 = 73 \\ 8x + 7x &\quad \text{becomes} \quad 8(5) + 7(5) = 40 + 35 = 75 \end{aligned}$$

However, if we replace x by 3, then

$$\begin{aligned} 5 + x &\quad \text{becomes} \quad 5 + 3 = 8 \\ x - 2 &\quad \text{becomes} \quad 3 - 2 = 1 \\ 12x + 13 &\quad \text{becomes} \quad 12(3) + 13 = 36 + 13 = 49 \\ 8x + 7x &\quad \text{becomes} \quad 8(3) + 7(3) = 24 + 21 = 45 \end{aligned}$$

It should be noted that the number represented by an algebraic expression depends on the replacement chosen for the letter that appears in it. Both letters and numerals are symbols that take the place of numbers. When a letter or numeral can represent only a single number, we shall call it a *constant.* For example, 5, 7, and $\frac{2}{3}$ are constants; x is also a constant if only one number is allowed as a replacement for it.

EXAMPLES

Find the numerical value of each of the following expressions for the indicated value of each variable.

1 $5x + 2$, when $x = 3$.

SOLUTION

$$5x + 2 = 5(3) + 2 = 15 + 2 = 17$$

2 $3x + 7y - 3$, when $x = 5$ and $y = 4$.

SOLUTION

$$\begin{aligned} 3x + 7y - 3 &= 3(5) + 7(4) - 3 \\ &= 15 + 28 - 3 \\ &= 40 \end{aligned}$$

3 $\dfrac{3x - 5y}{x + y}$, when $x = 7$ and $y = 1$.

SOLUTION

$$\frac{3x - 5y}{x + y} = \frac{3(7) - 5(1)}{7 + 1}$$

$$= \frac{21 - 5}{7 + 1} = \frac{16}{8}$$

$$= 2$$

PROBLEM SET 1.1

In problems 1–4, write each statement in symbols.

1 Ten equals three plus seven. 10 = 3+7

2 Twelve is not equal to sixteen. 12 ≠ 16

3 Thirty equals six times five. 30 = 6×5

4 Fifteen equals ninety divided by six. 15 = 90 ÷ 6

In problems 5–30, use the order of operations to find the value of each expression.

5 $4 \times 7 + 8$

6 $5 \times 8 - 4$

7 $2 + 7 \times 5 - 3$

8 $17 - 3 \times 5 + 7$

9 $7 \times 2 + 5 \times 4$

10 $9 \times 6 + 11 - 2$

11 $5(8 + 5) - 2$

12 $8(9 - 2) + 6$

13 $4 + 4 \times 4 - 4 \div 4$

14 $5 + 5 \times 5 - 5 \div 5 + 5 \times 5 \div 5$

15 $4 \times 8 \times 3 \div 12 + 10 \times 2 - 8 \times 3$

16 $17 \times 3 - 3 \times 2 \times 5 \div 15 + 7 \times 2$

17 $8 + (16 + 4) \div 5$

18 $24 \div 2 \times 3 - 2$

19 $\dfrac{8 \times 4 - 10}{2} - \dfrac{2 + 2 \times 6}{7}$

20 $\dfrac{18 - 12 \div 2 + 7}{2 \times 4}$

21 $\dfrac{5(6 - 3)}{10} + \dfrac{2(8 - 2)}{8}$

22 $\dfrac{8 + 12 \div 3 - 2}{10 \div 2}$

23 $\dfrac{24 + 6 \div 3}{7 + 6 \times 3}$

24 $\dfrac{5(5 - 3)}{2} + \dfrac{18}{3 \times 2} + 5$

25 $\dfrac{3(3 + 2) - 3 \times 3 + 2}{5 \times 2 - 2(2 - 1)}$

26 $\dfrac{6 + 2 \times 3}{2 \times 4 - 2 \times 3} + \dfrac{2(12 - 8)}{4 - 2}$

27 $\dfrac{(6 + 24 \div 4) \div (8 \times 4 - 20)}{7 - 2 \times 2}$

28 $\dfrac{16 + 10 - 4 \times 3 \times 2}{1 \times 2} + \dfrac{9 + 16 \div 8}{3 \times 4 - 10}$

29 $8 \times 10 \div 2 \times 3 + \dfrac{6 \div 2 \times 100}{5}$

30 $\dfrac{25 + 18 \div 2 \times 3 + 8 \times 5}{2 \times 2}$

In problems 31–36, write an expression in the variable x as described.

31 Add 5 to x and multiply the result by 6.

32 Subtract four times x from 8.

33 Multiply 7 by 5 more than x.

34 Multiply x by the sum of 3 and twice x.

35 Subtract 5 times x from 10 times x plus 3.

36 Add 7 to the product of x and three more than x.

In problems 37–46, find the numerical value of the given expression when $x = 2$.

37 $x + 5$

38 $x + 13$

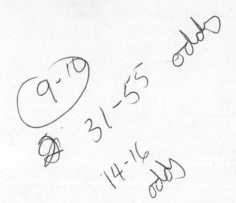

39 $4x - 3$

40 $3x - 1$

41 $3x$

42 $5x$

43 $3x + 2$

44 $7x + 1$

45 $7x - 3$

46 $9x - 4$

In problems 47–56, find the numerical value of the given expression for the indicated value of each variable.

47 $24 - 3x + 7; x = 4$

48 $5x - 3(x + 1) + 8; x = 2$

49 $x + 3y; x = 1, y = 4$

50 $5x - 3y; x = 7, y = 2$

51 $4x + 2y + 3; x = 3, y = 2$

52 $\dfrac{5x + 4y - 3}{2x}; x = 5, y = 7$

53 $\dfrac{2x + 7y}{x + y + 5}; x = 4, y = 2$

54 $\dfrac{6x - 3y + 9}{2x + y}; x = 7, y = 5$

55 $7(3x - 2y); x = 11, y = 8$

56 $11(5x + 6y - 1); x = 2, y = 9$

1.2 Sets

A *set* is a collection of objects. Other words, such as *collection, class, group,* or *aggregate,* can be used in nonmathematical situations, but the word set is usually used in mathematics. In order to better understand the use of sets, let us recall some nonmathematical uses; for instance, we often speak of a set of golf clubs, a chess set, the set of books in a library, and so forth. The reader will be able to think of other examples of sets that occur in everyday life.

The objects of sets are called *elements* or *members* of the set. We use lowercase letters, a, b, c, d, etc., to represent the elements of a set and capital letters, A, B, C, D, etc., to represent the sets. For example, a set A which has as its elements b, c, d, and e and no other elements can be written $A = \{b,c,d,e\}$. The elements are enclosed within braces to indicate that they belong to the set. The symbol used to show that an element belongs to a set is $\in$. In our example, $b \in A$, $c \in A$, $d \in A$, and $e \in A$ read "b belongs to set A," "c belongs to set A," "d belongs to set A," and "e belongs to set A." To indicate that a particular element does not belong to a set, we use the notation $\notin$: for example, $f \notin A$ and $g \notin A$.

We call this method of representing set A *enumeration*. For example, we can use enumeration to write the set B of all integers from 1 to 6 as

$$B = \{1,2,3,4,5,6\}$$

EXAMPLE

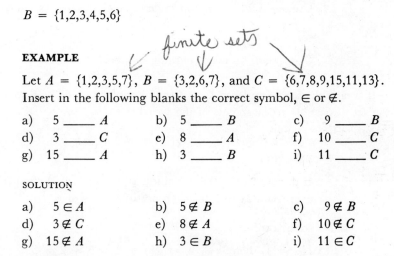

Let $A = \{1,2,3,5,7\}$, $B = \{3,2,6,7\}$, and $C = \{6,7,8,9,15,11,13\}$. Insert in the following blanks the correct symbol, $\in$ or $\notin$.

a) 5 ____ A	b) 5 ____ B	c) 9 ____ B
d) 3 ____ C	e) 8 ____ A	f) 10 ____ C
g) 15 ____ A	h) 3 ____ B	i) 11 ____ C

SOLUTION

a) $5 \in A$	b) $5 \notin B$	c) $9 \notin B$
d) $3 \notin C$	e) $8 \notin A$	f) $10 \notin C$
g) $15 \notin A$	h) $3 \in B$	i) $11 \in C$

If a set is so defined that it does not have any elements, we call it a *null set* or *empty set*. The symbol for the null set is $\emptyset$ or $\{\ \}$. For example, the set P of all presidents of the United States who died before their thirty-fifth birthday is an empty set, that is, $P = \emptyset$, since the requirements for being president of the United States include being at least 35 years old. It is important to note that 0 and $\emptyset$ do not mean the same thing. That is, $\emptyset$ is not equal to 0 or $\{0\}$, since $\{0\}$ is a set with the one element, 0, whereas $\emptyset$ (or $\{\ \}$) is the set with no elements; 0 represents the number of elements in $\emptyset$, although 0 is not an element of $\emptyset$.

A set is characterized by its elements. The order in which the elements appear is of no significance. For example, the set $B = \{1,2,3\}$ could be enumerated as follows:

$$B = \{1,3,2\} = \{2,1,3\} = \{2,3,1\} = \{3,1,2\} = \{3,2,1\}$$

There are two common methods used to describe sets. One is the enumeration method. The other is a set-description method that employs the identifying properties of the elements of the set. For example, the set B of all counting numbers between 1 and 100 could be described as follows. Let some variable, say x, represent an element of the set; then describe the set B as "the set of all x such that x is a counting number between 1 and 100." That is,

$$B = \{x | x \text{ is a counting number between 1 and 100}\}$$

This set description, called *set-builder notation*, takes the form $A = \{x | x \text{ has property } p\}$, which is read "$A$ is the set of all elements x such that x has the property p," where the vertical bar denotes "such that."

EXAMPLE

Use set-builder notation to describe C, the set of even counting numbers.

SOLUTION

$C = \{x | x \text{ is an even counting number}\}$

If a variable can be replaced by any member of a specified set, we call the set of possible replacements of the variable the *replacement set* or the *domain* of the variable.

EXAMPLE

If the domain of the variable x is the set $\{1,2,3\}$, find the value of the expression $3x + 2$ for each possible value of x.

SOLUTION

If $x = 1$, then $3x + 2 = 3(1) + 2 = 3 + 2 = 5$.
If $x = 2$, then $3x + 2 = 3(2) + 2 = 6 + 2 = 8$.
If $x = 3$, then $3x + 2 = 3(3) + 2 = 9 + 2 = 11$.

Finite and Infinite Sets

Consider the following three sets: A is the set of boys in a mathematics class, B is the set of positive counting numbers from 1 to 100, and C is the set of months in a year. Sets A, B, and C can be enumerated, because every element of each set can be named explicitly. Such sets are examples of finite sets. In general, a set M is said to be a *finite set* if the elements in M can be counted, in the usual way, using counting numbers 1, 2, 3, 4, . . . , and if this counting process eventually terminates, resulting in a specific number n (which is then said to be the number of elements in the set M). A set that is neither finite nor empty is called an *infinite set*. For example, the set N of all positive counting numbers is an infinite set, because enumeration is impossible in this case. It should be noted, however, that infinite sets whose elements form a general pattern can be described by enumeration. For example, the set N of all positive counting numbers can be described by enumeration as follows:

$$N = \{1,2,3,4, \ldots\}$$

where the three dots mean the same as "etc."

By contrast, the set B of all counting numbers from 1 through 100 can be described as

$$B = \{1,2,3, \ldots, 100\}$$

where the three dots represent the elements of B between 3 and 100.

EXAMPLES

Describe each of the following sets. Indicate if the set is finite or infinite.

1 A, the set of all positive odd counting numbers.

SOLUTION. The known pattern of the elements of A suggests that A can be written as $A = \{1,3,5, \ldots\}$. A is an infinite set.

2 E, the set of all even counting numbers from 4 through 300 inclusive.

SOLUTION. The set E can be enumerated as follows: $\{4,6,8, \ldots, 300\}$; therefore, E is a finite set.

Set Relations

Suppose that S is the set of students in a class and G is the set of girls in the same class. Clearly, all members of G are also members of S. We describe the relation between G and S by saying that G is a subset of S. Symbolically, we write $G \subseteq S$. This can be read as "G is a subset of S," or "G is contained in S." Specifically, a set A is a *subset* of a set B, written $A \subseteq B$, if every element of A is also an element of B. That is, if $A \subseteq B$ and $x \in A$, then $x \in B$. For example, the set $A = \{1,2,3\}$ is a subset of a set $B = \{1,2,3,5\}$, since every element of the set A is also an element of the set B.

We shall also agree that the empty set, $\emptyset$, is a subset of every set, including itself. This agreement implies that every element of $\emptyset$ must also be an element of any other set. Since $\emptyset$ contains no elements, this agreement does not contradict our definition of subsets.

EXAMPLE

List all possible subsets of $A = \{2,3\}$.

SOLUTION. $\emptyset$, $\{2\}$, $\{3\}$, and $\{2,3\}$ are the subsets of A.

Notice that the set of girls in a class and the set of boys in the same class have no members in common. Also, the set $A = \{9,3,8\}$ has no member in common with the set $B = \{4,6,5\}$. Such sets are called *disjoint sets.* In general, if two sets have no elements in common, they are disjoint sets.

Set Operations

Let $A = \{1,2,3,4,5\}$ and $B = \{3,4,5,6,7\}$. We can form a new set $\{1,2,3,4,5,6,7\}$ which contains each member of A as well as each member of B. The operation suggested by this example is called *set union.* More formally, the union of two sets, written as $A \cup B$, is the set of all elements in set A or in set B, or in both A and B. Symbolically,

$$A \cup B = \{x | x \in A \quad \text{or} \quad x \in B\}$$

EXAMPLE

If $A = \{3,5,1,7\}$ and $B = \{2,4,6,8\}$, find $A \cup B$.

SOLUTION. $A \cup B$ is the set that contains all the elements in both A and B. That is, $A \cup B = \{1,2,3,4,5,6,7,8\}$.

The set formed by including all the elements common to both A and B, where $A = \{1,2,3,4,5\}$ and $B = \{3,4,5,6,7\}$, is $\{3,4,5\}$. This set is called the *intersection* of A and B. More formally, the intersection of two sets, written as $A \cap B$, is the set of all elements common to both A and B. Symbolically,

$$A \cap B = \{x | x \in A \quad \text{and} \quad x \in B\}$$

EXAMPLE

If $A = \{2,3,5,6,12\}$ and $B = \{7,8,6,1,12,10\}$, find $A \cap B$.

SOLUTION. $A \cap B$ is the set whose elements are contained in both A and B. That is, $A \cap B = \{6,12\}$.

PROBLEM SET 1.2

1 Let $A = \{1,2,3,4,5,6,7\}$, $B = \{2,4,5,6,8\}$, and $C = \{6,8,9,10,12\}$. Insert in the following blanks the correct symbol, $\in$ or $\notin$.

4 ____ A	6 ____ B	9 ____ C	1 ____ C
10 ____ A	7 ____ B	8 ____ C	4 ____ C
6 ____ A	5 ____ B	10 ____ C	3 ____ C

2 Use enumeration to describe the set M of the letters in the word "Mississippi."

In problems 3–6, use set notation to describe each set.

3 The set of all days of the week whose names begin with the letter T.

4 The set of letters in the word "maximum."

5 The set of counting numbers greater than 1 and less than 10.

6 The set of even counting numbers less than 9.

In problems 7–10, use set-builder notation $\{x|x \text{ has the property } p\}$ to describe each set.

7 The set of positive counting numbers that are multiples of 5.

8 The set of odd positive numbers.

9 The set of counting numbers greater than 5 and less than 18.

10 The set of the first five letters of the English alphabet.

In problems 11–16, indicate whether the set is finite or infinite.

11 The set of even counting numbers between 1 and 11.

12 The set of people who pay income tax.

13 The set of odd counting numbers.

14 The set of the letters of the English alphabet.

15 The set $A = \{1,5,9,13, \ldots\}$.

16 The set of cars in the city of Boston.

In problems 17–20, indicate whether the statement is true or false.

17 $\{2\} \subseteq \{1,2\}$

18 $\{2,3\} \in \{2,3,4\}$

19 $\emptyset \subseteq \{1,2\}$

20 $\{1,2,3\} \in \{3,2,1\}$

21 If the domain of the variable is $\{0,1,2,3,4,5\}$, find the value of the expression $2x - 3$ for each possible value of the variable x.

22 List the first five elements of the set $\{x|x = 3n - 1, n \text{ is a counting number}\}$.

23 If the domain of the variable y is $\{0,1,2,3,4,5\}$, find the value of the expression $2y + 1$ for each possible value of the variable y.

24 Indicate the set relation that exists between the set of even counting numbers greater than 7 and the set $\{2,4,6\}$.

In problems 25–28, list all the subsets of each set.

25 $\{3\}$

26 $\{1,2\}$

27 $\{1,2,3\}$

28 $\{a,b,c,d\}$

In problems 29–46, let $A = \{1,2,4,6\}$, $B = \{2,3,5,6,7\}$, $C = \{5,6,4,8\}$, $D = \{1,4\}$, $E = \{6,7,9\}$, and $F = \{4\}$. Form each set.

29 $A \cup B$

30 $A \cup F$

31 $A \cup C$

32 $D \cup E$

33 $A \cap B$

34 $E \cap D$

35 $B \cap C$

36 $B \cup E$

37 $D \cup B$

38 $D \cap B$

39 $F \cap C$

40 $A \cup (F \cup D)$

41 $E \cap A$

42 $F \cap \emptyset$

43 $C \cup \emptyset$

44 $(A \cup B) \cup F$

45 $(A \cap B) \cap F$

46 $(F \cup E) \cap D$

1.3 Sets of Numbers

The language of sets can be used to describe some of the number sets of algebra. Historically, the first set of numbers developed were the *natural numbers*, 1, 2, 3, etc., also called the *counting numbers*. The letter N is used to designate the set of natural numbers; thus, $N = \{1,2,3,4, \ldots\}$. This set is also called the set of positive integers. The set N is an infinite set, because the process of counting its elements can never come to an end. However, there is one new "number" not obtainable by counting which is convenient to introduce at this point. This number is zero and is denoted symbolically by 0. The union of the set N and the set $\{0\}$ is called the *set of whole numbers* and is denoted by W. That is, $W = \{0,1,2,3, \ldots\}$.

The Number Line

The set of whole numbers W can be represented geometrically on a line by associating each whole number with a point on that line. Such a representation is called a *number line.* The number line can be constructed by starting with a line that extends endlessly in opposite directions. We use arrowheads to indicate such a line (Figure 1).

Figure 1

Now, choose a point on the line to associate with the number 0 (this point is called the *origin*) and another point to the right of the origin to associate with the number 1 (Figure 2).

Figure 2

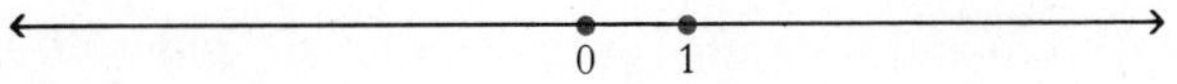

The distance between the two points labeled 0 and 1 is called the *unit length* or the *scale unit* of the number line. By measuring 1 unit length to the right of the point 1, we find the point to associate with 2. Repeating this process, other points can be found to associate with 3, 4, 5, . . . (Figure 3).

Figure 3

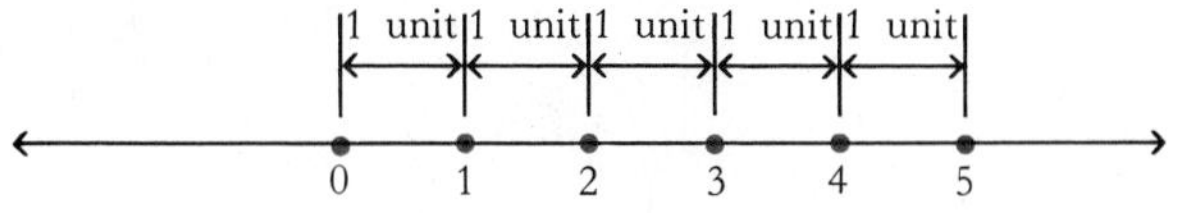

The side containing the point associated with 1 is called the *positive side* of the number line, and the direction from the origin to the point labeled 1 is called the *positive direction* on the number line. The point associated with a number on a number line is called the *graph* of that number, and the number is called the *coordinate* of that point. For example, the coordinates of points A, B, C, and D on the number line are 0, 2, 4, and 5, respectively (Figure 4).

Figure 4

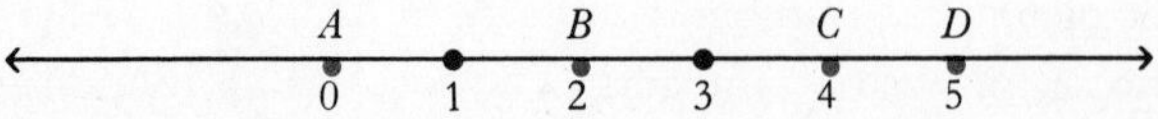

In order to assign coordinates to points to the left of the origin on the number line, we will introduce *negative numbers.* Thus, the coordinates of points E, F, G, and H on the number line (Figure 5) are -4, -3, -2, and -1 (which read "negative four," "negative three," "negative two," and "negative one," respectively). Note that a negative number is named by a numeral with a negative sign, or $-$, written to the left of an ordinary number symbol. The side of the line that contains the points with negative coordinates is called the *negative side* of the number line, and the direction from 0 to -1 is called the *negative direction.* The set $\{\ldots,-4,-3,-2,-1\}$ is called the *set of negative integers.*

Figure 5

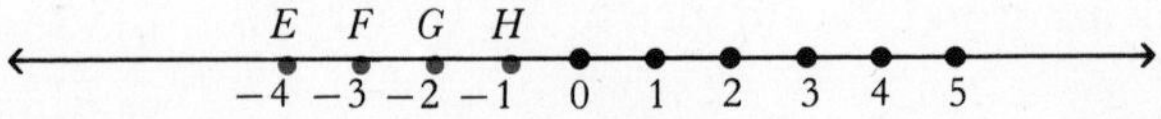

It should be pointed out that the negative integers are written with the negative sign affixed, and the positive integers are indicated by the absence of a sign. Hence, throughout this text, if a number has no sign affixed to it, it is assumed to be positive. The exception is the number zero, which is neither positive nor negative. Thus, the coordinate of a point that corresponds to the number $+5$ is written as 5. (*Note:* Some authors write the number names preceded by the raised symbol $+$ for the points to the right of zero and the raised symbol $-$ for the points to the left of zero. For example, a point corresponding to the number "positive five" is denoted by $^{+}5$ and a point corresponding to the number "negative five" is denoted by $^{-}5$. However, in this book we shall denote these numbers by 5 and -5.)

Real Numbers

The set W, together with the set of negative integers, is called the *set of integers* and is denoted by I. Thus, $I = \{\ldots,-4,-3,-2,-1,0,1,2,3,4,\ldots\}$. Investigating the graph of the set I (Figure 5), we note that there are other points not associated with an element of I. For example, the point halfway between 0 and 1 is not associated with an integer. Instead, the fraction $\frac{1}{2}$ can be associated with this point (Figure 6). We could go on marking more points because there is at least one more point between any two that we marked. For example, to find a point between A and C, we could take 0 and $\frac{1}{2}$ and find a number between the two, say $\frac{1}{4}$.

Figure 6

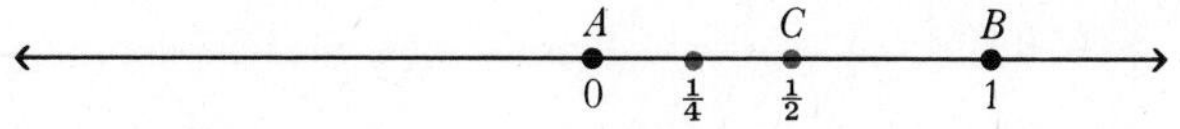

Note that there are infinitely many other points on the number line that require fractions to indicate their location with respect to the origin. Numbers such as those described above are called *rational numbers.* A rational number, then, is a number that can be expressed as a quotient of two integers. That is, a rational number can be expressed in the form $\frac{a}{b}$, where a and b are integers and $b \neq 0$. Examples of rational numbers are $\frac{3}{4}$, $\frac{21}{11}$, $-\frac{2}{3}$, $-\frac{11}{7}$, etc. The set of rational numbers is denoted by Q. Figure 7 shows the location of a few points which correspond to some rational numbers on the number line.

Figure 7

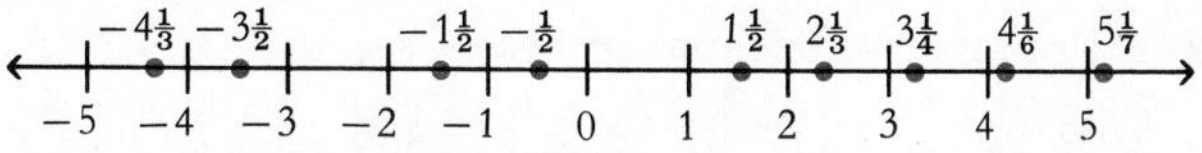

Since any integer can be expressed as the quotient of itself and 1 (that is, $3 = \frac{3}{1}$ and $-5 = -\frac{5}{1}$), the set I is a subset of the set Q. Symbolically, it is written $I \subseteq Q$.

EXAMPLES

Represent the following sets on the number line.

1 $\{-3, -2, -1, 2, 3\}$

SOLUTION. The members of the set are located on a number line (Figure 8).

Figure 8

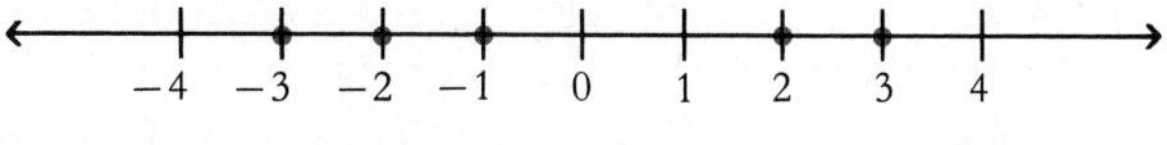

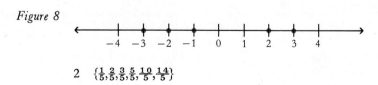

2 $\{\frac{1}{5}, \frac{2}{5}, \frac{3}{5}, \frac{5}{5}, \frac{10}{5}, \frac{14}{5}\}$

SOLUTION. The members of the set are located on a number line (Figure 9).

Figure 9

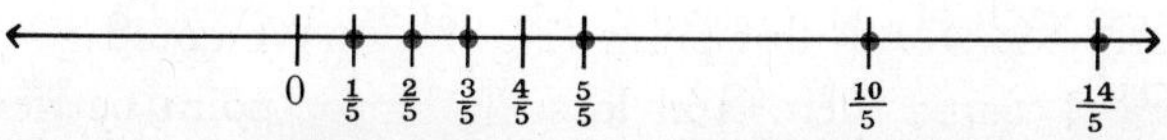

The rational numbers can also be described by investigating their decimal representations. For example, rational numbers such as 4, $\frac{1}{2}$, and $\frac{5}{4}$, which can be written with an integral number of decimal places as

$$4 = \frac{4}{1} = 4.0 \qquad \frac{1}{2} = 0.5 \qquad \text{and} \qquad \frac{5}{4} = 1.25$$

are sometimes called *terminating decimals*. In general, any terminating decimal can be written in the form $\frac{a}{b}$, where a and b are integers and $b \neq 0$. For example,

$$3.57 = 3 + \frac{5}{10} + \frac{7}{100} = \frac{357}{100}$$

EXAMPLES

1 Express each of the following rational numbers in decimal notation.

a) $\frac{1}{4}$ b) $-\frac{3}{5}$ c) $\frac{3}{8}$

SOLUTION. The rational numbers $\frac{1}{4}$, $-\frac{3}{5}$, and $\frac{3}{8}$ can be written with an integral number of decimal places:

a) $\frac{1}{4} = 0.25$ b) $-\frac{3}{5} = -0.6$ c) $\frac{3}{8} = 0.375$

2 Express each of the following terminating decimals in the form $\frac{a}{b}$, where $b \neq 0$.

a) 0.7 b) 3.7 c) 5.64

SOLUTION

a) $0.7 = \frac{7}{10}$
b) $3.7 = 3 + \frac{7}{10} = \frac{37}{10}$
c) $5.64 = 5 + \frac{6}{10} + \frac{4}{100} = \frac{564}{100}$

The rational numbers can also be expressed as repeating decimals. For example, the rational number $\frac{2}{3}$ is written as 0.6666666 . . . . The shortest complete repeating pattern is indicated by drawing a horizontal bar over it: $\frac{2}{3} = 0.\overline{6}$. Notice that the repeating block is the digit 6. The rational number $\frac{17}{99}$ is written 0.17171717 . . . or $0.\overline{17}$, with the repeating block 17. In general, every rational number can be represented by an eventually repeating decimal, and conversely, every eventually repeating decimal represents a rational number.

EXAMPLE

Express each of the following rational numbers as a decimal.

a) $\frac{10}{3}$ b) $\frac{5}{6}$ c) $\frac{7}{9}$

SOLUTION

a) $\frac{10}{3} = 3.\overline{3}$ has a repeating block (the digit 3).
b) $\frac{5}{6} = 0.8\overline{3}$ has a repeating block (the digit 3).
c) $\frac{7}{9} = 0.\overline{7}$ has a repeating block (the digit 7).

Although the set of rational numbers is the set of numbers represented by repeating decimals, there are decimals that do not repeat. For example, the decimal 1.01001000100001 . . . , where there is one more 0 after each 1 than there is before the 1, is a nonrepeating decimal. Another example of a nonrepeating decimal is a number used in plane geometry called π, where $\pi = 3.14159265358\ldots$. Also, numbers such as $\sqrt{2}$, $\sqrt{3}$, and $\sqrt{5}$, which can be approximated by $\sqrt{2} = 1.14142136\ldots$, $\sqrt{3} = 1.7320508\ldots$, and $\sqrt{5} = 2.23606797\ldots$, have nonrepeating-decimal representations. Such numbers are called *irrational numbers.*

The set of all numbers used to represent all the points on a number line is called the set of *real numbers.* That is, the set of real numbers includes integers, fractions, decimals, square roots, and all other roots. The number line that results from associating each of its points with a real number is called a *real line* or *real axis.*

PROBLEM SET 1.3

In problems 1–8, represent each set on a number line.

1 $\{-3,-2,-1,0,1,2,3\}$

2 $\{-4,-3,-2,-1,0\}$

3 $\{-4,-3,0,3\}$

4 $\{-6,-5,0,5,6\}$

5 $\{-\frac{2}{3},-\frac{1}{3},0,\frac{1}{3},\frac{2}{3}\}$

6 $\{\frac{3}{8},\frac{4}{8},\frac{5}{8},\frac{7}{8},\frac{8}{8},\frac{9}{8}\}$

7 $\{-2,-1,0,\frac{1}{2},\frac{3}{2},\frac{5}{2}\}$

8 $\{-\frac{5}{9},-\frac{4}{9},-\frac{3}{9},-\frac{2}{9},-\frac{1}{9}\}$

In problems 9–36, express each rational number in decimal notation.

9 $\dfrac{3}{10}$

10 $\dfrac{7}{10}$

11 $\dfrac{3}{2}$

12 $-\frac{7}{2}$

13 $\frac{15}{100}$

14 $-\frac{67}{100}$

15 $\frac{91}{1,000}$

16 $-\frac{875}{1,000}$

17 $-\frac{4}{5}$

18 $-\frac{5}{4}$

19 $\frac{3}{4}$

20 $\frac{6}{5}$

21 $-\frac{5}{8}$

22 $\frac{7}{8}$

23 $-\frac{2}{3}$

24 $-\frac{5}{6}$

25 $\frac{7}{20}$

26 $-\frac{13}{25}$

27 $-\frac{11}{12}$

28 $\frac{27}{20}$

29 $\frac{17}{25}$

30 $-\frac{7}{12}$

31 $\frac{7}{35}$

32 $\frac{6}{7}$

33 $-\frac{6}{25}$

34 $\frac{4}{3}$

35 $\frac{7}{15}$

36 $-\frac{18}{12}$

In problems 37–56, express each decimal in the form $\frac{a}{b}$, where a and b are integers.

37 0.43

38 0.54

39 0.72

40 0.06

41 0.27

42 1.72

43 2.64

44 -0.125

45 -0.008

46 -0.0098

47 -40.08

48 -36.63

49 -0.582

50 0.694

51 -0.209

52 0.918

53 -0.0527

54 0.00329

55 -0.00052

56 -0.000171

1.4 Numbers and Their Opposites—Absolute Value

Consider the representation of the set $\{-5,-4,-3,-2,-1,0,1,2,3,4,5\}$ on the number line as shown in Figure 1. If we select any positive number, say 4, which lies 4 units away from the origin, we can find exactly one number on the line the same distance from 0 as 4 but on the opposite side of 0. It is apparent from Figure 1 that the other number is -4. The numbers 4 and -4 are called *opposite numbers* or *additive inverses.*

Figure 1

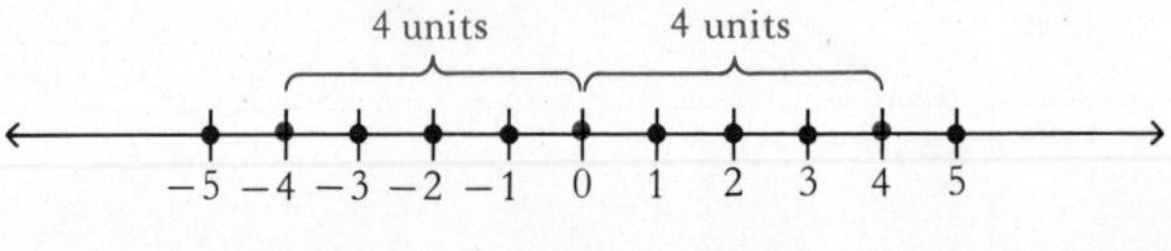

Specifically, if x is a real number, the additive inverse of x is that number which we denote by $-x$ and which is the same distance from 0 on the number line as x but on the opposite side of 0 and has the property that $x + (-x) = 0$ (Figure 2). Note that the additive inverse of 0 is 0, since $0 + (-0) = 0$.

Figure 2

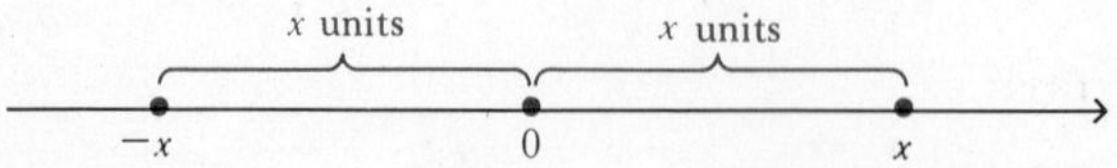

EXAMPLE

Find the additive inverse of each of the following numbers.

a) 2 b) 3 c) -2 d) -5 e) $-(-7)$ f) -20

SOLUTION

	Number	*Additive inverse*	*Reason*
a)	2	-2	$2 + (-2) = 0$
b)	3	-3	$3 + (-3) = 0$
c)	-2	2	$-2 + (2) = 0$
d)	-5	5	$-5 + (5) = 0$
e)	$-(-7)$	-7	$-(-7) + (-7) = 0$
f)	-20	20	$-20 + (20) = 0$

Absolute Value

We indicated in Figure 1 that the distance from -4 to 0 is the same as the distance from 0 to 4, so 4 and -4 are the additive inverses of each other. To express this idea, we say that -4 and 4 have the same *absolute value*, which is expressed symbolically as $|-4| = 4$ and $|4| = 4$, where the vertical bars, $|\ |$, denote the absolute value of a number. For example, the distance from -6 to 0 on the number line is 6 units, so $|-6| = 6$. Also, the distance from 0 to 6 is 6 units, so $|6| = 6$ (Figure 3). The absolute value of 0 is defined to be 0; that is, $|0| = 0$. It should be noted that since the distance is a physical measurement that is never negative, the absolute value of a number can never be negative.

Figure 3

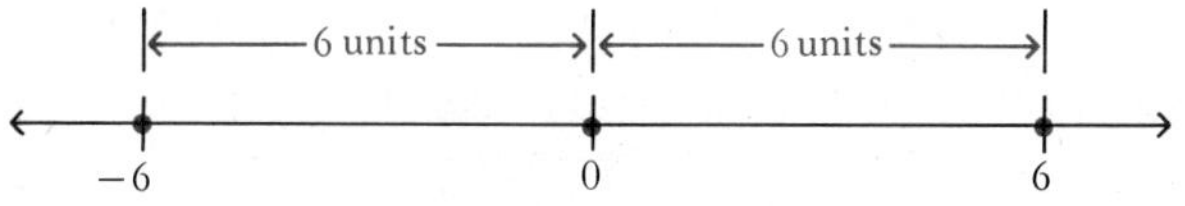

EXAMPLE

Find the value of the following expressions.

a) $|8|$ b) $|-7|$ c) $-|9|$ d) $|-3| + |-7|$

SOLUTION

a) $|8| = 8$, since 8 is a positive number that lies 8 units from the origin.
b) $|-7| = 7$, since -7 is a negative number that lies 7 units from the origin.
c) $-|9| = -(9) = -9$
d) $|-3| = 3$; also, $|-7| = 7$, so $|-3| + |-7| = 3 + 7 = 10$.

PROBLEM SET 1.4

In problems 1–16, give the additive inverse of each number.

1 9

2 12

3 -19

4 3

5 -13

6 0

7 -3

8 -11

9 4

10 -63

11 $-(-17)$

12 $-(-3)$

13 -100

14 3,000

15 179

16 -382

In problems 17–28, find the value of each expression.

17 $|9|$

18 $|28|$

19 $-|-6|$

20 $|104|$

21 $|-31|$

22 $|-79|$

23 $|-41|$

24 $|-75|$

25 $-|-35|$

26 $-|-29|$

27 $-|-65|$

28 $-|62|$

In problems 29–38, compute the value of each expression. Express the results without absolute-value symbols.

29 $|-3| - |3|$

30 $|-8| + |8|$

31 $|5 - 2|$

32 $|-7 + 7|$

33 $|-12| - |3|$

34 $|-2| + |5|$

35 $|-5| - |-3|$

36 $|-2| - |2|$

37 $|-2| + |-3|$

38 $|-5| - |-3|$

In problems 39–46, complete the statements by filling in the blanks.

39 $|-8| = |\_\_\_\_|$

40 $|3 + (-3)| = |\_\_\_\_|$

41 $|15| - |\_\_\_\_| = 0$

42 $-|21 + 34| = \_\_\_\_ + |51|$

43 $|-6| - |-5| = \_\_\_\_$

44 $\_\_\_\_ + |-3| = 0$

45 $|3 + 2| = |\_\_\_\_| + 2$

46 $-|-7| + \_\_\_\_ = 0$

In problems 47–50, determine which statements are true and which are false.

47 $|3| + |-3| = 2|3|$

48 $|7 - 5| = |5| + |-7|$

49 $|-4| - |4| = 0$

50 $|-5| + |5| = 0$

1.5 Addition and Subtraction of Signed Numbers

The rules for adding and subtracting positive numbers were discussed in arithmetic. However, in order to extend the notion of addition and subtraction to include negative numbers, we consider the meaning of adding and subtracting positive numbers on a number line.

Addition of Signed Numbers

In this section we shall use the number line to extend addition of positive numbers to addition of real numbers. For example, to add 2 and 3 on the number line (Figure 1), we start at the origin and move 2 units to the right; thus, the number 2 is represented by an arrow from 0 to 2. We then start at the number 2 and move 3 units to the right, so the number 3 is represented by an arrow between 2 and 5. Together, the sum of the two directed moves is $2 + 3 = 5$ (Figure 1*a*). It can also be seen that $3 + 2 = 5$ (Figure 1*b*).

Figure 1

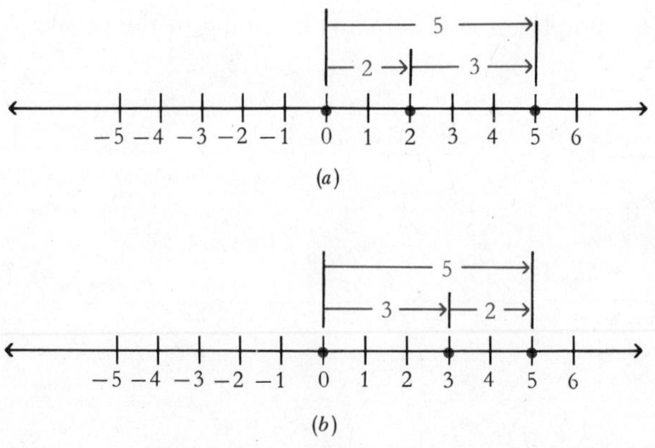

Now let us consider the sum $(-2) + (-3)$. This sum can be found by first moving 2 units to the left of the origin on the number line, then moving 3 units to the left of -2; that is, $(-2) + (-3) = -5$ (Figure 2).

Figure 2

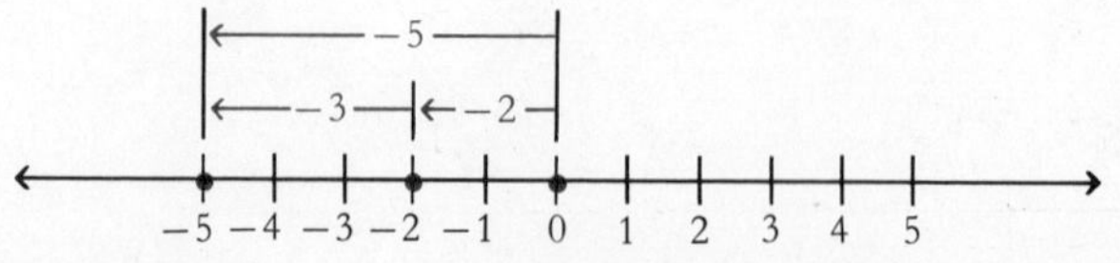

Note that in both cases illustrated above, the sums can be found by adding the absolute values of the numbers and retaining the common sign of the numbers. In general, we have the following rule: To add like signed numbers, add their absolute values and keep their common sign. For example,

$$3 + 4 = |3| + |4| = 7$$
$$(-4) + (-6) = -(|-4| + |-6|) = -(4 + 6) = -10$$

The addition of unlike signed numbers can also be illustrated on a number line. For example, the sum $(-3) + 8$ can be interpreted as a movement of 8 units to the right of the number -3 (Figure 3*a*). The sum $(-3) + 8$ can also be interpreted as a movement of 3 units to the left of 8 (Figure 3*b*). The same sum is obtained in both approaches; that is, $(-3) + 8 = 5$.

Figure 3

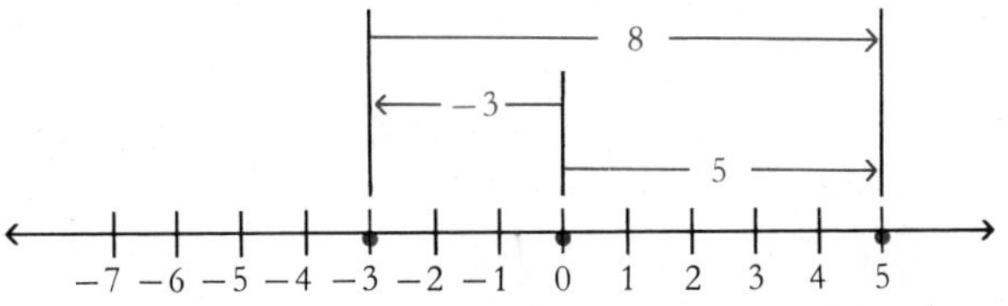

(*a*)

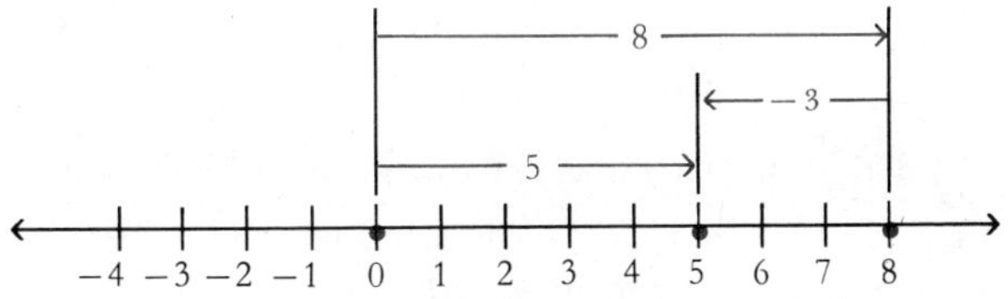

(*b*)

Now consider the sum $(-7) + 5$. By use of the number line (Figure 4) we see that $(-7) + 5 = -2$. This sum can also be found as follows:

$$\begin{aligned}(-7) + 5 &= (-2) + (-5) + 5 \\ &= (-2) + [(-5) + 5] \\ &= (-2) + 0 \\ &= -2\end{aligned}$$

Figure 4

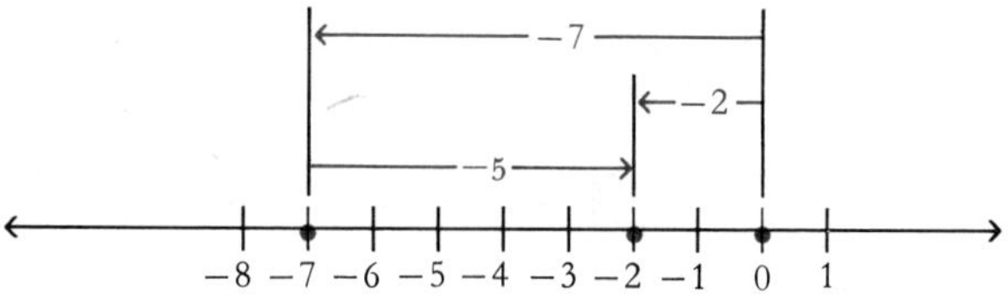

Note that in both examples, that is, $(-3) + 8 = 5$ and $(-7) + 5 = -2$, the sum can be found by subtracting the absolute values of the numbers and retaining the sign of the number with the largest absolute value. In general, we have the following rule: To add two unlike signed numbers, subtract their absolute values and retain the sign of the number with the greatest absolute value. For example,

$$3 + (-3) = 0 \qquad \text{(Additive inverse)}$$
$$(-3) + 5 = +(|5| - |3|) = +(5 - 3) = 2$$
$$(-7) + 2 = -(|7| - |2|) = -(7 - 2) = -5$$

EXAMPLE

Find the following sums.

a) $8 + 6$ b) $(-4) + (-7)$ c) $11 + (-11)$
d) $(-4) + 21$ e) $8 + (-23)$

SOLUTION

a) $8 + 6 = 14$
b) $(-4) + (-7) = -(4 + 7) = -11$
c) $11 + (-11) = 0$ (Additive inverse)
d) $(-4) + 21 = +(21 - 4) = 17$
e) $8 + (-23) = -(23 - 8) = -15$

Subtraction of Signed Numbers

To subtract two signed numbers, we appeal to the following definition: The *differences* of two real numbers a and b, denoted by $a - b$, is defined by $a + (-b)$. That is, to subtract b from a, add the additive inverse of b to a. This definition, together with the rules for adding signed numbers, provides a method for subtracting signed numbers. For example, the difference $(-7) - (3)$ can be found as follows:

$$(-7) - (3) = (-7) + (-3)$$
$$= -10$$

We can illustrate this subtraction on a number line by interpreting the subtractions of 3 from (-7) as a movement of 3 units to the left of -7; that is, $(-7) - (3) = -10$ (Figure 5).

Figure 5

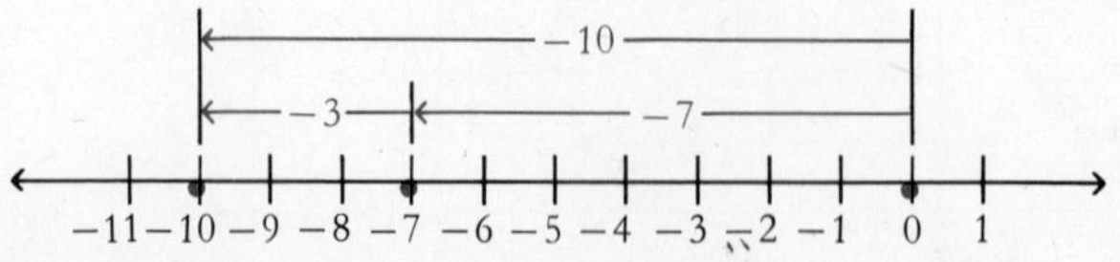

The rule for subtracting signed numbers can be stated as follows: To subtract signed numbers, change the sign of the number to be subtracted, then add by following the rules for adding signed numbers. For example,

$$9 - 4 = 9 + (-4) = 5$$
$$4 - 9 = 4 + (-9) = -5$$
$$(-9) - (6) = (-9) + (-6) = -15$$

EXAMPLE

Find the following differences.

a) $5 - 3$ b) $(-11) - 8$ c) $12 - (-4)$ d) $(-16) - (-19)$

SOLUTION

a) $5 - 3 = 5 + (-3) = 2$
b) $(-11) - 8 = (-11) + (-8) = -19$
c) $12 - (-4) = 12 + 4 = 16$
d) $(-16) - (-19) = -16 + 19 = 3$

PROBLEM SET 1.5

In problems 1–57, perform the additions and subtractions.

1 $9 + 7$

2 $13 + 16$

3 $17 + 21$

4 $19 + 35$

5 $7 + (-3)$

6 $7 + (-8)$

7 $27 + (-39)$

8 $65 + (-19)$

9 $62 + (-31)$

10 $71 + (-29)$

11 $10 + (-101)$

12 $65 + (-111)$

13 $75 + (-118)$

14 $89 + (-321)$

15 $39 + (-39)$

16 $65 + (-65)$

17 $(-23) + 14$

18 $(-381) + 379$

19 $(-10) + 4$

20 $(-120) + 113$

21 $(-11) + 20$

22 $(-25) + 77$

23 $(-100) + 43$

24 $(-83) + 83$

25 $(-165) + 165$

26 $(-79) + 79$

27 $(-8) + (-6)$

28 $(-79) + (-13)$

29 $(-11) + (-7)$

30 $(-87) + (-253)$

31 $(-13) + (-7) + 11$

32 $(-113) + (-127) + 201$

33 $(-5) + 7 + (-3)$

34 $(-18) + (-17) + (-59)$

35 $(-2) + (-3) + 8$

36 $|4| + (-300) + 161$

37 $(-21) + (-18) + (-39)$

38 $(-119) + (-105) + 216$

39 $18 - 13$

40 $65 - 39$

41 $16 - 7$

42 $19 - 13$

43 $49 - 37$

44 $41 - 29$

45 $7 - 13$

46 $19 - 27$

47 $18 - (-9)$

48 $27 - (-8)$

49 $31 - (-5)$

50 $37 - (-9)$

51 $(-25) - 4$

52 $(-49) - 7$

53 $(-65) - 7$

54 $(-57) - 16$

55 $(-17) - (-43)$

56 $(-22) - (-8)$

57 $(-25) - (-45)$

1.6 Multiplication and Division of Signed Numbers

In this section we develop a procedure for multiplying and dividing signed numbers.

Multiplication of Signed Numbers

Multiplication of two positive integers is often described in arithmetic as repeated addition. For instance, the product 4×3 is interpreted as $4 + 4 + 4$ or $3 + 3 + 3 + 3$. In both cases, we have $4 \times 3 = 12$. The same approach can be used to define the product of any real number by a positive number. For instance, to find $4 \times (-1)$, we first consider the following pattern of products:

$$\begin{aligned} 4 \times 3 &= 12 \\ 4 \times 2 &= 8 \\ 4 \times 1 &= 4 \\ 4 \times 0 &= 0 \end{aligned}$$

We observe that as the numbers on the left of the equals sign decrease by 1 each time, the products on the right decrease by 4.

The pattern that we have illustrated shows that the product of two positive numbers is a positive number. In symbols, if a and b are positive numbers, then $a \cdot b = +(ab)$. In particular, if a or b is zero, then $a \cdot 0 = 0$ and $0 \cdot b = 0$. Now, we note that this pattern will be preserved if the next number on the right is 4 less than 0. That is, if the number is -4. Therefore, we have $4 \times (-1) = -4$.

Continuing this pattern, we have

$$\begin{aligned} 4 \times (-1) &= -4 \\ 4 \times (-2) &= -8 \\ 4 \times (-3) &= -12 \end{aligned}$$

and so forth. In a similar manner, we find that

$$\begin{aligned} (-4) \times 3 &= -12 \\ (-4) \times 2 &= -8 \\ (-4) \times 1 &= -4 \\ (-4) \times 0 &= 0 \end{aligned}$$

Specifically, the product of a negative number and a positive number is a negative number. In symbols, if a and b are positive numbers, then

$$a(-b) = -(ab)$$

or

$$(-a)b = -(ab)$$

Let us now consider the product $(-4) \times (-1)$. Using the latter pattern, we notice that as the numbers on the left of equal signs decrease each time by 1, the products on the right increase by 4. To preserve such a pattern, we agree that $(-4) \times (-1)$ is 4 more than $(-4) \times 0$. That is, we must have $(-4) \times (-1) = 4$. The above pattern will then be continued as follows:

$$(-4) \times (-1) = 4$$
$$(-4) \times (-2) = 8$$
$$(-4) \times (-3) = 12$$

and so on.

This shows that the product of two negative numbers is a positive number. In symbols, if a and b are positive numbers, then $(-a) \cdot (-b) = ab$.

Summarizing the above rules for multiplying signed numbers we have:

1 If two numbers have like signs, their product is positive.
2 If two numbers have unlike signs, their product is negative.

EXAMPLES

1 Find the indicated products.

a) $(-9) \times 6$ b) $5 \times (-8)$ c) $6 \times (-15)$

SOLUTION

a) $(-9) \times 6 = -(9 \times 6) = -54$
b) $5 \times (-8) = -(5 \times 8) = -40$
c) $6 \times (-15) = -(6 \times 15) = -90$

2 Find the indicated products.

a) 9×3 b) $(-4) \times (-2)$ c) $(-5) \times (-3)$

SOLUTION

a) $9 \times 3 = 27$
b) $(-4) \times (-2) = 4 \times 2 = 8$
c) $(-5) \times (-3) = 5 \times 3 = 15$

When multiplying more than two signed numbers, the factors can be paired to determine the sign of the product. For example,

$$(-3) \times 2 \times (-5) \times 6$$

can be written

$$[(-3) \times (-5)] \times [2 \times 6] = 15 \times 12 = 180$$

Here, because there are an even number of negative factors, the product is positive. In the case of the product $(-4) \times (-7) \times (-2) \times 3$, we have

$$[(-4) \times (-7)] \times [(-2) \times 3] = 28 \times (-6) = -168$$

Here, because there are an odd number of negative factors, the product is negative. Thus, we can state a more general rule for finding the product of signed numbers: The product of signed numbers is positive if there is an even number of negative factors, and the product is negative if there is an odd number of negative factors.

EXAMPLE

Determine the indicated products.

a) $3 \times (-2) \times 6 \times (-7) \times (-8) \times 2 \times (-4)$
b) $(-1) \times 7 \times (-3) \times (-8) \times (-4) \times (-7)$

SOLUTION

a) By counting, we find there are four negative factors. Since four is an even number, the product is positive.
$3 \times (-2) \times 6 \times (-7) \times (-8) \times 2 \times (-4) = 16{,}128$
b) Since there are five negative factors, an odd number, the product is negative.
$(-1) \times 7 \times (-3) \times (-8) \times (-4) \times (-7) = -4{,}704$

Division of Signed Numbers

Division of signed numbers, denoted by the symbol $\div$, is the inverse operation of multiplication; that is, if

$$15 \div 5 = 3 \qquad \text{then } 3 \times 5 = 15$$

and if

$$(-28) \div (-7) = 4 \qquad \text{then } 4 \times (-7) = -28$$

In general, if a and b are real numbers, with $b \neq 0$, and if $a \div b = c$, then $c \cdot b = a$.

$$
\begin{aligned}
12 \div 4 &= 3 && \text{because } 3 \times 4 = 12 \\
12 \div (-4) &= (-3) && \text{because } (-3) \times (-4) = 12 \\
(-12) \div 4 &= (-3) && \text{because } (-3) \times 4 = -12 \\
(-12) \div (-4) &= 3 && \text{because } 3 \times (-4) = -12
\end{aligned}
$$

It should be noted here that division by zero is not permitted, because $-\frac{1}{0}$, $\frac{1}{0}$, and $\frac{0}{0}$ are not numbers. However, $-\frac{0}{1} = 0$ and $\frac{0}{1} = 0$.

In general, we have

1 If two numbers have like signs, their quotient is positive.
2 If two numbers have unlike signs, their quotient is negative.

EXAMPLE

Perform the following divisions.

a) $45 \div (-9)$ b) $(-42) \div 3$ c) $(-65) \div (-13)$

SOLUTION

a) $45 \div (-9) = -(45 \div 9) = -5$
b) $(-42) \div 3 = -(42 \div 3) = -14$
c) $(-65) \div (-13) = +(65 \div 13) = 5$

The numeral $\frac{a}{b}$ may be used to express the quotient of two numbers; that is, if $b \neq 0$, then we write

$$a \div b = \frac{a}{b}$$

The numeral $\frac{a}{b}$ is also called a *fraction*, where a is called the *numerator* and b is called the *denominator*. Accordingly, the sign of the fraction, $\frac{a}{b}$, will be positive when both a and b have like signs and negative when both a and b have unlike signs, that is, if $b \neq 0$:

1 $\dfrac{-a}{b} = (-a) \div b = -\left(\dfrac{a}{b}\right)$

2 $\dfrac{a}{-b} = a \div (-b) = -\left(\dfrac{a}{b}\right)$

3 $\dfrac{-a}{-b} = (-a) \div (-b) = \dfrac{a}{b}$

EXAMPLE

Find the following quotients.

a) $\frac{21}{-7}$ b) $\frac{-72}{24}$ c) $\frac{-91}{-13}$

SOLUTION

a) $\frac{21}{-7} = -\left(\frac{21}{7}\right) = -3$

b) $\frac{-72}{24} = -\left(\frac{72}{24}\right) = -3$

c) $\frac{-91}{-13} = \frac{91}{13} = 7$

PROBLEM SET 1.6

In problems 1–30, perform the multiplications.

1 3×5

2 7×13

3 19×7

4 135×6

5 $(-7) \times 2$

6 $(-13) \times 7$

7 $(-31) \times 5$

8 $(-29) \times 4$

9 $8 \times (-14)$

10 $17 \times (-3)$

11 $5 \times (-18)$

12 $6 \times (-25)$

13 $(-1) \times (-5)$

14 $(-2) \times (-3)$

15 $(-4) \times (-5)$

16 $(-13) \times (-31)$

17 $(-11) \times (-6)$

18 $(-79) \times (-4)$

19 $-(-2) \times (-2)$

20 $-(-4) \times (-7)$

21 $(-1) \times (-1) \times (-1)$

22 $3 \times (-2) \times (-5)$

23 $(-3) \times (-5) \times (-1)$

24 $(-1) \times (-2) \times (-5)$

25 $(-2) \times (-5) \times (-1) \times (-3)$

26 $(-2) \times (-3) \times (-4) \times (-5) \times (-1)$

27 $7 \times (-2) \times (-1) \times (-1) \times 3$

28 $3 \times (-2) \times (-2) \times (-10) \times (-4) \times (-1)$

29 $(-2) \times (-9) \times (-3) \times 5 \times (-1) \times (-1)$

30 $(-7) \times (-2) \times 7 \times 4 \times (-1) \times (-1) \times (-1)$

In problems 31–60, perform the divisions.

31 $10 \div 5$

32 $18 \div 9$

33 $\dfrac{34}{17}$

34 $\dfrac{65}{13}$

35 $(-18) \div 9$

36 $(-57) \div 19$

37 $(-28) \div 4$

38 $(-90) \div 15$

39 $\dfrac{-49}{7}$

40 $\dfrac{-39}{13}$

41 $36 \div (-12)$

42 $72 \div (-9)$

43 $51 \div (-3)$

44 $125 \div (-25)$

45 $\dfrac{216}{-6}$

46 $\dfrac{81}{-27}$

47 $(-32) \div (-8)$

48 $(-24) \div (-6)$

49 $(-36) \div (-18)$

50 $(-250) \div (25)$

51 $(-75) \div (-15)$

52 $0 \div (-5)$

53 $0 \div 17$

54 $9 \div 0$

55 $\dfrac{-42}{-14}$

56 $\dfrac{-625}{-25}$

57 $\dfrac{-243}{-3}$

58 $\dfrac{-289}{-17}$

59 $\dfrac{-7}{0}$

60 $\dfrac{0}{0}$

1.7 Basic Properties of Real Numbers

In Section 1.3 we discussed the procedure of locating real numbers on the number line. Here, we shall list a set of properties of real numbers. These properties serve as a foundation for justifying the algebraic steps in later chapters.

1 CLOSURE PROPERTY

a) *Closure property for addition:* The sum of two real numbers is always a real number. That is, if a and b are real numbers, then $a + b$ is a real number. For example, the sum of real numbers 3 and 7 is a real number 10. That is,

$$3 + 7 = 10$$

Also,

$$(-3) + 5 = 2$$

b) *Closure property for multiplication:* The product of two real numbers is a real number. That is, if a and b are real numbers, then $a \cdot b$ is a real number. For example, the product of the real numbers 8 and 5 is the real number 40. That is,

$$8 \times 5 = 40$$

Also,

$$5 \times (-7) = -35$$

It should be noted that while the set of real numbers is closed under the operations of addition and multiplication, not all sets are closed under these operations. To illustrate, we give the following example.

EXAMPLE

Consider the set $A = \{1,2\}$. Does the set A possess the closure properties for addition and multiplication?

SOLUTION. Since $1 + 2 = 3$ and 3 does not belong to A, the set A does not possess the closure property for addition. Also, since $2 \times 2 = 4$, and 4 does not belong to the set A, then A does not possess the closure property for multiplication.

2 COMMUTATIVE PROPERTY

a) *Commutative property for addition:* For all real numbers a and b,

$$a + b = b + a$$

For example, $6 + 11 = 11 + 6$, since

$$6 + 11 = 17$$
$$11 + 6 = 17$$

Also,

$$(-85) + (-19) = (-19) + (-85) = -104$$

b) *Commutative property for multiplication:* For all real numbers a and b,

$$a \cdot b = b \cdot a$$

For example, $5 \times 3 = 3 \times 5$, since

$$5 \times 3 = 15$$
$$3 \times 5 = 15$$

Also,

$$(-6) \times 4 = 4 \times (-6) = -24$$

EXAMPLES

Verify the commutative property by performing actual computation.

1 $73 + (-39)$ and $(-39) + 73$

SOLUTION

$$73 + (-39) = 34$$
$$(-39) + 73 = 34$$

Therefore,

$$73 + (-39) = (-39) + 73$$

2 $(-12) \times (-4)$ and $(-4) \times (-12)$

SOLUTION

$$(-12) \times (-4) = 48$$
$$(-4) \times (-12) = 48$$

Therefore,

$$(-12) \times (-4) = (-4) \times (-12)$$

3 ASSOCIATIVE PROPERTY

a) *Associative property for addition:* For all real numbers a, b, and c,

$$a + (b + c) = (a + b) + c$$

For example, $5 + (3 + 8) = (5 + 3) + 8$, since

$$5 + (3 + 8) = 5 + 11 = 16$$
$$(5 + 3) + 8 = 8 + 8 = 16$$

b) *Associative property for multiplication:* For all real numbers, a, b, and c,

$$a \cdot (b \cdot c) = (a \cdot b) \cdot c$$

For example, $4 \times (6 \times 2) = (4 \times 6) \times 2$, since

$$4 \times (6 \times 2) = 4 \times 12 = 48$$
$$(4 \times 6) \times 2 = 24 \times 2 = 48$$

EXAMPLES

Verify the associative property by performing the actual computations.

1 $77 + [(-9) + 2]$ and $[77 + (-9)] + 2$

SOLUTION

$$77 + [(-9) + 2] = 77 + (-7) = 70$$
$$[77 + (-9)] + 2 = 68 + 2 = 70$$

Therefore,

$$77 + [(-9) + 2] = [77 + (-9)] + 2$$

2 $4 \times [(-13) \times (-7)]$ and $[4 \times (-13)] \times (-7)$

SOLUTION

$$4 \times [(-13) \times (-7)] = 4 \times 91 = 364$$
$$[4 \times (-13)] \times (-7) = (-52) \times (-7) = 364$$

Therefore,

$$4 \times [(-13) \times (-7)] = [4 \times (-13)] \times (-7)$$

4 DISTRIBUTIVE PROPERTY

If a and b are real numbers, then

a) $a \cdot (b + c) = a \cdot b + a \cdot c$
b) $(a + b) \cdot c = a \cdot c + b \cdot c$

For example, $5 \times (4 + 6) = (5 \times 4) + (5 \times 6)$, since

$$5 \times (4 + 6) = 5 \times 10 = 50$$
$$(5 \times 4) + (5 \times 6) = 20 + 30 = 50$$

Also, $[(-5) + (-7)] \times 2 = [(-5) \times 2] + [(-7) \times 2]$, since

$$[(-5) + (-7)] \times 2 = (-12) \times 2 = -24$$
$$[(-5) \times 2] + [(-7) \times 2] = (-10) + (-14) = -24$$

EXAMPLES

Verify the distributive property by performing the actual computations.

1 $5 \times [(-8) + 2]$ and $[5 \times (-8)] + (5 \times 2)$

SOLUTION

$$5 \times [(-8) + 2] = 5 \times (-6) = -30$$
$$[5 \times (-8)] + (5 \times 2) = (-40) + 10 = -30$$

Therefore,

$$5 \times [(-8) + 2] = [5 \times (-8)] + (5 \times 2)$$

2 $[(-4) + (-7)] \times 8 = [(-4) \times 8] + [(-7) \times 8]$

SOLUTION

$$[(-4) + (-7)] \times 8 = (-11) \times 8 = -88$$
$$[(-4) \times 8] + [(-7) \times 8] = (-32) + (-56) = -88$$

Therefore,

$$[(-4) + (-7)] \times 8 = [(-4) \times 8] + [(-7) \times 8]$$

5 IDENTITY PROPERTY

a) *Identity property for addition:* If a is a real number, then

$$a + 0 = 0 + a = a$$

For example,

$$4 + 0 = 0 + 4 = 4$$

b) *Identity property for multiplication:* If a is a real number, then

$$a \cdot 1 = 1 \cdot a = a$$

For example,

$$5 \times 1 = 1 \times 5 = 5$$

6 INVERSE PROPERTY

a) *Additive inverse:* For each real number a, there is a real number, called the *additive inverse* and denoted by $-a$, such that

$$a + (-a) = (-a) + a = 0$$

For example,

$$19 + (-19) = (-19) + 19 = 0$$

For more examples of the additive inverse, see Section 1.4.

b) *Multiplicative inverse:* For each real number a, where $a \neq 0$, there is a real number a, called the *multiplicative inverse* or *reciprocal*, denoted by $\frac{1}{a}$, such that

$$a \cdot \frac{1}{a} = \frac{1}{a} \cdot a = 1$$

For example,

$$5 \times \tfrac{1}{5} = \tfrac{1}{5} \times 5 = 1$$

EXAMPLE

Find the multiplicative inverse of the following numbers.

a) 5 b) -4 c) $\frac{1}{7}$ d) $-\frac{1}{8}$ e) 0 f) x g) $\frac{1}{y}$

SOLUTION

	Number	Multiplicative inverse	Reason
a)	5	$\frac{1}{5}$	$5 \times \frac{1}{5} = 1$
b)	-4	$\frac{1}{-4}$	$(-4) \times \frac{1}{-4} = 1$
c)	$\frac{1}{7}$	7	$\frac{1}{7} \times 7 = 1$
d)	$-\frac{1}{8}$	-8	$(-\frac{1}{8}) \times (-8) = 1$
e)	0	None	
f)	x	$\frac{1}{x}, x \neq 0$	$x \cdot \frac{1}{x} = 1$
g)	$\frac{1}{y}, y \neq 0$	y	$\frac{1}{y} \cdot y = 1$

7 EQUALITY

The properties and definitions introduced so far have all involved the use of "equality," which is denoted by the symbol $=$. The statement $a = b$ means that a and b are two names for the same number. We shall list additional properties for a, b, and c elements of the set of real numbers R.

a) *Reflexive Property:* $a = a$.
b) *Symmetric Property:* If $a = b$, then $b = a$.
c) *Transitive Property:* If $a = b$ and $b = c$, then $a = c$.

Another important and frequently used property is the substitution property.

8 THE SUBSTITUTION PROPERTY

If $a = b$, then a can be substituted for b in any statement involving b without affecting the truthfulness of the statement. For example, if $a = b$ and $b + 5 = 11$, then $a + 5 = 11$.

The properties of the real numbers and the substitution property can be combined to form other true statements about the real numbers. These statements have applications to topics introduced later in the book, for example, solving equations.

Consider a, b, c, and d real numbers. Then:

a) *Addition Property of Equality:* If $a = b$ and $c = d$, then $a + c = b + d$.
b) *Multiplication Property of Equality:* If $a = b$ and $c = d$, then $ac = bd$.

9 THE CANCELLATION PROPERTIES

a) *Cancellation Property for Addition:* If a and b are real numbers and if $a + c = b + c$, then $a = b$.

b) *Cancellation Property for Multiplication:* If a and b are real numbers and if $ac = bc$, with $c \neq 0$, then $a = b$.

Now, we list additional properties that can be useful throughout the text.

10 PROPERTY (MULTIPLICATION BY ZERO)

$a \cdot 0 = 0 \cdot a = 0.$

11 PROPERTY

a) $-(-a) = a.$
b) $(-a)(b) = a(-b) = -(ab).$
c) $(-a)(-b) = ab.$

12 PROPERTY

If $a \cdot b = 0$, then either $a = 0$ or $b = 0$ or both $a = 0$ and $b = 0$.

Other properties of real numbers will be discussed as we encounter them.

EXAMPLE

State the properties that justify each of the following equalities.

a) $(-5) + (-7) = (-7) + (-5)$
b) $9 + [(-8) + 2] = [9 + (-8)] + 2$
c) $(-7) \times [3 \times (-2)] = [(-7) \times 3] \times (-2)$
d) $2 \times (-5) = (-5) \times 2$
e) $14 \times [2 + (-3)] = (14 \times 2) + [14 \times (-3)]$
f) $(-7) + 7 = 0$
g) $(-5) \times \dfrac{1}{-5} = 1$
h) $(-13) \times 0 = 0$
i) $(-2) \times (-3) = 6$

SOLUTION

a) $(-5) + (-7) = (-7) + (-5)$ (Commutative property for addition)

b) $9 + [(-8) + 2] = [9 + (-8)] + 2$ (Associative property for addition)

c) $(-7) \times [3 \times (-2)] = [(-7) \times 3] \times (-2)$ (Associative property for multiplication)

d) $2 \times (-5) = (-5) \times 2$ (Commutative property for multiplication)

e) $14 \times [2 + (-3)] = (14 \times 2) + [14 \times (-3)]$ (Distributive property)

f) $(-7) + 7 = 0$ (Additive inverse)

g) $(-5) \times \frac{1}{-5} = 1$ (Multiplicative inverse)

h) $(-13) \times 0 = 0$ (Property 10)

i) $(-2) \times (-3) = 6$ (Property 11c)

PROBLEM SET 1.7

In problems 1–4, indicate which sets are closed under the operations of addition and multiplication.

1 {1,3}

2 {−1,1}

3 {even positive integers}

4 {odd negative integers}

In problems 5–12, verify the commutative properties of addition and multiplication.

5 $12 + (-3)$ and $(-3) + 12$

6 $(-17) + (-13)$ and $(-13) + (-17)$

7 12×7 and 7×12

8 $(-16) \times 3$ and $3 \times (-16)$

9 $(-5) \times (-9)$ and $(-9) \times (-5)$

10 $(-5) \times (-4)$ and $(-4) \times (-5)$

11 $(-5) + (-79)$ and $(-79) + (-5)$

12 $(-17) + (-39)$ and $(-39) + (-17)$

In problems 13–20, verify the associative properties of addition and multiplication.

13 $(35 + 62) + 111$ and $35 + (62 + 111)$

14 $77 + (19 + 2)$ and $(77 + 19) + 2$

15 $[9 + (-8)] + (-13)$ and $9 + [(-8) + (-13)]$

16 $[8 + (-11)] + 21$ and $8 + [(-11) + 21]$

17 $(5 \times 7) \times 11$ and $5 \times (7 \times 11)$

18 $6 \times (8 \times 21)$ and $(6 \times 8) \times 21$

19 $[(-9) \times 13] \times 2$ and $(-9) \times (13 \times 2)$

20 $[(-7) \times (-5)] \times 11$ and $(-7) \times [(-5) \times 11]$

In problems 21–28, verify the distributive property.

21 $8 \times (5 + 11) = (8 \times 5) + (8 \times 11)$

22 $14 \times (95 + 3) = (14 \times 95) + (14 \times 3)$

23 $(-14) \times (91 + 3) = [(-14) \times 91] + [(-14) \times 3]$

24 $(-7) \times [(-5) + (-3)] = [(-7) \times (-5)] + [(-7) \times (-3)]$

25 $(17 + 22) \times 6 = (17 \times 6) + (22 \times 6)$

26 $(64 + 31) \times 8 = (64 \times 8) + (31 \times 8)$

27 $(-2) \times [(-37) + (-8)] = [(-2) \times (-37)] + [(-2) \times (-8)]$

28 $[(-13) + (-2)] \times (-6) = [(-13) \times (-6)] + [(-2) \times (-6)]$

In problems 29–36, determine the additive and the multiplicative inverses of each number.

29 12

30 $\frac{1}{3}$

31 -3

32 $-\frac{1}{2}$

33 $\frac{1}{5}$

34 0

35 $-\frac{1}{6}$

36 $-\frac{3}{4}$

In problems 37–56, name the property that justifies each statement.

37 $(-5) + 8 = 8 + (-5)$

38 $(-17) \times 2 = 2 \times (-17)$

39 $(-13) \times (-5) = (-5) \times (-13)$

40 $(-7) + (-3) = -10$

41 $(-7) \times (-3) = 21$

42 $9 + (-9) = 0$

43 $[(-5) \times (-4)] \times 6 = (-5) \times [(-4) \times 6]$

44 $(-5) + [(-2) + 3] = [(-5) + (-2)] + 3$

45 $(-16) \times [4 + (-3)] = [(-16) \times 4] + [(-16) \times (-3)]$

46 $(-19) + 0 = -19$

47 $(-13) \times 1 = -13$

48 $(-17) \times \dfrac{1}{-17} = 1$

49 If $(-5) \cdot x = 0$, then $x = 0$.

50 $a \cdot (b - c) = a \cdot b - a \cdot c$

51 If $a = b$, then $3a = 3b$.

52 If $a + b = a + c$, then $b = c$.

53 If $x = 4$, then $x + 7 = 4 + 7$.

54 $x \cdot \left(\dfrac{1}{x}\right) = 1, x \neq 0$

55 $0 \cdot a = 0$

56 $a + 0 = a$

1.8 Symbols of Inclusion

We have seen in Section 1.1 that the value of an expression involving addition, subtraction, multiplication, or division may depend on the order in which the operations are performed. Thus, by using the rules of order of operations we simplify the following expression: $7 \times 3 + 5 = 21 + 5 = 26$. To make the meaning of this expression clear and avoid any ambiguity, we use a pair of parentheses as a *grouping symbol* or as *symbols of inclusion*. Thus,

$$(7 \times 3) + 5 = 21 + 5 = 26$$

The symbols, ( ) and [ ], used earlier in the chapter to group numbers, are called parentheses and brackets, respectively. An additional symbol, which we denote by { }, is called braces and will be used in this section.

In a fraction such as $\dfrac{19 - 5}{7 - 3}$, the horizontal-bar symbol is often used as a grouping symbol, as well as a division sign, since it groups $19 - 5$ as well as $7 - 3$. The bar indicates that the number $(19 - 5)$ is to be divided by the number $(7 - 3)$. Thus,

$$\frac{19 - 5}{7 - 3} = \frac{14}{4} = \frac{7}{2}$$

EXAMPLES

Simplify each of the following expressions.

1 $3[40 - (5 + 3)]$

SOLUTION

$$\begin{aligned} 3[40 - (5 + 3)] &= 3[40 - 8] \qquad \text{(Why?)} \\ &= 3[32] \\ &= 96 \end{aligned}$$

2 $4[(7 + 2) \times (9 \div 3)]$

SOLUTION

$$\begin{aligned} 4[(7 + 2) \times (9 \div 3)] &= 4[9 \times 3] \qquad \text{(Why?)} \\ &= 4[27] \\ &= 108 \end{aligned}$$

3 $4 + 2\{3 + 2[5 + 6(3 - 1)] + 4\}$

SOLUTION

$$\begin{aligned} &4 + 2\{3 + 2[5 + 6(3 - 1)] + 4\} \\ &\quad = 4 + 2\{3 + 2[5 + 12] + 4\} \qquad \text{(Why?)} \\ &\quad = 4 + 2\{3 + 10 + 24 + 4\} \\ &\quad = 4 + 2\{41\} \\ &\quad = 4 + 82 \\ &\quad = 86 \end{aligned}$$

4 $5\{3 - 5[18 - 2(6 + 5) - (-3 + 7)] + 4\}$

SOLUTION

$$\begin{aligned} &5\{3 - 5[18 - 2(6 + 5) - (-3 + 7)] + 4\} \\ &\quad = 5\{3 - 5[18 - 12 - 10 + 3 - 7] + 4\} \\ &\quad = 5\{3 - 5[-8] + 4\} = 5\{3 + 40 + 4\} \\ &\quad = 5\{47\} = 235 \end{aligned}$$

PROBLEM SET 1.8

In problems 1–26, simplify each expression by removing all grouping symbols.

1 $[40 + (-6)] + [(-13) + 2]$

2 $3 + [(-7) + 3(2 + 5)]$

3 $[(-17) + 8] - (7 - 9)$

4 $-[(-8) - 15] - [(-5) - 12]$

5 $3 + 2\{(5 - 3) + 2(1 + 3)\}$

6 $(7 - 19) - \{3 - (15 - 6) - 2(4 - 11)\}$

7 $7[2 + \{8 + 3(5 - 2)\}]$

8 $-(12 - 25) + [4 - 3\{18 - (7 - 14)\} + 23]$

9 $3 + \{4(2 - 3) - 6[2 + 3(5 - 3)]\}$

10 $13 - ((-17) + 5) - [(-8) - (27 - 38)]$

11 $4[3\{10 + 2(5 - 2) + 5\} + 2]$

12 $16[(16 - 6) + 2\{24 - (20 - 3)\} + 5]$

13 $8[5 + 2\{6 + (16 - 6) + 3\} + 5]$

14 $14 - \{11 - [(5 - 18) - (-10 + 7)]\}$

15 $-[(23 - 31) - 2] - \{(-8) - [15 - (9 - 21) - (27 - 32)]\}$

16 $-[(36 - 53) - 67] - \{(-25) - [8 - (-7 - 15) - (62 - 75)]\}$

17 $-(17 - 13) - [8 - \{35 - (36 - 10)\} + 7]$

18 $-3[2 - (5 - 2) + 7\{3 - 2(6 - 3)\} + 81]$

19 $3 - [5 + 2(1 + 3) + 1] + 2$

20 $5 + [2 - \{3 - 2(5 + 2) + 4\} + 8]$

21 $7 - \{5 - 3(3 + 4) - 6\} + 13$

22 $6 - 2[3 - 2\{1 - 2(2 + 7) + 2\} - 7]$

23 $5 - 7[2 - 3\{3 - 2(3 - 2) + 5\} - 2]$

24 $3(5 + 6) - 4[3 - 5(2 + 4)]$

25 $8 - 7\{(3 - 2) + 3 - 2\} + 8$

26 $14 - \{2 - [3 + 11 - (7 - 3 + 5)] + 21\}$

REVIEW PROBLEM SET

In problems 1–8, use the order of operations to find the value of each expression.

1 $7 + 6 \times 2$

2 $6 \times 3 - 2 \times 5$

3 $5(12 - 6) + 13$

4 $3(2 + 8) - 3 \times 4$

5 $\frac{8 - 5}{3} + 3 \times 7$

6 $64 \div 8(4 + 3)$

7 $\dfrac{8 + 6\left(\frac{5 + 3}{8 - 4}\right) - 3}{8 - 4\left(\frac{6 - 3}{4 - 1}\right) + 6}$

8 $\dfrac{16 - 2 \times (4 + 6) \div 5 \times 2 + 3}{13 - 3 \times 2 + 8}$

In problems 9–14, find the value of each expression for the indicated value of each variable.

9 $5x + 13;\ x = 2$

10 $7x + 21;\ x = 1$

11 $5x + 7y;\ x = 3,\ y = 4$

12 $25x - 5y;\ x = 2,\ y = 3$

13 $5xy + 1;\ x = 7,\ y = 2$

14 $3xy + 6x;\ x = 5,\ y = 4$

In problems 15–18, use the variable x to write expressions for each statement.

15 The number that is 3 more than twice some number. x^2+3

16 The perimeter of a square whose side is a given number of units.

17 The number obtained by adding 10 to some number.

18 The number of inches in a given number of feet.

In problems 19–24, determine whether the statements are true or false if the sets A, B, C, and D are given by $A = \{0,1,2,3\}$, $B = \{0,1,2\}$, $C = \{1,2,3\}$, and $D = \{1,2,3,4\}$.

19 $B \subseteq A$

20 $C \subseteq B$

21 $B \subseteq D$

22 $C \subseteq D$

23 B has eight subsets

24 $\emptyset \subseteq B$

In problems 25–28, use set-builder notation to describe each set.

25 $A = \{0,2,4,6,8\}$

26 $B = \{17,18,19,20,21\}$

27 $C = \{0,5,10,15,20,25,30\}$

28 $D = \{1,3,5,7,9,11,13,15\}$

In problems 29–38, let $A = \{1,2,3\}$, $B = \{2,3,4,5\}$, and $C = \{2,5,7,8\}$. Find each set.

29 $A \cup B$

30 $(B \cup A) \cup C$

31 $A \cap B$

32 $B \cap C$

33 $A \cap C$

34 $A \cup C$

35 $B \cup C$

36 $(C \cup B) \cap A$

37 $A \cup \emptyset$

38 $B \cap \emptyset$

In problems 39 and 40, write the first five members of each set.

39 $A = \{2x | x \text{ is a positive integer}\}$

40 $B = \{2x + 1 | x \text{ is a positive integer}\}$

In problems 41–46, graph each set on the number line.

41 $A = \{0,1,4,7,10\}$

42 $B = \{-4,-2,0,2,4\}$

43 $C = \{-\frac{4}{7},-\frac{2}{7},0,\frac{2}{7},\frac{4}{7}\}$

44 $D = \{x|x \text{ is an odd positive integer less than } 9\}$

45 $E = \{-3,-1,1,4,6\}$

46 $F = \{0,\frac{1}{6},\frac{5}{6},1,\frac{7}{6}\}$

In problems 47–54, express each rational number in decimal representation.

47 $\frac{11}{40}$

48 $-\frac{5}{11}$

49 $-\frac{17}{25}$

50 $\frac{13}{40}$

51 $\frac{7}{9}$

52 $-\frac{2}{7}$

53 $\frac{5}{14}$

54 $-\frac{6}{13}$

In problems 55–60, express each decimal as a rational number in the form $\frac{a}{b}$, where $b \neq 0$.

55 0.97

56 0.113

57 -0.75

58 0.15471

59 0.1111

60 -0.6752

In problems 61–92, perform the indicated operations.

61 $7 + (-3)$

62 $(-8) + (-11)$

63 $41 + (-25)$

64 $341 + 788$

65 $(-475) + (-807)$

66 $(-185) + (47)$

67 $500 - 302$

68 $(-99) - (-67)$

69 $(-16) - (-545)$

70 $(-312) - (-505)$

71 $0 - (-81)$

72 $(-2{,}001) - (-999)$

73 $2 \times (-73)$

74 $(-3) \times 18$

75 $(-14) \times (-2)$

76 $(-173) \times (207)$

77 $(-95) \times (-105)$

78 $0 \times (4{,}632)$

79 $(-591) \times (-3{,}002)$

80 $(-14) \times (-1{,}717)$

81 $(-16) \div 8$

82 $(-279) \div (-9)$

83 $(-256) \div (-16)$

84 $0 \div (-117)$

85 $2{,}632 \div (-47)$

86 $(-58{,}183) \div (-83)$

87 $\frac{-32}{-8}$

88 $\frac{0}{17}$

89 $\frac{-56}{7}$

90 $\frac{-512}{32}$

91 $\frac{-196}{-14}$

92 $\frac{0}{-8}$

In problems 93–102, simplify each expression.

93 $|-13|$

94 $|-(-81)|$

95 $|-4| \cdot |-3|$

96 $|7 - 2| + |-3|$

97 $|-31| + |-79|$

98 $|-5| \cdot |7 - 3|$

99 $|5 - 2| + |8 - 3|$

100 $|-6| \cdot |10 - 4|$

101 $|10 - 1| + |10 + 1|$

102 $|14 - 5| - |8 - 5|$

In problems 103–120, justify the statements by giving the appropriate property. Assume that all variables represent real numbers.

103 $6 + (-4) = (-4) + 6$

104 $[(-10) + 2] + 7 = (-10) + (2 + 7)$

105 $(-13) \times (-2) = (-2) \times (-13)$

106 $(7 \times 8) \times (-5) = 7 \times [8 \times (-5)]$

107 $6 + (-6) = 0$

108 $x \cdot 1 = x$

109 If $5a = 0$, then $a = 0$.

110 $2(xy) = (2x)y$

111 $(-2) \times (-3) = 2 \times 3$

112 If $x = y$, then $2x = 2y$.

113 If $x = 2$, then $x + 3 = 2 + 3$.

114 $a \cdot \left(\frac{1}{a}\right) = 1, a \neq 0$

115 $(-6) \times (-7) = 6 \times 7$

116 $7(xy) = (7x)y$

117 If $x + z = y + z$, then $x = y$.

118 $a(3b + 2c) = a(3b) + a(2c)$

119 If $x + y = 5$, then $4(x + y) = 4 \times 5$.

120 $(-5)x = -(5x)$

In problems 121–124, simplify each expression by removing all grouping symbols.

121 $-3[2 - 5\{3 - (7 + 8) + 1\} + 2]$

122 $-2\{5 - [6 - (-3 - 4)]\} + 71$

123 $3 - \{2 - 2[1 - (2 - 5)]\}$

124 $-3\{1 - [1 - (2 - 7)]\}$

CHAPTER 2

Algebraic Operations of Polynomials

2 ALGEBRAIC OPERATIONS OF POLYNOMIALS

Introduction

In Chapter 1 we discussed the basic properties of numbers and their use in the four fundamental arithmetic operations. Here we shall apply these properties to special algebraic expressions called *polynomials*. Such expressions consist of any combination of variables or numbers or both, with appropriate operations. With this information, we shall learn how to add, subtract, multiply, and divide polynomials. First, it would be helpful to consider positive integral exponents.

2.1 Positive Integral Exponents

If a given product consists of two or more identical numbers, an abbreviation can be used to represent this repetition of numbers, and each number is called a *factor* of the product. For example, to represent the expression $2 \times 2 \times 2 \times 2$, we can write the abbreviated form 2^4, where the raised number 4 indicates that the number 2 is used as a repeated factor 4 times. Similarly, we can write 3^2, read "the second power of 3" or "3 squared" as an abbreviated form of 3×3. Also, x^3, read "the third power of x" or "x cubed," can be used to represent $x \cdot x \cdot x$, and y^5, read "the fifth power of y," represents $y \cdot y \cdot y \cdot y \cdot y$. Generalizing, we have the following definition:

DEFINITION 1 POSITIVE INTEGRAL EXPONENTS

The notation x^n represents a product, wherein x occurs as a factor n times. That is, if x is any real number, then

$$x^n = \underbrace{x \cdot x \cdot x \ldots x}_{n \text{ factors}}, \qquad \text{for } n \text{ a positive integer}$$

In the notation x^n, n is called the *exponent*, x is called the *base*, and x^n is called the nth *power of* x.

Using Definition 1, we have

$$\begin{aligned} x^4 &= x \cdot x \cdot x \cdot x \\ 7^3 &= 7 \times 7 \times 7 \\ (-3)^6 &= (-3) \times (-3) \times (-3) \times (-3) \times (-3) \times (-3) \end{aligned}$$

Notice that parentheses are used in the expression $(-3)^6$ to indicate that -3 is the base. However, when no parentheses are used, the expression -3^6 is written as $-3^6 = -(3 \times 3 \times 3 \times 3 \times 3 \times 3)$.

An exponent applies only to the base to which it is attached. For example,

$$2x^3 = 2 \cdot x \cdot x \cdot x \qquad \text{and} \qquad x^3y^4 = x \cdot x \cdot x \cdot y \cdot y \cdot y \cdot y$$

In the preceding definition of x^n, if $n = 1$, we have $x^1 = x$. Thus, when no exponent is shown, it is assumed to be the number 1. For example, $2 = 2^1$, $-5 = (-5)^1$, and $x^3y = x^3y^1$.

EXAMPLES

1 Rewrite the following expressions, using exponents.

a) $5 \times 5 \times 5 \times 5 \times 5$
b) $(-4) \times (-4) \times (-4)$
c) $x \cdot x \cdot x \cdot y \cdot y$

SOLUTION

a) $5 \times 5 \times 5 \times 5 \times 5 = 5^5$
b) $(-4) \times (-4) \times (-4) = (-4)^3$
c) $x \cdot x \cdot x \cdot y \cdot y = x^3y^2$

2 Find the value of each of the following expressions. Also, indicate the base and the exponent.

a) 4^3 b) $(-2)^5$ c) -3^2 d) $-(-3)^3$

SOLUTION. Using Definition 1, we have

a) $4^3 = 4 \times 4 \times 4 = 64$
The base is 4; the exponent is 3.

b) $(-2)^5 = (-2) \times (-2) \times (-2) \times (-2) \times (-2) = -32$
The base is -2; the exponent is 5.

c) $-3^2 = -(3 \times 3) = -(9) = -9$
The base is 3; the exponent is 2.

d) $-(-3)^3 = -[(-3) \times (-3) \times (-3)]$
$= -(-27) = 27$
The base is -3; the exponent is 3.

3 Evaluate $3^4 + 4(3)$.

SOLUTION

$$\begin{aligned} 3^4 + 4(3) &= (3 \times 3 \times 3 \times 3) + (3 + 3 + 3 + 3) \\ &= 81 + 12 = 93 \end{aligned}$$

To find the product of the expressions 2^3 and 2^2, we make use of Definition 1, so that

$$\begin{aligned} 2^3 \cdot 2^2 &= (2 \times 2 \times 2) \cdot (2 \times 2) \\ &= 2 \times 2 \times 2 \times 2 \times 2 = 2^5 \end{aligned}$$

Observe that the same result could have been obtained by adding the exponents of the factors 2^3 and 2^2. That is, $2^3 \cdot 2^2 = 2^{3+2} = 2^5$. The exponent 5 in the result is the sum of the exponents 3 and 2.

In a similar manner, the product $x^3 \cdot x^4$ can be found by adding the exponents 3 and 4, so that

$$x^3 \cdot x^4 = x^{3+4} = x^7$$

In general, we have the following property:

1 PROPERTY OF MULTIPLICATION OF LIKE BASES

Let a be a real number and m and n positive integers; then

$$a^m \cdot a^n = a^{m+n}$$

To verify this property, we have

$$\begin{aligned} a^m \cdot a^n &= \underbrace{(a \cdot a \cdot a \cdot a \ldots a)}_{m \text{ factors}}\underbrace{(a \cdot a \cdot a \ldots a)}_{n \text{ factors}} \qquad \text{(Definition)} \\ &= \underbrace{a \cdot a \cdot a \cdot a \ldots a}_{m+n \text{ factors}} \\ &= a^{m+n} \end{aligned}$$

EXAMPLE

Use Property 1 to find the products of the following expressions. Write each answer in exponential form.

a) $3^4 \cdot 3^7$ b) $(-3)^6 \cdot (-3)^{12}$ c) $x^3 \cdot x$ d) $(-a)^3 \cdot (-a)^4$

SOLUTION. Applying Property 1, we have

a) $3^4 \cdot 3^7 = 3^{4+7} = 3^{11}$

b) $$\begin{aligned} (-3)^6 \cdot (-3)^{12} &= (-3)^{6+12} \\ &= (-3)^{18} \\ &= 3^{18} \qquad \text{(Why?)} \end{aligned}$$

c) $x^3 \cdot x = x^3 \cdot x^1 = x^{3+1} = x^4$

d) $$\begin{aligned} (-a)^3 \cdot (-a)^4 &= (-a)^{3+4} \\ &= (-a)^7 \\ &= -a^7 \qquad \text{(Why?)} \end{aligned}$$

Next, we shall consider a procedure of simplifying expressions such as $(5^4)^3$. The exponent 3 tells us that the base 5^4 is a factor three times, so $(5^4)^3 = 5^4 \cdot 5^4 \cdot 5^4$. We know that $5^4 = 5 \times 5 \times 5 \times 5$, and so we can write

$$\begin{aligned} (5^4)^3 &= (5 \times 5 \times 5 \times 5)(5 \times 5 \times 5 \times 5)(5 \times 5 \times 5 \times 5) \\ &= 5^{12} \end{aligned}$$

Since $3 \times 4 = 12$, we have

$$(5^4)^3 = 5^{4\times 3} = 5^{12}$$

Generalizing, we have the following property:

2 PROPERTY OF THE POWER OF A POWER

Let a be a real number and m and n positive integers; then

$$(a^m)^n = a^{mn}$$

The verification of this property is straightforward and will be left as an exercise (see problem 71).

EXAMPLE

Use Property 2 to simplify the following expressions. Express the results in exponential form.

a) $(4^2)^6$ b) $[(-2)^3]^5$ c) $(x^{10})^8$ d) $[(-a)^2]^3$

SOLUTION. Applying Property 2, we have

a) $(4^2)^6 = 4^{2\times 6} = 4^{12}$
b) $[(-2)^3]^5 = (-2)^{3\times 5} = (-2)^{15} = -2^{15}$
c) $(x^{10})^8 = x^{10\times 8} = x^{80}$
d) $[(-a)^2]^3 = (-a)^{2\times 3} = (-a)^6 = a^6$

Now, we can use the properties of numbers in Section 1.7 to develop an additional rule for exponents. Consider the expression $(3y)^4$. By Definition 1 of exponents, we have

$$\begin{aligned}(3y)^4 &= (3y)(3y)(3y)(3y)\\ &= 3\times 3\times 3\times 3\cdot y\cdot y\cdot y\cdot y \qquad \text{(Why?)}\\ &= 3^4\cdot y^4\end{aligned}$$

so that

$$(3y)^4 = 3^4\cdot y^4$$

Note that the same result could be obtained by simply raising each factor of the base to the indicated power. In general, we have the following property:

3 PROPERTY OF THE POWER OF A PRODUCT

Let a and b be real numbers and n a positive integer; then

$$(ab)^n = a^nb^n$$

This property can be verified in a similar manner as Property 1 (see problem 72).

EXAMPLES

1 Use Property 3 to simplify the following expressions. Express the results in exponential form.

a) $(5x)^2$ b) $(xy)^3$ c) $(-x)^5$ d) $(xyz)^4$

SOLUTION. Applying Property 3, we have

a) $(5x)^2 = 5^2x^2 = 25x^2$

b) $(xy)^3 = x^3y^3$

c) $(-x)^5 = [(-1)x]^5$
$= (-1)^5x^5$
$= (-1)x^5$
$= -x^5$

d) $(xyz)^4 = [x(yz)]^4$
$= x^4(yz)^4$
$= x^4y^4z^4$

2 Use Property 1, 2, or 3 to rewrite the following expressions in a simpler form.

a) $(2x^3)^4$ b) $(2x^2y^3)^3$ c) $(-3xy^2)^2(2x^3y)^3$

SOLUTION

a) $(2x^3)^4 = 2^4(x^3)^4$ (Property 3)
$= 16x^{12}$ (Property 2)

b) $(2x^2y^3)^3 = 2^3(x^2)^3(y^3)^3$ (Property 3)
$= 8x^6y^9$ (Property 2)

c) $(-3xy^2)^2(2x^3y)^3 = (-3)^2 \cdot x^2 \cdot (y^2)^2 \cdot 2^3(x^3)^3 \cdot y^3$ (Property 3)
$= 9x^2y^4 \cdot 8x^9y^3$ (Property 2)
$= 9 \cdot 8 \cdot x^{2+9}y^{4+3}$ (Property 1)
$= 72x^{11}y^7$

The preceding properties of exponents are summarized as follows: If a and b are real numbers and m and n are positive integers, then:

1 *Multiplication-of-Like-Bases Property:* $a^m \cdot a^n = a^{m+n}$.
2 *Power-of-a-Power Property:* $(a^m)^n = a^{mn}$.
3 *Power-of-a-Product Property:* $(ab)^n = a^nb^n$.

PROBLEM SET 2.1

In problems 1–8, identify the base and the exponent in each expression.

1 3^{16}

2 -7^4

3 $(-4)^5$

4 $-(-11)^{13}$

5 $(-1)^{25}$

6 x^{17}

7 $(2x)^5$

8 $(-3y)^{31}$

In problems 9–18, find the value of each expression.

9 3^4

10 $(-4)^3$

11 $-(-5)^3$

12 $(-1)^{11}$

13 $(-1)^4 + 3^2$

14 $2^5 + 5(2)$

15 $(-3)^2 - 3^2$

16 $2^3 + (-2)^3$

17 $2^2 - 2(-1)(3) + 3^2$

18 $(-3)^2 + 2(-3)(-2) + (-2)^2$

In problems 19–28, write each expression in exponential form.

19 $5 \times 5 \times 5 \times 5 \times 5 \times 5$

20 $(-7) \times (-7) \times (-7) \times (-7)$

21 $(-1) \times (-1) \times (-1) \times (-1) \times (-1)$

22 $(-2) \times (-2) \times (-2) \times (3) \times (3)$

23 $3 \times 3 + (-6) \times (-6) \times (-6)$

24 $7 \times 5 \times 5 + 3 \times (-2) \times (-2)$

25 $a \cdot a \cdot a \cdot a \cdot a \cdot a \cdot a \cdot a$

26 $a \cdot a \cdot b \cdot b \cdot b \cdot b \cdot b$

27 $-x \cdot x \cdot x \cdot x \cdot x \cdot x \cdot y \cdot y \cdot y \cdot y$

28 $3 \cdot x \cdot y \cdot z \cdot x \cdot y \cdot z \cdot z$

In problems 29–40, use Property 1 to find the products of the expressions. Write each answer in exponential form.

29 $2^3 \cdot 2^2$

30 $5^2 \cdot 5^4$

31 $7^5 \cdot 7$

32 $3^4 \cdot 3^5$

33 $(-2)^3(-2)^5$

34 $(-5)^4(-5)^6$

35 $(-4)^2(-4)^3$

36 $(-6)^3(-6)^2$

37 x^4x^8

38 y^5y^{13}

39 $(-b)^2(-b)$

40 $(-a)^2(-a)^3$

In problems 41–50, use Property 2 to simplify the expressions. Express the results in exponential form.

41 $(3^2)^3$

42 $(4^3)^8$

43 $[(-5)^3]^4$

44 $[(-3)^4]^{15}$

45 $(x^4)^2$

46 $[(-1)^3]^5$

47 $(z^9)^5$

48 $[(-b)^2]^7$

49 $[(x^2)^3]^4$

50 $(x^{3n})^2$, where n is a positive integer

In problems 51–60, use Property 3 to simplify the expressions. Express the results in exponential form.

51 $(2x)^3$

52 $(-3x)^2$

53 $(uv)^2$

54 $(-yz)^3$

55 $(-5x)^4$

56 $[-5(-y)]^4$

57 $[-3(-x)]^3$

58 $(2tuv)^5$

59 $[-2(-r)(-s)]^4$

60 $[(-x)(-y)]^6$

In problems 61–70, use Property 1, 2, or 3 to write the expressions in simpler form.

61 $(x^2y^4)^3$

62 $(-3x^4)^5$

63 $(-5x^3)^4$

64 $(7ab^2)^3$

65 $(4x^2y^3)^2(-xy^2)^3$

66 $(-x^2y^3z^4)^3(-x^2y^3z^4)^2$

67 $(2x^3y^5)^4(x^3y^7a)^2$

68 $(4a^2b^3)^2(ac^2)^3$

69 $(-3rs)^2(r^4s^3)^3$

70 $(-x^4y^5)(x^2y^3)^2$

71 Verify Property 2.

72 Verify Property 3.

2.2 Polynomials

Algebraic expressions such as 7, $3x$, $2x + 1$, $\frac{3x - 5}{7}$, $y^2 + 3y - 4$, and $z^4 - 3z^3 - z^2 + z + 1$ are called polynomials. In general, a *polynomial* in one variable is an algebraic expression in which the exponents of the variables are nonnegative integers and all the variables are in the numerator of the expression. Algebraic expressions such as $4x^{1/2} + 3x^{1/4} + 1$ and $y^3 - 5y - \frac{1}{y^2}$ are not polynomials, since the exponents of the first expression are not integers and the variable y^2 appears in the denominator of the second expression.

EXAMPLE

Which of the following expressions is a polynomial in one variable?

a) $3x^3 - 5x^2 + 7x + 2$
b) $t^4 - \frac{2}{3}t^3 + 6t^2 - \frac{1}{2}t + 4$
c) $2x^3 - \dfrac{3}{x^2} + \dfrac{1}{x} - 8$

SOLUTION

a) The expression $3x^3 - 5x^2 + 7x + 2$ is a polynomial in one variable.
b) The expression $t^4 - \frac{2}{3}t^3 + 6t^2 - \frac{1}{2}t + 4$ is a polynomial in one variable.
c) The expression $2x^3 - \dfrac{3}{x^2} + \dfrac{1}{x} - 8$ is not a polynomial in one variable, since the variables x^2 and x appear in the denominator.

Consider the algebraic expression $5x^3 - 7x^2 + 8x + \dfrac{13}{x}$. The quantities $5x^3$, $7x^2$, $8x$, and $\dfrac{13}{x}$, separated by the signs + and −, are called the *terms* of the expression. Polynomials containing one, two, or three terms are given special names. That is, a polynomial with exactly one term is called a *monomial*, a polynomial with exactly two terms is called a *binomial*, and a polynomial with exactly three terms is called a *trinomial*. For example, $-\frac{3}{2}x^2$ is a monomial, $1 - 3y$ is a binomial, and $x^2 - 3x + 2$ is a trinomial.

EXAMPLE

Give the terms of the following polynomials and identify each as a monomial, binomial, or trinomial.

a) $3x^2 + 2x$ b) $-7x$ c) $4y^3 - 2y + 1$

SOLUTION

a) The terms are $3x^2$ and $2x$. Since there are two terms, $3x^2 + 2x$ is a binomial.
b) The only term is $-7x$, so $-7x$ is a monomial.
c) The terms are $4y^3$, $-2y$, and 1. Since there are three terms, $4y^3 - 2y + 1$ is a trinomial.

The *degree* of a term with one variable is the exponent on the variable. The *degree* of a polynomial in one variable is the highest exponent in any nonzero term of the polynomial. For example, the degree of $7x - 1$ is 1, the degree of $-3y^3 + 2y^2 + 6$ is 3, and the degree of $z^7 - 3z^4 + 2z + 10$ is 7. A nonzero real number, such as 5, is called a *polynomial of degree zero*, while the number zero is not assigned a degree. Instead, it is called the *zero polynomial*. Any factor (or factors) of a product is called the *coefficient* of the others. Thus, in the product xy, x is the coefficient of y and y is the coefficient of x.

The numerical factor of any term of a polynomial is called its *numerical coefficient*. For example, the numerical coefficients of $2x^2 - 3x + 6$ are 2, -3, and 6. If any term of a polynomial does not contain a numerical factor, we assume that its numerical coefficient is 1, since for any real number a, $a = 1 \cdot a$. Thus, the numerical coefficients of $y^2 - y + 2$ are 1, -1, and 2, since $y^2 - y + 2 = 1 \cdot y^2 - 1 \cdot y + 2$.

EXAMPLE

Find the degree and the numerical coefficients of each of the following polynomials.

a) $6x + 2$
b) $y^2 + 3y - 4$
c) $-z^3 + 6z^2 - 5z + 7$
d) $4x^5 + x^4 - 3x^3 + 2x^2 - x + 7$

SOLUTION

a) The degree is 1 and the numerical coefficients are 6 and 2.
b) The degree is 2 and the numerical coefficients are 1, 3, and -4.
c) The degree is 3 and the numerical coefficients are -1, 6, -5, and 7.
d) The degree is 5 and the numerical coefficients are 4, 1, -3, 2, -1, and 7.

Evaluation of a Polynomial

In Section 1.1 we discussed the fact that in evaluating an algebraic expression, operations must be performed in a particular order. We indicated that multiplication and division must be performed before addition and subtraction. We must now modify our previous agreement to include the operation of raising to a power. For example, to find the value of the polynomial $z^2 - 3z + 5$ for $z = -3$, we use the following order of operations:

1 Replace z by -3 and raise the number -3 to a power so that $(-3)^2 - 3(-3) + 5 = 9 - 3(-3) + 5$.
2 Multiply powers of -3 by their numerical coefficients so that $9 - 3(-3) + 5 = 9 + 9 + 5$.
3 Perform the additions or subtractions so that $9 + 9 + 5 = 23$. Therefore, the value of the polynomial $z^2 - 3z + 5$ for $z = -3$ is 23.

EXAMPLES

1 Evaluate $5x + 2$ for

a) $x = -2$ b) $x = 0$ c) $x = 4$

SOLUTION

a) For $x = -2$, we have $5x + 2 = 5(-2) + 2 = -10 + 2 = -8$.
b) For $x = 0$, we have $5x + 2 = 5(0) + 2 = 0 + 2 = 2$.
c) For $x = 4$, we have $5x + 2 = 5(4) + 2 = 20 + 2 = 22$.

2 Evaluate each of the following polynomials for $y = -2$.

a) $2y + 3$
b) $y^2 + 3y - 5$
c) $-y^3 + 2y^2 + 3y - 6$

SOLUTION. Replacing y by -2 in parts (a), (b), and (c), we have

a) $2y + 3 = 2(-2) + 3 = -4 + 3 = -1$

b) $$\begin{aligned} y^2 + 3y - 5 &= (-2)^2 + 3(-2) - 5 \\ &= 4 - 6 - 5 \\ &= -7 \end{aligned}$$

c) $$\begin{aligned} -y^3 + 2y^2 + 3y - 6 &= -(-2)^3 + 2(-2)^2 + 3(-2) - 6 \\ &= -(-8) + 2(4) + 3(-2) - 6 \\ &= 8 + 8 - 6 - 6 \\ &= 4 \end{aligned}$$

3 Evaluate $2x^2y + 3xy^2 - 4$ for $x = -1$ and $y = 2$.

SOLUTION. Replacing x by -1 and y by 2, we have

$$\begin{aligned} 2x^2y + 3xy^2 - 4 &= 2(-1)^2(2) + 3(-1)(2)^2 - 4 \\ &= 2(1)(2) + 3(-1)(4) - 4 \\ &= 4 - 12 - 4 \\ &= -12 \end{aligned}$$

PROBLEM SET 2.2

In problems 1–10, determine if the given expression is a polynomial in one variable.

1 $x^2 - 2x + 4$

2 $y^3 + 3y^2 - 5$

3 $\frac{1}{5}x^2 - 2x + \frac{1}{3}$

4 $x^4 + 3x^2 + \frac{7}{3}x - 17$

5 $\frac{1}{2}y - \frac{2}{y}$

6 0

7 6

8 $3x^2 + \frac{7}{x}$

9 $\frac{3z + 2}{z^2}$

10 $4x^2 + 3xy - 7y^2$

In problems 11–24, identify the given polynomial as a monomial, binomial, trinomial, or none of these. Find the degree of the polynomial and list the numerical coefficients.

11 $x^2 - 5x + 6$

12 $3x^2 - 7$

13 $2x^3 + 6x^2 - 5x + 4$

14 $24z^3$

15 $-5x$

16 $4y^2 - 4y + 1$

17 $15 - 8z^5$

18 $-x^4 + 3x^3 + \frac{1}{2}x^2 - \frac{3}{4}x + \frac{2}{3}$

19 11

20 0

21 $-\frac{1}{2}x$

22 $\frac{x + 5}{3}$

23 $\frac{-7x + 3}{9}$

24 $\frac{1}{3}y^2$

In problems 25–34, evaluate the polynomials for $x = 2$.

25 $4x - 1$

26 $3x + 2$

27 $2x^2 - x + 4$

28 $3x^3 - 2x^2 + x - 4$

29 $x^3 - 2x + x^2 - 1$

30 $7x^3 + 3x^2 - 11$

31 $3x^4 - 15$

32 $-2x^4 - 3x^2 + 20$

33 $x^4 - 3x^2 + 17x + 23$

34 $x^5 - x^4 + x^3 - x^2 + x - 1$

In problems 35–40, evaluate the expression $3x^2 + xy - z^2y$ for the given values of x, y, and z.

35 $x = 1, y = -1, z = 2$

36 $x = 0, y = 1, z = 2$

37 $x = -1, y = 1, z = -1$

38 $x = 3, y = 1, z = 2$

39 $x = 2, y = -3, z = 1$

40 $x = -2, y = 3, z = 2$

2.3 Addition and Subtraction of Polynomials

Since polynomials are expressions that represent real numbers, then the addition properties of real numbers discussed in Chapter 1 also apply to polynomials. Assume that P, Q, and R represent polynomial expressions. Then:

1 COMMUTATIVE PROPERTY OF ADDITION

$$P + Q = Q + P$$

2 ASSOCIATIVE PROPERTY OF ADDITION

$$P + (Q + R) = (P + Q) + R$$

3 DISTRIBUTIVE PROPERTIES

a) $P(Q + R) = (PQ) + (PR)$
b) $(P + Q)R = (PR) + (QR)$

Addition of Polynomials

We can use the preceding properties to add polynomials. For example, suppose that we are to add the monomials $5x^2$ and $2x^2$. By the distributive property, we have $5x^2 + 2x^2 = (5 + 2)x^2 = 7x^2$. On the other hand, the sum of the monomials $2x$ and $3y$ can only be expressed as $2x + 3y$ or $3y + 2x$, since the distributive property does not apply in this case. The terms $5x^2$ and $2x^2$ are called similar terms. In general, *similar terms* or *like terms* are those whose variables are exactly alike but which may have different numerical coefficients, that is, those terms involving the same variables, with each variable having the same exponent. For example, the expressions $3x^2y$ and $2x^2y$ are similar terms and so are $-3xy^2$ and $5xy^2$. However, $2x^2$ and $3x$ are not similar terms, since the exponents of the variables are not alike. Also, $6x^3y$ and $-7xy^2$ are not similar terms.

It can be readily observed from the preceding example that in the addition of similar monomial terms we find the algebraic sum of the numerical coefficients and attach to that sum the variables common to each of the addends. This is further illustrated as follows:

EXAMPLE

Perform the addition of the following monomial expressions.

a) $7x + 8x$
b) $3x^2 + (-6x^2)$
c) $4y^3 + 9y^3 + (-2y^3)$
d) $12x^2y + 8x^2y + 3xy^2$

SOLUTION

a) $7x + 8x = (7 + 8)x$ (Distributive property)
$= 15x$

b) $3x^2 + (-6x^2) = [3 + (-6)]x^2$ (Distributive property)
$= -3x^2$

c) $4y^3 + 9y^3 + (-2y^3)$
$= [4 + 9 + (-2)]y^3$ (Distributive property)
$= 11y^3$

d) $12x^2y + 8x^2y + 3xy^2$
$= (12x^2y + 8x^2y) + 3xy^2$ (Associative property)
$= (12 + 8)x^2y + 3xy^2$ (Distributive property)
$= 20x^2y + 3xy^2$

Additions of polynomials of more than one term can be reduced to simply adding monomials. For example, the sum of polynomials

$$(5x^2 + 3x + 2) + (3x^2 + 2x + 7)$$

can be performed by arranging the terms into groups of similar terms and by using the commutative and associative properties. Thus,

$$(5x^2 + 3x + 2) + (3x^2 + 2x + 7)$$
$$= (5x^2 + 3x^2) + (3x + 2x) + (2 + 7)$$

Applying the distributive property to the right-hand side, we have

$$\begin{aligned}(5x^2 + 3x + 2) + (3x^2 + 2x + 7)& \\ &= (5 + 3)x^2 + (3 + 2)x + (2 + 7) \\ &= 8x^2 + 5x + 9\end{aligned}$$

Often the rearrangement and association of similar terms when adding polynomials is done by using the "vertical scheme." For example, the polynomials $5x^2 + 3x + 2$ and $3x^2 + 2x + 7$ can be added as follows:

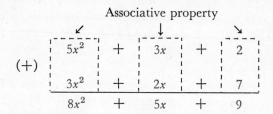

Associative property

$$\begin{array}{lccccc} & \swarrow & & \downarrow & & \searrow \\ & \boxed{5x^2} & + & \boxed{3x} & + & \boxed{2} \\ (+) & & & & & \\ & 3x^2 & + & 2x & + & 7 \\ \hline & 8x^2 & + & 5x & + & 9 \end{array}$$

It is important to note that the vertical arrangement implicitly applies the same properties that were used to add polynomials "horizontally."

EXAMPLES

Use the vertical scheme to perform the following additions.

1 $(4x^3 + 6x^2 - 3x - 7) + (5x - 2x^2 + x^3 - 2)$.

SOLUTION. First, arrange similar terms in columns:

$$\begin{array}{r} 4x^3 + 6x^2 - 3x - 7 \\ x^3 - 2x^2 + 5x - 2 \\ \hline \end{array}$$

Then, add those terms column by column, so that

$$\begin{array}{lr} & 4x^3 + 6x^2 - 3x - 7 \\ (+) & x^3 - 2x^2 + 5x - 2 \\ \hline & 5x^3 + 4x^2 + 2x - 9 \end{array}$$

2 $(-8x^5 + 5x^3 + 2x^2 + 4) + (10x^5 + 7x^4 - x^3 - 3x)$.

SOLUTION. Arranging and combining similar terms, we have

$$\begin{array}{lr} & -8x^5 + 0x^4 + 5x^3 + 2x^2 + 0x + 4 \\ (+) & 10x^5 + 7x^4 - x^3 + 0x^2 - 3x + 0 \\ \hline & 2x^5 + 7x^4 + 4x^3 + 2x^2 - 3x + 4 \end{array}$$

3 $(4x^2y - 3xy^2 + 2xy - 3) + (5xy^2 + xy - 6x^2y + 11)$.

SOLUTION. By arranging similar terms in columns and combining, we have

$$\begin{array}{rrrrr} & 4x^2y & -\ 3xy^2 & +\ 2xy & -\ 3 \\ (+) & -6x^2y & +\ 5xy^2 & +\ xy & +\ 11 \\ \hline & -2x^2y & +\ 2xy^2 & +\ 3xy & +\ 8 \end{array}$$

4 $(x^2 - 2x + 1) + (3x^2 + 4x - 3) + (5x^2 - 3x + 4)$.

SOLUTION. Arranging similar terms in columns and combining, we have

$$\begin{array}{rrrr} & x^2 & -\ 2x & +\ 1 \\ & 3x^2 & +\ 4x & -\ 3 \\ (+) & 5x^2 & -\ 3x & +\ 4 \\ \hline & 9x^2 & -\ x & +\ 2 \end{array}$$

Subtraction of Polynomials

To subtract polynomials, we make use of the definition of subtraction introduced in Chapter 1. That is, for any real numbers a and b, $a - b = a + (-b)$. Subtraction, like addition, can be performed only with similar terms. For example, to subtract the monomial $3x^2y$ from $9x^2y$, we have

$$\begin{aligned} 9x^2y - 3x^2y &= 9x^2y + (-3x^2y) \\ &= [9 + (-3)]x^2y \\ &= 6x^2y \end{aligned}$$

Note that the same result could be obtained by simply subtracting the numerical coefficients of the second term from the first, so that

$$\begin{aligned} 9x^2y - 3x^2y &= (9 - 3)x^2y \\ &= 6x^2y \end{aligned}$$

EXAMPLE

Perform the following subtractions of monomials.

a) $7x^3 - 2x^3$
b) $-14x^4y^2 - 8x^4y^2$
c) $12m^3n - (-3m^3n)$
d) $-3z^5 - (-6z^5)$

SOLUTION

a) $7x^3 - 2x^3 = (7 - 2)x^3 = 5x^3$
b) $-14x^4y^2 - 8x^4y^2 = (-14 - 8)x^4y^2 = -22x^4y^2$
c) $12m^3n - (-3m^3n) = [12 - (-3)]m^3n = 15m^3n$
d) $-3z^5 - (-6z^5) = [-3 - (-6)]z^5 = 3z^5$

Subtraction of polynomials of more than one term can be accomplished just as addition is. That is, we first group similar terms by applying the commutative and associative properties, and then perform the subtraction of the similar monomial terms as illustrated above. For example, to subtract $x^2 + 3x + 4$ from $3x^2 + 7x + 8$, we write

$$\begin{aligned}(3x^2 + 7x + 8) - (x^2 + 3x + 4) \\ &= (3x^2 - x^2) + (7x - 3x) + (8 - 4) \\ &= (3 - 1)x^2 + (7 - 3)x + (8 - 4) \\ &= 2x^2 + 4x + 4\end{aligned}$$

Using the vertical scheme, we arrange similar terms in columns, so that

$$\begin{array}{r} 3x^2 + 7x + 8 \\ \underline{x^2 + 3x + 4} \end{array}$$

Then we use the rules for subtracting signed numbers and combine similar terms:

$$\begin{array}{rr} & 3x^2 + 7x + 8 \\ (-) & \underline{x^2 + 3x + 4} \\ & 2x^2 + 4x + 4 \end{array}$$

EXAMPLES

In examples 1–4, use the vertical scheme to perform the subtraction.

1 Subtract $4x^2 - 3x + 5$ from $6x^2 - x - 3$.

SOLUTION. Arranging similar terms in columns and using the rules for subtracting signed numbers when combining similar terms, we have

$$\begin{array}{rr} & 6x^2 - x - 3 \\ (-) & \underline{4x^2 - 3x + 5} \\ & 2x^2 + 2x - 8 \end{array}$$

2 Subtract $2xy^2 - 3x^2y + 7$ from $-5x^2y + 6xy^2 - 11$.

SOLUTION. Arrange similar terms in columns, then proceed as in example 1.

$$\begin{array}{rr} & -5x^2y + 6xy^2 - 11 \\ (-) & -3x^2y + 2xy^2 + \ \ 7 \\ \hline & -2x^2y + 4xy^2 - 18 \end{array}$$

3 Subtract $3x^2 + x$ from $x^3 - 2x^2 - 3$.

SOLUTION

$$\begin{array}{rr} & x^3 - 2x^2 + 0x - 3 \\ (-) & 0x^3 + 3x^2 + \ \ x + 0 \\ \hline & x^3 - 5x^2 - \ \ x - 3 \end{array}$$

4 Subtract $5x^2 - 3x + 6$ from the sum of $2x^2 - 5x + 7$ and $x^2 + 3x - 10$.

SOLUTION. First, we perform the addition of $2x^2 - 5x + 7$ and $x^2 + 3x - 10$:

$$\begin{array}{rr} & 2x^2 - 5x + \ \ 7 \\ (+) & x^2 + 3x - 10 \\ \hline & 3x^2 - 2x - \ \ 3 \end{array}$$

Then, we subtract $5x^2 - 3x + 6$ from the resulting sum:

$$\begin{array}{rr} & 3x^2 - 2x - 3 \\ (-) & 5x^2 - 3x + 6 \\ \hline & -2x^2 + \ \ x - 9 \end{array}$$

PROBLEM SET 2.3

In problems 1–20, perform the additions of the given polynomial expressions.

1 $10x + 7x$

2 $12xy + 7xy$

3 $-12y + 20y$

4 $-11m^2 + (-9m^2)$

5 $5xy + (-3xy)$

6 $8uv + (-10uv)$

7 $4ab + 6ab + 3ab$

8 $3yz^2 + (-6yz^2) + (-7yz^2)$

9 $7x^2y^3 + (-4x^2y^3) + 2x^3y^3$

10 $13x^2y + 5xy^2 + (-5x^2y)$

11 $(3x + 8) + (5x + 3)$

12 $(4y - 7) + (-6y + 1)$

13 $(2x^2 + 5x - 1) + (x^2 - 2x + 3)$

14 $(3x^2 - 5x + 7) + (5 + 9x - 2x^2)$

15 $(4x^3 + 2x^2 - 3x + 2) + (x^3 - 5x^2 + 6x + 3)$

16 $(-x^5 + 3x^3 + x - 12) + (2x^5 + 3x^4 - 7x^2 + 3x + 8)$

17 $(5y^4 + 3y^2 - 6y^3 + 7) + (-2y^4 + y^3 - 3y^2 + 7y)$

18 $(14a^3b^2 - 12a^2b + 10a) + (-11a^3b^2 + 10a^2b - 8a)$

19 $(3x^2y^2 - 5xy^2 + 8x^2y) + (7xy^2 - 2x^2y^2 - 3x^2y)$

20 $(25y^4z - 15y^3z^2 + 8y^2z^3 - 7yz^4) + (-9y^4z + 10y^3z^2 + 6yz^4 - 8y^2z^3)$

In problems 21–40, perform the subtractions of the given polynomial expressions.

21 $7x - 5x$

22 $11y - 8y$

23 $19y - (-6y)$

24 $24a - (-16a)$

25 $-13z^3 - (-7z^3)$

26 $-18xy - (-18xy)$

27 $3uv - 11uv$

28 $4x^5b - 13x^5b$

29 $-12x^2y - (-27x^2y)$

30 $-9xy^3 - (-17xy^3)$

31 $(5y + 8) - (2y + 6)$

32 $(-11x + 10) - (3x - 4)$

33 $(3x^2 + 2x + 5) - (2x^2 + x + 1)$

34 $(4y^2 + 7y - 8) - (y^2 - 5y + 2)$

35 $(7x^2 - 5xy - 3y^2) - (4y^2 - 3xy + 2x^2)$

36 $(10m^3n - 8m^2n^2 - 9mn^3) - (-5m^3n + 4mn^3 - 8m^2n^2)$

37 $(x^3 - 2x^2 + 7x - 8) - (-2x^3 + 4x^2 + 5x - 6)$

38 $(3x^5 - 11x^3 + 5x) - (x^5 - 2x^4 + 6x^2 + x - 3)$

39 $(-x^4 + 5x^2 - 3x + 11) - (-4x^3 + 2x^2 - 3x + 9)$

40 $(3y^3z^2 + 5y^2z^3 - 8yz + 11) - (2y^2z^3 - 5y^3z^2 + 2yz)$

In problems 41–50, perform the indicated operations.

41 $(2x^2 - 3x + 11) + (x^2 + 5x - 2) + (3x^2 - 7x + 1)$

42 $(x^2 + xy - y^2) + (2x^2 - 3xy - y^2) + (3x^2 + 4xy - y^2)$

43 $(8y^3 - 7y + 6) + (3y^3 + 2y - 4) - (7y^3 + y + 5)$

44 $(14yz - 11y - 3z) + (-12yz + 8y - 2z) - (yz + y - 4z)$

45 $(x^3 + x^2 - x + 1) + (2x^3 - 3x^2 + 2x + 3) + (3x^3 + 5x^2 - 7x + 6)$

46 $(y^3 - 3y^2 + 6y - 8) + (-2y^3 + 7y - 2 + y^2) + (1 - 3y + 3y^2 - y^3)$

47 $(4z^4 - 7z^2 + 6) - (z^4 + z^2 - 2) - (2z^4 - 3z^2 + 5)$

48 $(11x^2y^3 - 8x^3y + 2xy^2) - (7x^3y - 3xy^2 + 8x^2y^3) - (xy^2 + x^2y^3 - 2x^3y)$

49 $(3x^5 + 4x^4 + 6x^3 - 5x^2 + x - 7) - (2x^4 + 3x^2 - 2) - (x^5 - x^3 - 3x)$

50 $(3y^2 + y - 2) - (y^2 - y + 3) + (2y^2 + 3y - 1) - (-y^2 - 2y + 3)$

2.4 Multiplication of Polynomials

The multiplication properties of real numbers that were discussed in Chapter 1 can be applied to polynomial expressions. Suppose that P, Q, and R are polynomial expressions. Then:

1 COMMUTATIVE PROPERTY OF MULTIPLICATION

$$PQ = QP$$

2 ASSOCIATIVE PROPERTY OF MULTIPLICATION

$$P(QR) = (PQ)R$$

Multiplication of Monomials

To multiply monomial expressions, we use the rules for exponents and the commutative and associative properties (page 81). For example, to multiply $5x^3$ by $4x^4$, we have

$$\begin{aligned}(5x^3)(4x^4) &= (5 \times 4)(x^3 \cdot x^4) \quad \text{(Commutative property)}\\ &= 20x^{3+4}\\ &= 20x^7\end{aligned}$$

In the preceding example, the product was essentially found by multiplying the numerical coefficients and applying the rule for multiplying like bases by adding exponents; that is, $a^m \cdot a^n = a^{m+n}$.

EXAMPLES

Find the following products of monomials.

1 $(9x^4)(4x)$

SOLUTION

$$\begin{aligned}(9x^4)(4x) &= (9 \times 4)(x^4 \cdot x)\\ &= 36x^{4+1}\\ &= 36x^5\end{aligned}$$

2 $(-2x^3y^4)(11xy^2)$

SOLUTION

$$\begin{aligned}(-2x^3y^4)(11xy^2) &= (-2 \times 11)(x^3 \cdot x)(y^4 \cdot y^2)\\ &= -22x^4y^6\end{aligned}$$

3 $(5y^3)(2y^4)(3y)$

SOLUTION

$$\begin{aligned}(5y^3)(2y^4)(3y) &= (5 \times 2 \times 3)(y^3 \cdot y^4 \cdot y)\\ &= 30y^8\end{aligned}$$

4 $(-xy^2)(-2x^2y)(-3y^3)$

SOLUTION

$$\begin{aligned}(-xy^2)(-2x^2y)(-3y^3) &= (-1xy^2)(-2x^2y)(-3y^3)\\ &= [(-1) \times (-2) \times (-3)](x \cdot x^2)(y^2 \cdot y \cdot y^3)\\ &= -6x^3y^6\end{aligned}$$

Multiplication of a Monomial and a Polynomial

To multiply a monomial by a polynomial, we use the distributive property applied to polynomials. The property states that if P, Q, and R are polynomial expressions, then

a) $P(Q + R) = PQ + PR$.
b) $(P + Q)R = PR + QR$.

For example, to find the product of $4x^3$ and $3x + 2$, we have

$$\begin{aligned} 4x^3(3x + 2) &= (4x^3)(3x) + (4x^3)(2) \\ &= (4 \times 3)(x^3 \cdot x) + (4 \times 2)x^3 \\ &= 12x^4 + 8x^3 \end{aligned}$$

The procedure is further illustrated by the following examples.

EXAMPLES

Find the following products.

1 $2x(3x^3 + 5x)$

SOLUTION

$$\begin{aligned} 2x(3x^3 + 5x) &= (2x)(3x^3) + (2x)(5x) \\ &= (2 \times 3)(x \cdot x^3) + (2 \times 5)(x \cdot x) \\ &= 6x^4 + 10x^2 \end{aligned}$$

2 $-4y^3(-3y^2 + y - 2)$

SOLUTION

$$\begin{aligned} -4y^3(-3y^2 + y - 2) &= (-4y^3)(-3y^2) + (-4y^3)(y) + (-4y^3)(-2) \\ &= 12y^5 - 4y^4 + 8y^3 \end{aligned}$$

3 $5x^2y(2xy^2 - 4x + 5y^2)$

SOLUTION

$$\begin{aligned} 5x^2y(2xy^2 - 4x + 5y^2) &= (5x^2y)(2xy^2) + (5x^2y)(-4x) + (5x^2y)(5y^2) \\ &= 10x^3y^3 - 20x^3y + 25x^2y^3 \end{aligned}$$

Multiplication of Polynomials

The distributive property can also be used to reduce the multiplication of a polynomial by another polynomial to a multiplication of monomials. For example, to multiply $x + 3$ by $x + 4$, we have

$$\begin{aligned}(x+3)(x+4) &= x(x+4) + 3(x+4) && \text{(Distributive property)}\\ &= x^2 + 4x + 3x + 12 && \text{(Distributive property)}\\ &= x^2 + 7x + 12\end{aligned}$$

EXAMPLES

Determine the following products.

1 $(2x + y)(x - 3y)$

SOLUTION

$$\begin{aligned}(2x+y)(x-3y) &= 2x(x-3y) + y(x-3y)\\ &= 2x^2 - 6xy + xy - 3y^2\\ &= 2x^2 - 5xy - 3y^2\end{aligned}$$

2 $(x + 1)(x^2 - 2x + 1)$

SOLUTION

$$\begin{aligned}(x+1)(x^2-2x+1) &= x(x^2-2x+1) + 1(x^2-2x+1)\\ &= x^3 - 2x^2 + x + x^2 - 2x + 1\\ &= x^3 - x^2 - x + 1\end{aligned}$$

3 $(x^2 - xy + y^2)(x - xy + y)$

SOLUTION

$$\begin{aligned}(x^2-xy+y^2)(x-xy+y) &= x^2(x-xy+y) + (-xy)(x-xy+y)\\ &\quad + y^2(x-xy+y)\\ &= x^3 - x^3y + x^2y - x^2y + x^2y^2 - xy^2\\ &\quad + xy^2 - xy^3 + y^3\\ &= x^3 - x^3y + x^2y^2 - xy^3 + y^3\end{aligned}$$

Since in many multiplications involving polynomials, the actual computations can become quite tedious, as in the example above, any device to help perform the computations and at the same time simplify the work is desirable. One such device, the vertical scheme, is illustrated by the following example. To find the product of $x + 2$ and $x^3 - x^2 + 3x - 4$, we arrange the polynomial in a vertical scheme:

$$\begin{array}{rrrrrl}
x^3 & - \;x^2 & + \;3x & -\;4 & & \\
x & +\;2 & & & & \\
\hline
x^4 & -\;x^3 & +\;3x^2 & -\;4x & & \longleftarrow [x(x^3 - x^2 + 3x - 4)] \\
 & 2x^3 & -\;2x^2 & +\;6x & -\;8 & \longleftarrow [2(x^3 - x^2 + 3x - 4)] \\
\hline
x^4 & +\;x^3 & +\;x^2 & +\;2x & -\;8 & \longleftarrow \text{product}
\end{array}$$

This shortcut involves arranging the partial products so that similar terms are in the same column ready for the final step of addition. As with the vertical scheme for addition, it is important to note that the "mechanics" of this method are based on the properties. The vertical arrangement facilitates the application of the properties.

EXAMPLES

Use the vertical scheme to find the following products.

1 $(3x - 4y)(2x + 3y)$

SOLUTION

$$\begin{array}{rrr}
3x & -\;4y & \\
2x & +\;3y & \\
\hline
6x^2 & -\;8xy & \\
 & 9xy & -\;12y^2 \\
\hline
6x^2 & +\;xy & -\;12y^2
\end{array}$$

2 $(x + 2y)(x^2 - xy + y^2)$

SOLUTION

$$\begin{array}{rrrr}
x^2 & -\;xy & +\;y^2 & \\
x & +\;2y & & \\
\hline
x^3 & -\;x^2y & +\;xy^2 & \\
 & 2x^2y & -\;2xy^2 & +\;2y^3 \\
\hline
x^3 & +\;x^2y & -\;xy^2 & +\;2y^3
\end{array}$$

3 $(x^2 - 2x + 1)(x^2 + 4x + 4)$

SOLUTION

$$\begin{array}{r} x^2 + 4x + 4 \\ x^2 - 2x + 1 \\ \hline x^4 + 4x^3 + 4x^2 \qquad\qquad \\ - 2x^3 - 8x^2 - 8x \qquad \\ x^2 + 4x + 4 \\ \hline x^4 + 2x^3 - 3x^2 - 4x + 4 \end{array}$$

PROBLEM SET 2.4

In problems 1–10, find the products of the monomials.

1 $(2x^5)(4x^2)$

2 $(10y^3)(11y)$

3 $(-3y^4)(7y^3)$

4 $(8z^8)(-5z^3)$

5 $(-5x^2y)(-3xy^4)$

6 $(-9yz)(-7y^2z^3)$

7 $(4x^3)(2x)(-5x^4)$

8 $(12y^2)(-2y^3)(3y)$

9 $(-xy^2)(-2x^3)(-7x^2y)$

10 $(7m^2n)(-3mn^3)(-5m^3n^2)$

In problems 11–20, find the products of the monomials and polynomials.

11 $x(5x + 2)$

12 $y(2y + 3)$

13 $2x(3x - 6)$

14 $-3z(4z - 5)$

15 $3x(x^2 - x + 5)$

16 $-yz(3y^2 - 7yz + 9z^2)$

17 $-4ab^2(-a^2b + 2ab - 3b^2)$

18 $5xz(4x^2 + 3y^2 - 8z^2)$

19 $-8xy(2x^2y - 3xy^2 + x^2y^3)$

20 $3x^3y^2(xy^2 - 2x^2y^3 + 4x^3y)$

In problems 21–40, find the products of the polynomials by use of the vertical scheme.

21 $(x + 1)(x + 5)$

22 $(-2y + 7)(3y + 5)$

23 $(3x - 5)(x + 3)$

24 $(2x + 3)(5x - 2)$

25 $(2x - 3)(3x + 2)$

26 $(5x - 7)(2x - 3)$

27 $(4x^2 + 1)(x^2 - 1)$

28 $(2y^2 + 5)(y^2 - 1)$

29 $(2x - 1)(x^2 - 3x + 4)$

30 $(3y + 1)(2y^3 - 5y + 7)$

31 $(x + 3y)(x^2 - 4xy + 5y^2)$

32 $(4x - 3y)(3x^2 - xy - 9y^2)$

33 $(x^2 + 2)(x^2 - 2x + 3)$

34 $(2y^2 - 1)(3y^2 + 3y + 5)$

35 $(3y^2 - 2z^2)(4y^2 - yz + 5z^2)$

36 $(5 - 3x^2)(7 - 2x - 4x^2)$

37 $(x^2 + 2x + 1)(x^2 - 4x + 3)$

38 $(2x^2 + 3xy - y^2)(3x^2 - xy + 2y^2)$

39 $(y^2 + 2xy + 3x^2)(x - xy + y)$

40 $(3x^2 + 4x - 1)(5x^3 - 2x^2 + 3x - 2)$

2.5 Multiplication of Binomials

In Section 2.4 we learned to multiply binomial expressions by use of the distributive properties or the vertical scheme. For example, to find the product $(x + 2)(x + 7)$ by applying the distributive properties, we have

$$\begin{aligned}(x + 2)(x + 7) &= x(x + 7) + 2(x + 7)\\ &= x^2 + 7x + 2x + 14\\ &= x^2 + 9x + 14\end{aligned}$$

This product can be accomplished by using a device that simplifies the computation much more quickly and easily without writing all the steps. The following diagram provides such a shortcut method:

First term: $(x+2)(x+7) = \boxed{x^2} + 9x + 14$

$(x)(x)$

Second term: $(x+2)(x+7) = x^2 + \boxed{9x} + 14$

$\oplus$ $(2)(x) + (7)(x)$

Third term: $(x+2)(x+7) = x^2 + 9x + \boxed{14}$

$(2) \times (7)$

It should be noted that the computations used by the scheme above prove to be convenient provided that the first and second terms of the two binomials are similar.

EXAMPLES

Find the following products by the method described above.

1 $(x+2)(2x+1)$

SOLUTION

First term: $(x+2)(2x+1) = 2x^2 + \_\_\_ + \_\_\_$

$(x)(2x)$

Second term: $(x+2)(2x+1) = 2x^2 + 5x + \_\_\_$

$\oplus$ $(1)(x) + 2(2x)$

Third term: $(x+2)(2x+1) = 2x^2 + 5x + 2$

$(2) \times (1)$

Therefore, $(x+2)(2x+1) = 2x^2 + 5x + 2$.

2 $(x-3)(x+4)$

SOLUTION

$(-3) \times (4)$

$(x)(x)$

$(x-3)(x+4) = x^2 + x - 12$

$\oplus$ $(4)(x) + (-3)(x)$

Therefore, $(x-3)(x+4) = x^2 + x - 12$.

3 $(3x + 2)(4x - 5)$

SOLUTION

$(2) \times (-5)$

$(3x)(4x)$

$$(3x + 2)(4x - 5) = 12x^2 - 7x - 10$$

$\oplus$ $(-5)(3x) + (2)(4x)$

4 $(2x + 3y)(3x - 5y)$

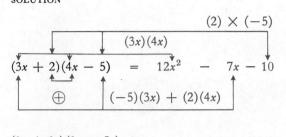

SOLUTION. We can mentally obtain this product, using the following scheme:

$$(2x + 3y)(3x - 5y) = 6x^2 - xy - 15y^2$$

First term: $(2x)(3x) = 6x^2$

Second term: $(2x)(-5y) + (3y)(3x) = -xy$

Third term: $(3y)(-5y) = -15y^2$

5 $(5x - 4y)(2x - y)$

SOLUTION

$$(5x - 4y)(2x - y) = 10x^2 - 13xy + 4y^2$$

First term: $(5x)(2x) = 10x^2$

Second term: $(5x)(-y) + (-4y)(2x) = -13xy$

Third term: $(-4y)(-y) = 4y^2$

The product of two binomials in all cases above may be found according to the following formulas, if desired:

(1) $$(x + p)(x + q) = x^2 + (p + q)x + pq$$

(2) $$(sx + p)(rx + q) = (sr)x^2 + (sq + pr)x + pq$$

The following examples will illustrate this procedure.

EXAMPLES

Use the preceding formulas to find the following products.

1 $(x + 3)(x + 7)$

SOLUTION. Letting $p = 3$ and $q = 7$ in the formula

$$(x + p)(x + q) = x^2 + (p + q)x + pq$$

we have

$$\begin{aligned}(x + 3)(x + 7) &= x^2 + (3 + 7)x + (3) \times (7) \\ &= x^2 + 10x + 21\end{aligned}$$

2 $(y - 2)(y - 3)$

SOLUTION. Letting $p = -2$ and $q = -3$ in formula (1) above, and replacing x by y, we have

$$(y + p)(y + q) = y^2 + (p + q)y + pq$$

so that

$$\begin{aligned}(y - 2)(y - 3) &= y^2 + [(-2) + (-3)]y + (-2) \times (-3) \\ &= y^2 - 5y + 6\end{aligned}$$

3 $(3x + 2)(4x + 5)$

SOLUTION. Letting $s = 3$, $p = 2$, $r = 4$, and $q = 5$ in the formula

$$(sx + p)(rx + q) = (sr)x^2 + (sq + pr)x + pq$$

we have

$$\begin{aligned}(3x + 2)(4x + 5) &= (3)(4)x^2 + [(3)(5) + (2)(4)]x + (2) \times (5) \\ &= 12x^2 + (15 + 8)x + 10 \\ &= 12x^2 + 23x + 10\end{aligned}$$

The procedure of finding the product of two binomials using formulas (1) and (2) will prove to be helpful in Chapter 3 when we develop a method for factoring second-degree polynomials.

PROBLEM SET 2.5

In problems 1–30, find the products of binomials by using the first method illustrated in this section.

1 $(x + 1)(x + 2)$

2 $(y + 2)(y + 6)$

3 $(x - 3)(x + 5)$

4 $(z + 7)(z - 9)$

5 $(y - 2)(y - 4)$

6 $(x - 10)(x - 11)$

7 $(x + 7)(x + 8)$

8 $(x + 5y)(x + 13y)$

9 $(z + 11)(z - 11)$

10 $(y - 7)(y + 9)$

11 $(x - 6)(x - 6)$

12 $(2x + 3)(2x + 3)$

13 $(2x + 5y)(x + 3y)$

14 $(3x - 7y)(2x - 9y)$

15 $(3x - 1)(x + 4)$

16 $(7x + 2)(3x - 1)$

17 $(4y + 7z)(3y - 2z)$

18 $(8y - 3z)(3y + 2z)$

19 $(x + 4y)(x + 5y)$

20 $(5x + 2)(3x + 10)$

21 $(3x - 2y)(2x - 3y)$

22 $(5x + 3)(5x + 3)$

23 $(3x + 2y)(2x - 3y)$

24 $(9x - 4)(9x + 4)$

25 $(3x - 2y)(2x + 3y)$

26 $(11y - 2z)(3y + z)$

27 $(6y - 7)(9y + 8)$

28 $(15x + 7)(4x + 2)$

29 $(10x - 3y)(5x - 2y)$

30 $(12x - 9y)(9x + 7y)$

In problems 31–40, use the formula $(x + p)(x + q) = x^2 + (p + q)x + pq$ to find the products.

31 $(x + 7)(x + 1)$

32 $(x + 11)(x + 12)$

33 $(x - 8)(x - 10)$

34 $(y - 6)(y - 13)$

35 $(y - 13)(y + 2)$

36 $(z + 4)(z - 16)$

37 $(x + 2y)(x + 7y)$

38 $(x - 6y)(x - 5y)$

39 $(z + 5y)(z - 9y)$

40 $(y + 7z)(y + 9z)$

In problems 41–50, use the formula $(sx + p)(rx + q) = (sr)x^2 + (sq + pr)x + pq$ to find the products.

41 $(2x + 3)(x + 7)$

42 $(y + 4)(3y + 11)$

43 $(2x - 5)(3x + 8)$

44 $(4x + 3)(3x - 4)$

45 $(4y - 11)(3y - 9)$

46 $(5x - 8)(4x - 3)$

47 $(3x + y)(2x + 3y)$

48 $(11x - 4y)(10x + 3y)$

49 $(7y - 5z)(5y + 7z)$

50 $(13x + 4z)(3x + 5z)$

2.6 Division of Polynomials

In Section 2.1 we introduced properties of exponents that involved the multiplication operations. In the present section we shall consider the division of polynomials. We begin with the division of monomials, but to understand this, we must first look at two additional properties of exponents that involve division.

Division Properties of Exponents

The rule for dividing expressions of like bases is similar to the multiplication property. To find the quotient $\frac{a^m}{a^n}$, where $a \neq 0$ and m and n are positive integers, we must examine three cases. In these cases, m is greater than n, m is equal to n, and m is less than n. Before we state the general rule, let us illustrate these cases by some examples. To illustrate the case m is greater than n, consider the quotient $\frac{5^6}{5^4}$, which is written

$$\frac{5^6}{5^4} = \frac{5 \times 5 \times 5 \times 5 \times 5 \times 5}{5 \times 5 \times 5 \times 5} = 5 \times 5 = 5^2 = 5^{6-4}$$

Note that the difference $6 - 4$ gives the new exponent 2. Similarly, $\frac{x^5}{x^3}$ can be found by writing

$$\frac{x^5}{x^3} = \frac{x \cdot x \cdot x \cdot x \cdot x}{x \cdot x \cdot x} = x \cdot x = x^2$$

Or, this result could be obtained by simply subtracting exponents. That is,

$$\frac{x^5}{x^3} = x^{5-3} = x^2$$

To illustrate the case $m = n$, consider the quotient $\frac{2^6}{2^6}$, which is written

$$\frac{2^6}{2^6} = \frac{2 \times 2 \times 2 \times 2 \times 2 \times 2}{2 \times 2 \times 2 \times 2 \times 2 \times 2} = 1$$

However, by subtracting exponents, we have $\frac{2^6}{2^6} = 2^{6-6} = 2^0$. Therefore, we define $2^0 = 1$. In general, if a is any real number and $a \neq 0$, we define $a^0 = 1$.

The case where m is less than n can be illustrated as follows:

$$\frac{5^3}{5^7} = \frac{5 \times 5 \times 5}{5 \times 5 \times 5 \times 5 \times 5 \times 5 \times 5}$$

$$= \frac{1}{5 \times 5 \times 5 \times 5} = \frac{1}{5^4}$$

The same result could have been obtained by subtracting exponents: $\frac{5^3}{5^7} = \frac{1}{5^{7-3}} = \frac{1}{5^4}$. In general, if m is less than n, then for $a \neq 0$, $\frac{a^m}{a^n} = \frac{1}{a^{n-m}}$, where $n - m$ represents a positive integer; for example,

$$\frac{x^4}{x^7} = \frac{1}{x^{7-4}} = \frac{1}{x^3} \qquad \text{for } x \neq 0$$

The three cases just discussed can be expressed by the following property.

1 DIVISION PROPERTY

Let a be any real number except zero, and m and n positive integers; then

$$\frac{a^m}{a^n} = \begin{cases} a^{m-n} & \text{if } m \text{ is greater than } n \\ a^0 = 1 & \text{if } m = n \\ \dfrac{1}{a^{n-m}} & \text{if } m \text{ is less than } n \end{cases}$$

EXAMPLES

1 Use Property 1 to rewrite the following expressions and simplify the results where possible.

a) $\frac{3^6}{3^2}$ b) $\frac{x^{10}}{x^7}$ c) $\frac{(-y)^9}{(-y)^5}$

SOLUTION. Applying Property 1, we have

a) $\frac{3^6}{3^2} = 3^{6-2} = 3^4 = 81$

b) $\frac{x^{10}}{x^7} = x^{10-7} = x^3$

c) $\frac{(-y)^9}{(-y)^5} = (-y)^{9-5} = (-y)^4 = y^4$ (Why?)

2 Use Property 1, where m is not larger than n, to rewrite the following expressions and simplify the results when possible.

a) $\frac{4^3}{4^5}$ b) $\frac{z^5}{z^{11}}$ c) $\frac{x^5}{x^5}$ d) $\frac{-(-3)^{70}}{(-3)^{70}}$

SOLUTION

a) $\frac{4^3}{4^5} = \frac{1}{4^{5-3}} = \frac{1}{4^2} = \frac{1}{16}$

b) $\dfrac{z^5}{z^{11}} = \dfrac{1}{z^{11-5}} = \dfrac{1}{z^6}$

c) $\dfrac{x^5}{x^5} = x^{5-5} = x^0 = 1$

d) $\dfrac{-(-3)^{70}}{(-3)^{70}} = -(-3)^{70-70} = -(-3)^0 = -1$

Now consider the power of the quotient represented by $\left(\dfrac{x}{y}\right)^4$. Using Definition 1 of exponents, we can rewrite this expression as follows:

$$\begin{aligned}\left(\frac{x}{y}\right)^4 &= \frac{x}{y}\cdot\frac{x}{y}\cdot\frac{x}{y}\cdot\frac{x}{y}\\ &= \frac{x\cdot x\cdot x\cdot x}{y\cdot y\cdot y\cdot y}\\ &= \frac{x^4}{y^4}\end{aligned}$$

Note that the same result could be obtained by simply raising each term of the base to the indicated power. In general, we have the following property.

2 POWER OF A QUOTIENT PROPERTY

Let a and b be real numbers where $b \neq 0$ and n is a positive integer; then

$$\left(\frac{a}{b}\right)^n = \frac{a^n}{b^n}$$

EXAMPLE

Use Property 2 to rewrite the following expressions and simplify the results when possible.

a) $\left(\dfrac{3}{4}\right)^2$ b) $\left(\dfrac{z}{y}\right)^6$ c) $\left(\dfrac{-x}{y}\right)^3$

SOLUTION. Applying Property 2, we have

a) $\left(\dfrac{3}{4}\right)^2 = \dfrac{3^2}{4^2} = \dfrac{9}{16}$

b) $\left(\dfrac{z}{y}\right)^6 = \dfrac{z^6}{y^6}$

c) $\left(\dfrac{-x}{y}\right)^3 = \dfrac{(-x)^3}{y^3} = \dfrac{-x^3}{y^3}$

Division of a Polynomial by a Monomial

To divide a monomial by another monomial, we apply Property 1. For example, to divide $6x^6y^3$ by $3x^4y^2$, we have

$$\frac{6x^6y^3}{3x^4y^2} = \left(\frac{6}{3}\right)\left(\frac{x^6}{x^4}\right)\left(\frac{y^3}{y^2}\right)$$
$$= 2x^{6-4}y^{3-2}$$
$$= 2x^2y$$

However, to divide a polynomial by a monomial, we use the fact that $\frac{a}{b} = a\left(\frac{1}{b}\right)$, for $b \neq 0$ and the distributive property. For example, to divide $x^3 + 5x^2 + 3x$ by x, we have

$$\frac{x^3 + 5x^2 + 3x}{x} = (x^3 + 5x^2 + 3x)\frac{1}{x} \qquad \text{(Definition of division)}$$
$$= x^3\left(\frac{1}{x}\right) + 5x^2\left(\frac{1}{x}\right) + 3x\left(\frac{1}{x}\right) \qquad \text{(Distributive property)}$$
$$= \frac{x^3}{x} + \frac{5x^2}{x} + \frac{3x}{x} \qquad \text{(Definition of division)}$$
$$= x^2 + 5x + 3$$

From the example above, we can see that the division of a polynomial by a monomial is accomplished by dividing each term of the polynomial by the monomial divisor. The following examples will illustrate this procedure.

EXAMPLES

1 Divide $x^4 + x^5$ by x^2.

SOLUTION

$$\frac{x^4 + x^5}{x^2} = \frac{x^4}{x^2} + \frac{x^5}{x^2}$$
$$= x^2 + x^3$$

2 Divide $4y^5z^6 - 12y^4z^5 + 8y^3z^4$ by $2yz^2$.

SOLUTION

$$\frac{4y^5z^6 - 12y^4z^5 + 8y^3z^4}{2yz^2} = \frac{4y^5z^6}{2yz^2} - \frac{12y^4z^5}{2yz^2} + \frac{8y^3z^4}{2yz^2}$$
$$= 2y^4z^4 - 6y^3z^3 + 4y^2z^2$$

3 Divide $x^4 + 3x^3 - 2x^2 + 5x + 2$ by x^2.

SOLUTION

$$\frac{x^4 + 3x^3 - 2x^2 + 5x + 2}{x^2} = \frac{x^4}{x^2} + \frac{3x^3}{x^2} - \frac{2x^2}{x^2} + \frac{5x}{x^2} + \frac{2}{x^2}$$
$$= x^2 + 3x - 2 + \frac{5}{x} + \frac{2}{x^2}$$

Notice that the result in example 3 is not a polynomial because of the expressions $\frac{5}{x}$ and $\frac{2}{x^2}$. Thus, we see that although the sum, difference, and the product of two polynomials always result in a polynomial, the quotient of two polynomials may not.

Division of a Polynomial by a Polynomial

To divide a polynomial by another polynomial, we use a method similar to the "long-division" method of arithmetic. For example, 6,741 ÷ 21. We usually do this long division in the following way:

```
      321
 21)6741
    63       [= 21 × 3]
     44
     42      [= 21 × 2]
      21
      21     [= 21 × 1]
       0
```

We can also perform this division by changing the dividend 6,741 and the divisor 21 to their expanded forms first:

$$21 = 2 \cdot 10 + 1$$

and

$$6{,}741 = 6 \cdot 10^3 + 7 \cdot 10^2 + 4 \cdot 10 + 1$$

and then dividing:

$$
\begin{array}{r@{}l}
 & \quad 3\cdot 10^2 + 2\cdot 10 \; + 1 \\
2\cdot 10 + 1 \,\big) & \overline{6\cdot 10^3 + 7\cdot 10^2 + 4\cdot 10 + 1} \\
 & \underline{6\cdot 10^3 + 3\cdot 10^2} \qquad\qquad [= (2\cdot 10 + 1)(3\cdot 10^2)] \\
 & \qquad\quad 4\cdot 10^2 + 4\cdot 10 \\
 & \qquad\quad \underline{4\cdot 10^2 + 2\cdot 10} \qquad [= (2\cdot 10 + 1)(2\cdot 10)] \\
 & \qquad\qquad\qquad 2\cdot 10 + 1 \\
 & \qquad\qquad\qquad \underline{2\cdot 10 + 1} \quad [= (2\cdot 10 + 1)(1)] \\
 & \qquad\qquad\qquad\qquad\quad 0
\end{array}
$$

Note that the expanded forms of the numbers in this example are polynomials, each term of which has a known base of 10. Let us change this problem to a similar one by changing the known base of 10 to one of base x. We then have the long division of $6x^3 + 7x^2 + 4x + 1$ by $2x + 1$.

$$
\begin{array}{r@{}l}
 & \quad 3x^2 + 2x \; + 1 \\
2x + 1 \,\big) & \overline{6x^3 + 7x^2 + 4x + 1} \\
 & \underline{6x^3 + 3x^2} \qquad\qquad [= (2x + 1)(3x^2)] \\
 & \qquad 4x^2 + 4x \\
 & \qquad \underline{4x^2 + 2x} \qquad [= (2x + 1)(2x)] \\
 & \qquad\qquad 2x + 1 \\
 & \qquad\qquad \underline{2x + 1} \quad [= (2x + 1)(1)] \\
 & \qquad\qquad\qquad 0
\end{array}
$$

In general, to divide a polynomial by another polynomial, we arrange both polynomials in descending powers of one of the variables and follow the procedure above.

EXAMPLES

1 Divide $x^3 + 2x^2 - 2x - 1$ by $x - 1$.

SOLUTION

$$
\begin{array}{r@{}l}
 & \quad x^2 + 3x \; + 1 \\
x - 1 \,\big) & \overline{x^3 + 2x^2 - 2x - 1} \\
 & \underline{x^3 - \;\; x^2} \\
 & \qquad 3x^2 - 2x \\
 & \qquad \underline{3x^2 - 3x} \\
 & \qquad\qquad x - 1 \\
 & \qquad\qquad \underline{x - 1} \\
 & \qquad\qquad\quad 0
\end{array}
$$

Therefore, when we divide $x^3 + 2x^2 - 2x - 1$ by $x - 1$, we obtain the quotient $x^2 + 3x + 1$ and the remainder 0. We check this by using multiplication.

Check:

$$(x - 1)(x^2 + 3x + 1) = x^3 + 2x^2 - 2x - 1$$

2 Divide $2x^4 + 5 - 3x + 3x^3$ by $2x - 1$.

SOLUTION. Arrange the dividend in descending powers of x:

$$
\begin{array}{r l}
 & x^3 + 2x^2 + x - 1 \\
\cline{2-2}
2x - 1 \,\big) & 2x^4 + 3x^3 + 0x^2 - 3x + 5 \\
 & 2x^4 - x^3 \\
\cline{2-2}
 & \qquad 4x^3 + 0x^2 \\
 & \qquad 4x^3 - 2x^2 \\
\cline{2-2}
 & \qquad\qquad 2x^2 - 3x \\
 & \qquad\qquad 2x^2 - x \\
\cline{2-2}
 & \qquad\qquad\qquad -2x + 5 \\
 & \qquad\qquad\qquad -2x + 1 \\
\cline{2-2}
 & \qquad\qquad\qquad\qquad 4
\end{array}
$$

Therefore, the quotient is $x^3 + 2x^2 + x - 1$ and the remainder is 4. The result is expressed as follows:

$$\frac{2x^4 + 2x^3 - 3x + 5}{2x - 1} = x^3 + 2x^2 + x - 1 + \frac{4}{2x - 1}$$

Check by multiplication:

$$
\begin{aligned}
(2x - 1)&\left[x^3 + 2x^2 + x - 1 + \frac{4}{2x - 1}\right] \\
&= (2x - 1)(x^3 + 2x^2 + x - 1) + 4 \\
&= (2x^4 + 3x^3 - 3x + 1) + 4 \\
&= 2x^4 + 3x^3 - 3x + 5
\end{aligned}
$$

3 Divide $x^3 + 2y^3 + 5x^2y + 7xy^2$ by $x + 2y$.

SOLUTION. In this case, we first arrange the dividend and the divisor in descending powers of x:

$$
\begin{array}{r l}
 & x^2 + 3xy + y^2 \\
x + 2y \,\big)\!\!\! & \overline{x^3 + 5x^2y + 7xy^2 + 2y^3} \\
 & \underline{x^3 + 2x^2y} \\
 & \qquad 3x^2y + 7xy^2 \\
 & \qquad \underline{3x^2y + 6xy^2} \\
 & \qquad\qquad xy^2 + 2y^3 \\
 & \qquad\qquad \underline{xy^2 + 2y^3} \\
 & \qquad\qquad\qquad 0
\end{array}
$$

Therefore, the quotient is $x^2 + 3xy + y^2$ and the remainder is 0.

Check:

$$(x + 2y)(x^2 + 3xy + y^2) = x^3 + 5x^2y + 7xy^2 + 2y^3$$

4 Divide $2x^4 - 3x^3 + 5x^2 + 2x + 7$ by $x^2 - x + 1$.

SOLUTION

$$
\begin{array}{r l}
 & 2x^2 - x + 2 \\
x^2 - x + 1 \,\big)\!\!\! & \overline{2x^4 - 3x^3 + 5x^2 + 2x + 7} \\
 & \underline{2x^4 - 2x^3 + 2x^2} \\
 & \quad -x^3 + 3x^2 + 2x \\
 & \quad \underline{-x^3 + x^2 - x} \\
 & \qquad\quad 2x^2 + 3x + 7 \\
 & \qquad\quad \underline{2x^2 - 2x + 2} \\
 & \qquad\qquad\quad 5x + 5
\end{array}
$$

Therefore, the quotient is $2x^2 - x + 2$ and the remainder is $5x + 5$, so

$$\frac{2x^4 - 3x^3 + 5x^2 + 2x + 7}{x^2 - x + 1} = 2x^2 - x + 2 + \frac{5x + 5}{x^2 - x + 1}$$

Check:

$$
\begin{aligned}
(x^2 - x + 1)&\left(2x^2 - x + 2 + \frac{5x + 5}{x^2 - x + 1}\right) \\
&= (x^2 - x + 1)(2x^2 - x + 2) + (5x + 5) \\
&= (2x^4 - 3x^3 + 5x^2 - 3x + 2) + 5x + 5 \\
&= 2x^4 - 3x^3 + 5x^2 + 2x + 7
\end{aligned}
$$

PROBLEM SET 2.6

In problems 1–16, use Property 1 to rewrite each expression and simplify the result when possible. 1–40

1 $\frac{2^4}{2^2}$

2 $\frac{3^8}{3^5}$

3 $\frac{(-3)^5}{(-3)^3}$

4 $\frac{(-5)^7}{(-5)^2}$

5 $\frac{x^9}{x^4}$

6 $\frac{z^{12}}{z^9}$

7 $\frac{y^{11}}{y^6}$

8 $\frac{(-x)^{10}}{(-x)^5}$

9 $\frac{(2x)^7}{(2x)^4}$

10 $\frac{(x^2)^5}{(x^2)^2}$

11 $\frac{5^4}{5^6}$

12 $\frac{(-2)^9}{(-2)^{13}}$

13 $\frac{x^7}{x^{11}}$

14 $\frac{(-z)^{15}}{(-z)^{15}}$

15 $\frac{y^{12}}{y^{12}}$

16 $\frac{(2x)^4}{(2x)^7}$

In problems 17–26, use Property 2 to rewrite the expressions and simplify the results when possible.

17 $\left(\frac{2}{3}\right)^3$

18 $\left(\frac{4}{5}\right)^4$

19 $\left(\frac{x}{y}\right)^5$

20 $\left(\frac{x}{y}\right)^9$

21 $\left(\frac{y}{-z}\right)^4$

22 $\left(\frac{-x}{2z}\right)^6$

23 $\left(\frac{-x}{z}\right)^3$

24 $\left(\frac{3x}{4y}\right)^2$

25 $\left(\frac{2x}{y^2}\right)^3$

26 $\left(\frac{-2x^3}{z}\right)^3$

In problems 27–40, divide as indicated.

27 $12x^4$ by $3x$

28 $20y^6$ by $5y^4$

29 $24x^5y^3$ by $8x^2y$

30 $42x^7y^2$ by $7x^2y^4$

31 $-15x^6y^3z^2$ by $3x^5yz^4$

32 $18xy^5z^2$ by $-6x^4y^2z^5$

33 $4x^3 + 10x^2$ by $2x$

34 $10y^4 - 15y^5$ by $5y^3$

35 $30x^3y^4 - 42x^5y^6 + 18x^7y^5$ by $6x^2y^3$

36 $75x^7y^5z^3 + 50x^6y^4z^2 - 25x^5y^3z^3$ by $-25x^4y^2z^2$

37 $6x^4z^4 - 9x^3z + 15xz^3$ by $3x^2z^2$

38 $33x^4y - 44xy^3 + 22x^2y^2$ by $-11x^2y^3$

39 $4z^2 - 8zy + 12y^2$ by $4z^3y$

40 $7x^5y - 14x^4y^2 - 21x^3y^3 + 28x^2y^4$ by $7x^2y^3$

In problems 41–62, divide by the long-division method.

41 $x^2 + 5x + 6$ by $x + 2$

42 $x^2 - x - 12$ by $x + 3$

43 $x^2 - 7x + 6$ by $x - 1$

44 $x^2 + x - 20$ by $x - 4$

45 $2y^2 + 3y + 1$ by $2y + 1$

46 $6z^2 - 5z - 4$ by $3z - 4$

47 $x^3 + 3x^2 + 3x + 1$ by $x + 1$

48 $y^3 - 6y^2 + 12y - 8$ by $y - 2$

49 $3x^3 + 8x^2 + 13x + 6$ by $3x + 2$

50 $x^4 + 2x^3 + 2x^2 + 2x + 1$ by $x + 1$

51 $x^3 - 2x^2 - 2x - 3$ by $x^2 + x + 1$

52 $6x^3 + x^2 + 7x + 6$ by $2x^2 - x + 3$

53 $3x^4 + 3x^3 - 6x^2 + x - 4$ by $x^2 - x + 2$

54 $4x^4 - 7x^3 + 2x^2 + x - 3$ by $x^2 - 2x + 1$

55 $x^3 + y^3 - 3xy^2 - 3x^2y$ by $x + y$

56 $2x^4 + 7x^3y + 9x^2y^2 + 5xy^3 + y^4$ by $2x + y$

57 $8x^3 - 27y^3$ by $2x - 3y$

58 $81x^4 - z^4$ by $3x - z$

59 $x^5 + y^5$ by $x + y$

60 $y^5 + 32z^5$ by $y + 2z$

61 $x^5 + x^4 + 5x^2 - 4x + 6$ by $x^3 + x^2 - 2x + 3$

62 $x^5 + 2x^4y - 2x^3y^2 - 2x^2y^3 + 3xy^4 - y^5$ by $x^2 + xy - y^2$

REVIEW PROBLEM SET

In problems 1–6, find the value of the expressions.

1 4^3

2 $(-3)^3$

3 $(-3)^2 + 2(-3)$

4 $2^3 + 3(2)$

5 $4^2 - 2(4)(-1) + (-1)^2$

6 $(-3)^2 + 2(-3)(1) + (1)^2$

In problems 7–12, write the expressions in equivalent exponential form.

7 $3 \cdot x \cdot x \cdot x \cdot y$

8 $-5x \cdot x \cdot y \cdot y \cdot z$

9 $-4 \cdot y \cdot y \cdot z + 7 \cdot y \cdot y \cdot z \cdot z \cdot z$

10 $8 \cdot a \cdot a \cdot b \cdot b \cdot b - 4 \cdot a \cdot a \cdot a \cdot a$

11 $y \cdot y \cdot y + 2 \cdot y \cdot x \cdot x - 3 \cdot x \cdot x \cdot x$

12 $2 \cdot x \cdot x \cdot y \cdot y \cdot z \cdot z \cdot z + 3 \cdot z \cdot z \cdot z \cdot z$

In problems 13–32, use the properties of exponents to rewrite the expressions.

13 $3^2 \cdot 3^4$

14 $(-2)^3(-2)^2$

15 x^2x^{11}

16 z^5z^3

17 y^4yy^2

18 $(-x)^2(-x)^3(-x)^4$

19 $(2^3)^2$

20 $[(-3)^3]^2$

21 $(x^4)^3$

22 $(z^4)^3$

23 $(-y^5)^4$

24 $[(-x)^2]^4$

25 $(3y)^2$

26 $(-2x)^3$

27 $(xy)^4$

28 $(5yz)^4$

29 $(x^3y^2)^4$

30 $(-u^2v^4)^3$

31 $(2xy^2)^3(-x^2y)^2$

32 $(x^2y)^3(yz^2)^2(xz)^3$

In problems 33–38, determine if the given expression is a polynomial in one variable.

33 $x^3 - 2x^2 + 1$

34 $\frac{3}{2}x^2 - \frac{1}{2}x + 3$

35 $\dfrac{2x^3 - 3}{x^2}$

36 $\dfrac{3}{2x^2} - \dfrac{1}{2x} + 3$

37 $y - 4y^2 + \frac{1}{2}y^3$

38 $\dfrac{1}{y^2} + \dfrac{2}{y} + 5$

In problems 39–46, identify the given polynomial as a monomial, binomial, or trinomial, if applicable. Find the degree of the polynomial and list the numerical coefficients.

39 $2x^2 - 7$

40 $2y^4$

41 $-7x^2$

42 $4x^3 - 3x - 7$

43 $y^2 + 4y - 3$

44 $8x^2 - 7x^3$

45 $-x^3 + 6x^2 - 2x + 4$

46 $y^5 - 3y^4 + 7y^3 - y$

In problems 47–50, evaluate the given polynomial for $x = 3$.

47 $-3x + 12$

48 $3x^2 - 4x - 5$

49 $x^3 - 2x^2 + x - 8$

50 $x^4 - 4x^2 + 3$

In problems 51–54, evaluate $2x^3 - 3xy + y$ for the given values of x and y.

51 $x = 1, y = -2$

52 $x = -1, y = -2$

53 $x = -2, y = 1$

54 $x = 3, y = 2$

In problems 55–66, perform the additions of the polynomials.

55 $3xy^2 + 17xy^2$

56 $18y^2 + (-11y^2)$

57 $-2xyz + 9xyz$

58 $-23z^3 + (-17z^3)$

59 $6xy^3 + 9xy^3 + 5xy^3$

60 $7z^2 + (-2z^2) + 11z^2$

61 $(27x + 13) + (-16x + 12)$

62 $(4y^2 - 3) + (-3y^2 + 1)$

63 $(8x^2 + x - 13) + (-5x^2 + 3x + 7)$

64 $(x^3 + 9x^2 - 11x - 13) + (-5x^3 - 7x^2 + 6x - 2)$

65 $(2y^3 + 4y^2 + 9y + 17) + (3y^3 - 12y^2 - 4y + 3)$

66 $(x^2y - xy^2 + xy + 3) + (3xy + 2xy^2 - 3x^2y)$

In problems 67–76, perform the subtractions of the polynomials.

67 $21x^2 - 12x^2$

68 $-18y - 11y$

69 $4y^3 - (-5y^3)$

70 $-11x^3 - 7x^3$

71 $(x^2 + x - 12) - (x^2 - x + 3)$

72 $(3y^2 - 8y + 4) - (-y^2 + 2y + 3)$

73 $(2x^3 - 10x^2 + 7x - 9) - (x^3 - 8x^2 - 4x - 4)$

74 $(5z^3 + 7z^2 - z + 2) - (-z^3 - 2z^2 + 3z - 13)$

75 $(3x^3y^2 + 4x^2y - 3xy^3) - (2x^2y + 2xy^3 - x^3y^2)$

76 $(13x^3y^3 - 11x^2y^2 + 7xy + 9) - (10 - 3xy - 7x^2y^2 + 8x^3y^3)$

In problems 77–80, perform the indicated operations.

77 $(3y^2 + 5y - 1) + (5y^2 - 2y + 2) + (-3y^2 + y - 7)$

78 $(x^3 + 2x - 7) + (x^2 - 3x + 6) + (3x^3 - 2x + 5)$

79 $(8x^2 - 11x + 7) + (2x^2 + 6x - 3) - (5x^2 + 3x + 13)$

80 $(4x^2y - 3xy^2 + 6xy) + (x^2y + 2xy^2 - 3xy) - (5xy^2 + xy - 2x^2y)$

In problems 81–90, find the products.

81 $(7x^3y^2z)(-3xz^2)$

82 $(9y^3z^4)(5y^2z)$

83 $3y(2y^3 + 5y - 4)$

84 $5xz^2(2x^3y - 3y^2z + xyz)$

85 $(x + 2)(x^2 - x + 1)$

86 $(y - 5)(2y^2 + 3y + 1)$

87 $(2x + 3y)(3x^2 - xy + y^2)$

88 $(3y - z)(y^2 - 2yz + z^2)$

89 $(x^2 + 2x + 1)(x^2 - 3x + 2)$

90 $(2x^2 - x - 1)(3x^2 + 2x + 1)$

In problems 91–100, find the products of binomials.

91 $(x + 6)(x + 7)$

92 $(x + 12)(x + 15)$

93 $(y - 1)(y - 9)$

94 $(z - 2)(z - 13)$

95 $(2x + 7)(x - 8)$

96 $(3x - 4)(7x + 6)$

97 $(y - 4z)(3y + 5z)$

98 $(2x + 7y)(3x + 5y)$

99 $(5x - 3y)(6x - 5y)$

100 $(9x - 8y)(7x + 6y)$

In problems 101–110, use properties of exponents to rewrite the expressions.

101 $\frac{7^6}{7^4}$

102 $\dfrac{5^8}{5^8}$

103 $\dfrac{x^3}{x^5}$

104 $\dfrac{y^{10}}{y^3}$

105 $\left(\dfrac{x}{y}\right)^7$

106 $\left(\dfrac{-z}{y^2}\right)^3$

107 $\dfrac{(-4)^5}{(-4)^3}$

108 $\dfrac{(5x)^4}{(5x)^7}$

109 $\dfrac{(3x)^6}{(3x)^4}$

110 $\left(\dfrac{x^3}{-2z}\right)^4$

In problems 111–116, divide as indicated.

111 $40x^2y^3$ by $8xy$

112 $50x^4y^3z^2$ by $25x^2z^2$

113 $24x^3y^5 - 30x^5y^3$ by $6x^2y$

114 $15x^8y^7z^5 - 35x^6y^5z^4$ by $5x^4yz^3$

115 $16x^7 + 12x^4 - 4$ by $4x^4$

116 $7z^3 - 14y^2z + 21y^4$ by $7y^2z^4$

In problems 117–124, divide by the long-division method.

117 $x^2 + 8x - 9$ by $x - 1$

118 $y^2 + 4y - 21$ by $y + 7$

119 $x^3 + 2x^2 - 5x + 12$ by $x + 4$

120 $6z^3 - 7z^2 + 3z + 2$ by $3z + 1$

121 $x^4 + 6x^3 - 3x^2 + 4x + 3$ by $x^2 - x + 1$

122 $2z^4 - z^3 - 6z^2 + 7z - 2$ by $2z^2 - 3z + 1$

123 $x^5 + y^5 + xy^4 + 4x^3y^2 + x^2y^3 + 4x^4y$ by $y^2 + x^2 + 2xy$

124 $x^6 - y^6$ by $x^2 - y^2$

CHAPTER 3

Special Products and Factoring Polynomials

3 SPECIAL PRODUCTS AND FACTORING POLYNOMIALS

Introduction

In Chapter 2 we discussed the operation of multiplication, which enabled us to obtain a product of two or more factors. In this chapter we shall "reverse" the multiplication operation, that is, start with a given polynomial and express it as the product of two or more polynomials. This process is called *factoring* and is considered a fundamental tool in algebra that will have many uses in later chapters. The chapter also includes "special products" which not only provide shortcuts for expanding frequently used product types but give clues and insights into factoring.

3.1 Common Factors

The distributive property, $P(Q + R) = PQ + PR$, provides a bridge for going back and forth between products and factoring. A polynomial is *factorable* if it can be expressed as the product of two or more polynomials. If a polynomial has no factors other than itself and 1, or its negative and -1, the polynomial is called *prime*. To factor a polynomial, we express it as the product of prime polynomials, or as the product of a monomial and prime polynomials. Once we accomplish these objectives, we refer to the final result as a *complete factorization* of the given polynomial. We should note that throughout the book, unless otherwise specified, we shall agree to accept as factors of any polynomial with integral coefficients only those factors which also contain integral coefficients.

Common Monomial Factors

The procedure of multiplying a monomial by a polynomial was discussed in Section 2.4. The result is a polynomial, in which each term contains a common monomial factor. The reverse process suggests a method of expressing such polynomials in a product form. For instance, the binomial $x^2 + 2x$ can be written

$$\begin{aligned} x^2 + 2x &= x \cdot x + x \cdot 2 \\ &= x(x + 2) \end{aligned}$$

In this case, x is a common factor and x and $x + 2$ are prime factors. Now consider the polynomial $2x^4y^5 + 4x^3y^4 - 8x^2y^3$. Inspection shows that each term of the polynomial is divisible by $2x^2y^3$. Thus, we write the polynomial as

$$\begin{aligned} &2x^4y^5 + 4x^3y^4 - 8x^2y^3 \\ &\qquad = (2x^2y^3)(x^2y^2) + (2x^2y^3)(2xy) + (2x^2y^3)(-4) \end{aligned}$$

We can see at a glance that $2x^2y^3$ is a common factor, so that

$$2x^4y^5 + 4x^3y^4 - 8x^2y^3 = 2x^2y^3(x^2y^2 + 2xy - 4)$$

The above examples suggest the following steps for factoring a polynomial containing a common factor:

1 Determine the common factor by inspecting the prime factors of each term.
2 Apply the distributive property by removing the product of the common prime factors in step 1. Or simply, divide the polynomial by the product of the common prime factors of step 1; the indicated product of the divisor and the quotient are the factors of the given polynomial.

EXAMPLES

Factor the given polynomials.

1 $4x + 12$

SOLUTION

Step 1. Write the factors of each term.

$$4x = (2)(2)(x)$$
$$12 = 2 \times 2 \times 3$$

We observe that the common factor is 4.

Step 2. Remove the common factor 4 by using the distributive property:

$$4x + 12 = 4(x + 3)$$

or, we might express the process this way:

$$4\overline{)4x + 12}^{\;x + 3}$$

Therefore, the factored form is $4x + 12 = 4(x + 3)$.

2 $6x^4 + 12x^2$

SOLUTION

Step 1. Write the factored form of each term:

$$6x^4 = 2(3) \cdot x \cdot x \cdot x \cdot x$$
$$12x^2 = 2(2)(3) \cdot x \cdot x$$

We observe that the common factor is $6x^2$.

Step 2. Apply the distributive property:

$$6x^4 + 12x^2 = 6x^2(x^2 + 2)$$

3 $-4xy^3 + 6xy^2 - 14xy$

SOLUTION

Step 1. Inspection shows that the monomial expression $-2xy$ is contained in each term of the polynomial. Thus, $-2xy$ is the common factor.

Step 2. Apply the distributive property:

$$-4xy^3 + 6xy^2 - 14xy = -2xy(2y^2 - 3y + 7)$$

4 $4x^7y^5 + 10x^3y^8 - 16x^4y^6 + 18x^5y^7$

SOLUTION

Step 1. Inspection shows that each term of the polynomial is divisible by $2x^3y^5$, and so the common factor is $2x^3y^5$.

Step 2. Apply the distributive property:

$$\begin{aligned} &4x^7y^5 + 10x^3y^8 - 16x^4y^6 + 18x^5y^7 \\ &\quad = 2x^3y^5(2x^4) + 2x^3y^5(5y^3) + 2x^3y^5(-8xy) + 2x^3y^5(9x^2y^2) \\ &\quad = 2x^3y^5(2x^4 + 5y^3 - 8xy + 9x^2y^2) \end{aligned}$$

PROBLEM SET 3.1

In problems 1–40, find the factors of the polynomials by removing the common monomials.

1 $x^2 - x$

2 $x^2 + 3x$

3 $y^2 - 7y$

4 $4t^2 + 2t$

5 $3t^3 + 6t^2$

6 $9x^2 + 3x$

7 $10x^2 - 5x$

8 $ax + ay$

9 $ax - ay$

10 $27x^4 - 18ax^2$

11 $a^2b - ab^2$

12 $a^2b^2 + ab^3$

13 $\pi r^2h + 2\pi rh$

14 $6lw + 3wh$

15 $6rs^2 + 30r^2s$

16 $9m^2n - 18mn^2$

17 $6p^2q + 24pq^2$

18 $5abc + 20abc^2$

19 $2x^2 + 4x + 6$

20 $3x^2 - 6x - 3$

21 $x^3 + 3x^2 + x$

22 $2a^3 - 8a^2 - 6a$

23 $x^5 + 2x^4 + 3x^3$

24 $y^6 + 7y^5 - 11y^3$

25 $3x^3 - 6x^5 + 15x^2$

26 $24x^2y^3 - 36xy^2 + 18xy$

27 $4x^2y + 12xy - 7xy^2$

28 $-4ax^2 + 6a^2x^3 - 8x$

29 $5x - 10x^2y^2 - 25y^2$

30 $12abc^2 - 8a^2b^2c + 4a^2b^2c^2$

31 $-m^2n - mn^2 - 3m^2n^2 - 6mn$

32 $2xyz^2 - 6x^2yz + 10xy^2z$

33 $121x^2y^3 + 55x^3y^2 + 77xy^4$

34 $16x^2y^2 + 64xy^3 + 48x^3y$

35 $8x^3y^6 - 16xy^5 + 24x^2y^3$

36 $8a^2b^2c^2 - 4ab^2c^4 - 3ab^3c$

37 $24a^5x^3 - 21a^2x^4 + 18a^2x^3$

38 $-4x^2y^4 - 8x^3y^3 + 2x^3y^4$

39 $-35r^3t - 25r^2t^2 - 15rt^3$

40 $3x^2b^3 - 6x^3b^2 + 12xb^4$

3.2 Special Products

Certain products of polynomials occur often enough in algebra to be worthy of special consideration. Learning these special products has the same advantage as learning the multiplication tables in arithmetic. In this section we develop special products that enable us to find many products in less time than would be required otherwise. These special products will prove valuable in factoring certain types of polynomials in Sections 3.3 and 3.6.

Square of a Binomial

For Special Products 1 and 2, we shall assume that A and B represent real numbers or *algebraic expressions.*

SPECIAL PRODUCT 1

The square of the sum of two terms in a binomial is equal to the sum of the squares of each term and twice the product of the two terms. In symbols,

$$(A + B)^2 = A^2 + 2AB + B^2$$

To verify Special Product 1, we have

$$\begin{aligned}(A + B)^2 &= (A + B)(A + B) = A(A + B) + B(A + B)\\ &= A^2 + AB + BA + B^2\\ &= A^2 + 2AB + B^2\end{aligned}$$

SPECIAL PRODUCT 2

The square of the difference of two terms in a binomial is equal to the sum of the squares of each term less twice the product of the two terms. In symbols,

$$(A - B)^2 = A^2 - 2AB + B^2$$

To verify Special Product 2, we have

$$\begin{aligned}(A - B)^2 &= (A - B)(A - B) = A(A - B) - B(A - B)\\ &= A^2 - AB - BA + B^2\\ &= A^2 - 2AB + B^2\end{aligned}$$

EXAMPLES

Use Special Products 1 and 2 to find the following products.

1 $(x + 4y)^2$

SOLUTION. Substituting x for A and $4y$ for B in Special Product 1, we have

$$\begin{aligned}(x + 4y)^2 &= x^2 + 2(x)(4y) + (4y)^2\\ &= x^2 + 8xy + 16y^2\end{aligned}$$

2 $(xy - 5)^2$

SOLUTION. Substituting xy for A and 5 for B in Special Product 2, we have

$$\begin{aligned}(xy - 5)^2 &= (xy)^2 - 2(xy)(5) + 5^2 \\ &= x^2y^2 - 10xy + 25\end{aligned}$$

3 $(8x + 3y)^2$

SOLUTION. Substituting $8x$ for A and $3y$ for B in Special Product 1, we have

$$\begin{aligned}(8x + 3y)^2 &= (8x)^2 + 2(8x)(3y) + (3y)^2 \\ &= 64x^2 + 48xy + 9y^2\end{aligned}$$

4 $(6x - 7y^2)^2$

SOLUTION. Substituting $6x$ for A and $7y^2$ for B in Special Product 2, we have

$$\begin{aligned}(6x - 7y^2)^2 &= (6x)^2 - 2(6x)(7y^2) + (7y^2)^2 \\ &= 36x^2 - 84xy^2 + 49y^4\end{aligned}$$

Special Product of the Form $(A + B)(A - B)$

Let A and B represent real numbers or algebraic expressions.

SPECIAL PRODUCT 3

The product of the sum and the difference of the same two terms is the difference between their squares. In symbols,

$$(A + B)(A - B) = A^2 - B^2$$

To verify Special Product 3, we have

$$\begin{aligned}(A + B)(A - B) &= A(A - B) + B(A - B) = A^2 - AB + BA - B^2 \\ &= A^2 - B^2\end{aligned}$$

EXAMPLES

Use Special Product 3 to find the following products.

1 $(x + 3)(x - 3)$

SOLUTION. Substituting x for A and 3 for B in Special Product 3, we have

$$\begin{aligned}(x + 3)(x - 3) &= x^2 - (3)^2 \\ &= x^2 - 9\end{aligned}$$

2 $(2x + 1)(2x - 1)$

SOLUTION. Letting $A = 2x$ and $B = 1$ in Special Product 3, we have

$$\begin{aligned}(2x + 1)(2x - 1) &= (2x)^2 - (1)^2 \\ &= 4x^2 - 1\end{aligned}$$

3 $(7 - 2z)(7 + 2z)$

SOLUTION. By Special Product 3, we have

$$\begin{aligned}(7 - 2z)(7 + 2z) &= (7)^2 - (2z)^2 \\ &= 49 - 4z^2\end{aligned}$$

4 $(3x - 2y)(3x + 2y)$

SOLUTION. By Special Product 3, we have

$$\begin{aligned}(3x - 2y)(3x + 2y) &= (3x)^2 - (2y)^2 \\ &= 9x^2 - 4y^2\end{aligned}$$

PROBLEM SET 3.2

In problems 1–24, use Special Products 1 and 2 to find the square of each binomial.

1 $(x + 5)^2$

2 $(x + 9)^2$

3 $(y + 6)^2$

4 $(z + 10)^2$

5 $(2x + 7)^2$

6 $(5x + 8)^2$

7 $(4 + 3yz)^2$

8 $(11 + 2z)^2$

9 $(3x + 2y)^2$

10 $(8xz + 9y)^2$

11 $(x - 2)^2$

12 $(x - 8)^2$

13 $(y - 7)^2$

14 $(z - 4)^2$

15 $(3x - 5)^2$

16 $(11 - 3y)^2$

17 $(x - 9y)^2$

18 $(8x - 5y)^2$

19 $(5x - 7y)^2$

20 $(7y - 6z)^2$

21 $(2xy - 9)^2$

22 $(10 - 3yz)^2$

23 $(6x - 7y)^2$

24 $(9y - 5z)^2$

In problems 25–40, use Special Product 3 to find the following products.

25 $(x + 4)(x - 4)$

26 $(z - 6)(z + 6)$

27 $(x + 10)(x - 10)$

28 $(11 - y)(11 + y)$

29 $(y - 8)(y + 8)$

30 $(x - 9)(x + 9)$

31 $(3x + y)(3x - y)$

32 $(2x + 7)(2x - 7)$

33 $(5 - x)(5 + x)$

34 $(13 - z)(13 + z)$

35 $(6x - 7y)(6x + 7y)$

36 $(3x + 11y)(3x - 11y)$

37 $(8y + 7z)(8y - 7z)$

38 $(7x + 15y)(7x - 15y)$

39 $(15x - 9y)(15x + 9y)$

40 $(20x - 13z)(20x + 13z)$

3.3 Factoring Polynomials by Special Products

In Section 3.2 we introduced special products of polynomials which occur often enough in algebra to be worth memorizing. Now we reverse the process of Section 3.2. That is, we factor those polynomials that fit the forms of special products.

Factoring a Perfect Trinomial

In the previous section we considered special products that result in trinomials when we square binomials. That is:

1 $(A + B)^2 = A^2 + 2AB + B^2$
2 $(A - B)^2 = A^2 - 2AB + B^2$

By reversing the process, we can factor this type of trinomial. Such expressions may be recognized if we remember the following facts:

1 Two of the terms must be perfect squares.
2 The other term must equal twice the product (disregarding the sign) of the square roots of the perfect-square terms.

The following examples will illustrate how these special products are used to factor certain second-degree polynomials.

EXAMPLES

Use Special Product 1 or 2 to factor the following trinomials.

1 $x^2 + 6x + 9$

SOLUTION

$$x^2 + 6x + 9 = x^2 + 2(3)x + 3^2$$

which is in the form of Special Product 1,

$$A^2 + 2AB + B^2 = (A + B)^2$$

where $A = x$ and $B = 3$, so

$$x^2 + 2(3)x + 3^2 = (x + 3)^2$$

or

$$x^2 + 6x + 9 = (x + 3)^2$$

2 $x^2 - 10x + 25$

SOLUTION

$$x^2 - 10x + 25 = x^2 - 2(5)x + 5^2$$

which is in the form of Special Product 2,

$$A^2 - 2AB + B^2 = (A - B)^2$$

where $A = x$ and $B = 5$, so that

$$x^2 - 2(5)x + 5^2 = (x - 5)^2$$

or

$$x^2 - 10x + 25 = (x - 5)^2$$

3 $4x^2 + 12x + 9$

SOLUTION

$$4x^2 + 12x + 9 = (2x)^2 + 2(2x)(3) + (3)^2$$

which is in the form of Special Product 1 with $A = 2x$ and $B = 3$, so that

$$4x^2 + 12x + 9 = (2x + 3)^2$$

4 $16z^2 - 8z + 1$

SOLUTION

$$16z^2 - 8z + 1 = (4z)^2 - 2(4z)(1) + (1)^2$$

which is in the form of Special Product 2 with $A = 4z$ and $B = 1$, so that

$$16z^2 - 8z + 1 = (4z - 1)^2$$

5 $9x^2 + 30xy + 25y^2$

SOLUTION

$$9x^2 + 30xy + 25y^2 = (3x)^2 + 2(3x)(5y) + (5y)^2$$

so that, by Special Product 1, we have

$$9x^2 + 30xy + 25y^2 = (3x + 5y)^2$$

Factoring the Difference of Two Squares

From Special Product 3,

$$(A - B)(A + B) = A^2 - B^2$$

we see at once that $A - B$ and $A + B$ are factors of $A^2 - B^2$. Since A and B represent any two algebraic expressions, we can always find the factors of the difference of two squares.

When the above special product is used, it is written as

$$A^2 - B^2 = (A - B)(A + B)$$

We illustrate by the following examples.

EXAMPLES

Use Special Product 3 to factor the given binomials.

1 $x^2 - 4$

SOLUTION

$$x^2 - 4 = x^2 - 2^2$$

which is in the form of Special Product 3,

$$A^2 - B^2 = (A - B)(A + B)$$

where $A = x$ and $B = 2$, so that

$$x^2 - 2^2 = (x - 2)(x + 2)$$

or

$$x^2 - 4 = (x - 2)(x + 2)$$

2 $9 - y^2$

SOLUTION

$$9 - y^2 = 3^2 - y^2$$

which is in the form of Special Product 3, where $A = 3$ and $B = y$, so that

$$9 - y^2 = (3 - y)(3 + y)$$

3 $16x^2 - 25$

SOLUTION

$$16x^2 - 25 = (4x)^2 - 5^2$$

so that, by Special Product 3, we have

$$16x^2 - 25 = (4x - 5)(4x + 5)$$

4 $4x^2 - 9z^2$

SOLUTION

$$4x^2 - 9z^2 = (2x)^2 - (3z)^2$$

so that, by Special Product 3, we have

$$4x^2 - 9z^2 = (2x - 3z)(2x + 3z)$$

5 $4x^2y^2 - 25z^2$

SOLUTION

$$4x^2y^2 - 25z^2 = (2xy)^2 - (5z)^2$$

so that, by Special Product 3, we have

$$4x^2y^2 - 25z^2 = (2xy - 5z)(2xy + 5z)$$

PROBLEM SET 3.3

In problems 1–24, use Special Product 1 or 2 to factor the polynomials.

1 $x^2 + 2x + 1$

2 $z^2 - 2z + 1$

3 $x^2 - 4x + 4$

4 $y^2 + 10y + 25$

5 $4x^2 + 20x + 25$

6 $x^2 + 8x + 16$

7 $x^2 - 6xy + 9y^2$

8 $4x^2 - 12xy + 9y^2$

9 $16y^2 + 40y + 25$

10 $36y^2 + 132y + 121$

11 $9z^2 + 42z + 49$

12 $9 - 12z + 4z^2$

13 $36x^2 - 12x + 1$

14 $81x^2 + 126x + 49$

15 $16y^2 + 24yz + 9z^2$

16 $25x^2 - 110x + 121$

17 $64x^2 - 48x + 9$

18 $49y^2 + 70yz + 25z^2$

19 $36x^2 + 60xy + 25y^2$

20 $100x^2 - 180xz + 81z^2$

21 $-6ab + a^2b^2 + 9$

22 $-4x + 4x^2 + 1$

23 $4a^4 + 12a^2 + 9$

24 $121b^4 - 22b^2 + 1$

In problems 25–50, use Special Product 3 to factor the polynomials.

25 $x^2 - 9$

26 $y^2 - 16$

27 $4 - x^2$

28 $36 - x^2$

29 $4x^2 - 1$

30 $4x^2 - 9$

31 $25 - 81y^2$

32 $64x^2 - 9y^2$

33 $4a^2 - b^2$

34 $9a^2 - 25$

35 $25 - 49a^2$

36 $4x^2 - 9a^2$

37 $4x^2 - 16$

38 $100 - 121y^2$

39 $a^2b^2 - 1$

40 $x^2y^2 - 9$

41 $x^2z^2 - 16$

42 $a^2b^2c^2 - 144$

43 $x^2y^2 - z^2$

44 $16x^2 - 49y^2$

45 $9m^2 - 144n^2$

46 $16a^2b^2 - 25c^2$

47 $9x^2 - 16z^2$

48 $25x^2 - 49y^2z^2$

49 $81a^2b^2 - 25c^2$

50 $64 - 49y^2z^2$

3.4 Factoring Trinomials of the Form $x^2 + bx + c$

In this section and the following one, we shall factor expressions of the form $ax^2 + bx + c$, where a, b, and c are integers and $a \neq 0$. Such expressions are called *second-degree trinomials* or *quadratic trinomials*. We first consider the case where $a = 1$, that is, trinomial expressions of the form

$x^2 + bx + c$. Examples of such trinomials are $x^2 + 5x + 6$, $x^2 - x - 12$, and $x^2 + 8x + 7$.

One way of factoring a trinomial such as $x^2 + 5x + 6$ is to write it as

$$x^2 + 5x + 6 = x^2 + 3x + 2x + 6$$

so that by the distributive property we have

$$x^2 + 5x + 6 = x(x + 3) + 2(x + 3)$$

This new expression has terms which contain a common binomial factor of $x + 3$, so that

$$x^2 + 5x + 6 = (x + 2)(x + 3)$$

To factor trinomials by this method can be tedious, even for the relatively simple expressions such as the example above. Thus, we shall consider an alternative procedure. If the binomials $x + p$ and $x + q$ are factors of the quadratic trinomial $x^2 + bx + c$, we can write

$$\begin{aligned} x^2 + bx + c &= (x + p)(x + q) \qquad \text{for some integers } p \text{ and } q \\ &= x^2 + px + qx + pq \\ &= x^2 + (p + q)x + pq \end{aligned}$$

so that

$$x^2 + bx + c = x^2 + (p + q)x + pq$$

The corresponding coefficients are indicated in the diagram.

Therefore, to factor a quadratic trinomial of the form $x^2 + bx + c$, we must be able to find integers p and q such that

$$b = p + q \qquad \text{and} \qquad c = pq$$

For example, in factoring the expression $x^2 + 5x + 6$, we try to find integers p and q so that

$$x^2 + 5x + 6 = (x + p)(x + q) = x^2 + (p + q)x + pq$$

where

$$5 = p + q \qquad \text{and} \qquad 6 = pq$$

That is, we attempt to find those factors of 6 whose sum is 5. Since both 6 and 5 are positive, we need only to examine the positive factors of 6. (Why?) The positive integral factors of 6 are 1 and 6, or 2 and 3. Since $1 + 6 = 7$ and $2 + 3 = 5$, replacing p by 2 and q by 3, we have

$$x^2 + 5x + 6 = (x + 2)(x + 3)$$

Similarly, to factor $x^2 - 4x + 3$, we try to find integers p and q such that

$$x^2 - 4x + 3 = (x + p)(x + q) = x^2 + (p + q)x + pq$$

where

$$-4 = p + q \qquad \text{and} \qquad 3 = pq$$

Since the factors of 3 are 1 and 3 or -1 and -3, and of these two possibilities, only $(-1) + (-3) = -4$, we have

$$x^2 - 4x + 3 = (x - 1)(x - 3)$$

In general, to factor a trinomial of the form $x^2 + bx + c$, we adopt the following steps:

1 Determine all pairs of integral factors that have c as their product, then identify the particular pair that has b as its sum.
2 Substitute one of the factors in step 1 for p and the other for q in the product $(x + p)(x + q)$ to obtain the binomial factors of the trinomial.

EXAMPLES

Factor the following trinomials.

1 $x^2 + 8x + 7$

SOLUTION

Step 1. The pair of integral factors of the constant term 7 whose sum is 8 are 1 and 7.

Step 2. Replace p by 1 and q by 7 in $(x + p)(x + q)$:

$$x^2 + 8x + 7 = (x + 1)(x + 7)$$

2 $x^2 - 8x + 12$

SOLUTION

Step 1. The pair of integral factors of the constant term 12 whose sum is -8 are -6 and -2.

Step 2. Replace p by -6 and q by -2 in $(x + p)(x + q)$ so that

$$x^2 - 8x + 12 = (x - 6)(x - 2)$$

3 $x^2 + 2x - 15$

SOLUTION

Step 1. The pair of integral factors of the constant term -15 whose sum is 2 are -3 and 5.

Step 2. Replace p by -3 and q by 5 in $(x + p)(x + q)$ so that

$$x^2 + 2x - 15 = (x - 3)(x + 5)$$

4 $x^2 - 7xy - 18y^2$

SOLUTION

Step 1. The pair of integral factors of the term -18 whose sum is -7 are 2 and -9.

Step 2. Replace p by $2y$ and q by $-9y$ in $(x + p)(x + q)$ so that

$$x^2 - 7xy - 18y^2 = (x + 2y)(x - 9y)$$

PROBLEM SET 3.4

In problems 1–40, factor the quadratic trinomials.

1 $x^2 + 3x + 2$

2 $y^2 + 4y + 3$

3 $x^2 + 6x + 5$

4 $z^2 + 7z + 6$

5 $y^2 - 10y + 21$

6 $x^2 - 9x + 20$

7 $z^2 - 7z + 10$

8 $x^2 - 12x + 35$

9 $x^2 + 2x - 3$

10 $y^2 - y - 20$

11 $x^2 - 4x - 12$

12 $z^2 + 6z - 27$

13 $y^2 - 11y + 30$

14 $x^2 - 7x - 8$

15 $x^2 + 15x + 56$

16 $y^2 - 13y + 40$

17 $x^2 + 3x - 40$

18 $x^2 + 15x + 54$

19 $z^2 - 12z + 20$

20 $x^2 + 2x - 35$

21 $x^2 - x - 30$

22 $z^2 - 5z - 50$

23 $y^2 + 2yz - 63z^2$

24 $x^2 + 14xy - 15y^2$

25 $x^2 - 4xy - 32y^2$

26 $y^2 + 7yz - 30z^2$

27 $x^2 + 13xz + 42z^2$

28 $x^2 + 15xz + 44z^2$

29 $y^2 - 4yz - 45z^2$

30 $x^2 - 17xy + 72y^2$

31 $x^2 - 18x + 81$

32 $x^2 - 20xz - 44z^2$

33 $x^2 + 8xy - 33y^2$

34 $x^2 + 16xy + 64y^2$

35 $x^2 + x - 110$

36 $y^2 - y - 132$

37 $x^2 + 11xy - 42y^2$

38 $y^2 + 2yz - 48z^2$

39 $x^2 - 9xz - 36z^2$

40 $z^2 - 15zx + 36x^2$

3.5 Factoring Trinomials of the Form $ax^2 + bx + c$, $a \neq 0$

In this section we shall generalize our factoring to include quadratic trinomials of the form $ax^2 + bx + c$, where $a \neq 0$ and $a \neq 1$. That is, we shall attempt to factor trinomials such as $2x^2 + 7x + 3$, $3x^2 - 5x + 2$, and $6x^2 + 5x - 6$. If such trinomials are factorable, then their factors are of the forms $sx + p$ and $rx + q$. For instance, to verify that $3x - 2$ and $2x + 3$ are factors of $6x^2 + 5x - 6$, we multiply the two factors $3x - 2$ and $2x + 3$ to obtain the polynomial $6x^2 + 5x - 6$. That is,

$$\begin{aligned}(3x - 2)(2x + 3) &= (3)(2)x^2 + [(3)(3) + (-2)(2)]x + (-2)(3)\\ &= 6x^2 + 5x - 6\end{aligned}$$

This is also true in general; that is, if the binomials $(sx + p)$ and $(rx + q)$ are factors of the quadratic trinomial $ax^2 + bx + c$, we can proceed as we did in Section 3.4 by writing the trinomial as

$$\begin{aligned}ax^2 + bx + c &= (sx + p)(rx + q)\\ &= srx^2 + sxq + prx + pq\\ &= srx^2 + (sq + pr)x + pq\end{aligned}$$

for some integers s, p, r, and q so that

$$ax^2 + bx + c = srx^2 + (sq + pr)x + pq$$

The corresponding coefficients are indicated in the diagram.

Therefore, to factor a quadratic trinomial of the form $ax^2 + bx + c$ we must be able to find integers s, p, r, and q such that

$$a = sr \qquad b = sq + pr \qquad \text{and} \qquad c = pq$$

For example, to factor $2x^2 + 7x + 3$, we try to find integers s, p, r, and q such that

$$2x^2 + 7x + 3 = (sx + p)(rx + q) = srx^2 + (sq + pr)x + pq$$

We start with the fact that $2 = sr$, $7 = sq + pr$, and $3 = pq$. From the statement $2 = sr$, we have either $s = 2$ and $r = 1$ or $s = 1$ and $r = 2$. This is because the only positive factors of 2 are 2 and 1. To complete the factoring of the trinomial, we must find values of p and q that are factors of 6 such that $sq + pr = 7$. There are no definite rules that give us help here. However, trial and error and insight gained by experience are the key to success in performing such factoring. We try $p = 3$ and $q = 1$ or $p = 1$ and $q = 3$. Since $7 = sq + pr = (2) \times (3) + (1) \times (1)$, we take $s = 2$, $q = 3$, $p = 1$, and $r = 1$. Therefore, the factors of the trinomial $2x^2 + 7x + 3$ are $2x + 1$ and $x + 3$. That is,

$$2x^2 + 7x + 3 = (2x + 1)(x + 3)$$

To factor the trinomial $3x^2 - 7x - 6$, we attempt to find integers, s, p, q, and r such that

$$3x^2 - 7x - 6 = (sx + p)(rx + q) = srx^2 + (sq + pr)x + pq$$

where $3 = sr$, $-7 = sq + pr$, and $-6 = pq$. The number -6 has several possible pairs of factors while 3 has only 3 and 1 as positive factors. We list below the possible combinations that satisfy the above requirements:

Possible combinations	*Products*
$(3x - 6)(x + 1)$	$3x^2 - 3x - 6$
$(3x - 1)(x + 6)$	$3x^2 + 17x - 6$
$(3x - 3)(x + 2)$	$3x^2 + 3x - 6$
$(3x - 2)(x + 3)$	$3x^2 + 7x - 6$
$(3x + 6)(x - 1)$	$3x^2 + 3x - 6$
$(3x + 1)(x - 6)$	$3x^2 - 17x - 6$
$(3x + 3)(x - 2)$	$3x^2 - 3x - 6$
$(3x + 2)(x - 3)$	$3x^2 - 7x - 6$

From the above possibilities, we see that the correct factorization is

$$3x^2 - 7x - 6 = (3x + 2)(x - 3)$$

With sufficient practice, one is usually able to find the correct factors by checking possible combinations mentally without actually listing all possibilities. However, an awareness of the distribution of signs in the trinomial can be quite helpful.

The following rules give a clue of the sign analysis:

1 If the sign of the last term of a factorable trinomial is positive, the signs of the second terms in both binomial factors are the same as that of the middle term of the trinomial.
2 If the sign of the last term of a factorable trinomial is negative, the signs of the second terms of the two binomial factors differ, and the sign of the middle term goes with the larger number used in the determined factors.

EXAMPLES

Factor the following trinomials.

1 $2x^2 + 11x + 12$

SOLUTION. There are two factors, each a binomial; hence we express $2x^2 + 11x + 12$ in the form of adjacent parentheses,

$$(2x + \_\_\_\_)(x + \_\_\_\_).$$

The rule of signs will confirm this possibility. Next, we consider the positive integral factors of 12, which are 1 and 12, 2 and 6, or 3 and 4. Trying different possible combinations of the factors 2 and 12 we find that $(2) \times (4) + (3) \times (1) = 11$, which is the coefficient of the middle term, so that $2x + 3$ and $x + 4$ are the factors of the trinomial. Hence,

$$2x^2 + 11x + 12 = (2x + 3)(x + 4)$$

2 $3x^2 - 10x + 8$

SOLUTION. By the rule of signs, we first have

$$3x^2 - 10x + 8 = (3x - \_\_\_\_)(x - \_\_\_\_)$$

The positive factors of 8 are 8 and 1 or 4 and 2. Trying different possible combinations in the expression above, we find that $(3) \times (-2) + (-4) \times (1) = -10$, which gives the coefficient of the middle term. So $3x - 4$ and $x - 2$ are factors of the trinomial. Hence,

$$3x^2 - 10x + 8 = (3x - 4)(x - 2)$$

3 $6x^2 + 7x - 3$

SOLUTION. Since the positive factors of 3 are 3 and 1, by the rule of signs we have either

$$6x^2 + 7x - 3 = (\_\_\_\_ x + 3)(\_\_\_\_ x - 1)$$

or

$$6x^2 + 7x - 3 = (\_\_\_\_ x - 3)(\_\_\_\_ x + 1)$$

The positive factors of 6 are 6 and 1 or 3 and 2. Trying different possible combinations in the expressions above, we find that $(2) \times (-1) + (3) \times (3) = 7$, which gives the coefficient of the middle term. Therefore, $2x + 3$ and $3x - 1$ are factors of the trinomial. Hence,

$$6x^2 + 7x - 3 = (2x + 3)(3x - 1)$$

4 $7x^2 - 9xy - 10y^2$

SOLUTION. Since the positive factors of 7 are 7 and 1, by the rule of signs, we have either

$$7x^2 - 9xy - 10y^2 = (7x - \_\_\_\_)(x + \_\_\_\_)$$

or

$$7x^2 - 9xy - 10y^2 = (7x + \_\_\_\_)(x - \_\_\_\_)$$

The positive factors of 10 are 10 and 1 or 5 and 2. Trying different possible combinations in the expression above, we find that $(7) \times (-2) + (5) \times (1) = -9$, which is the coefficient of the middle term, so that $7x + 5y$ and $x - 2y$ are factors of the trinomial. Therefore,

$$7x^2 - 9xy - 10y^2 = (7x + 5y)(x - 2y)$$

PROBLEM SET 3.5

In problems 1–40, factor the quadratic trinomials.

1 $2x^2 + 5x + 3$

2 $3y^2 + 4y + 1$

3 $3x^2 + 7x + 2$

4 $5x^2 + 14x + 8$

5 $2y^2 - 7y + 6$

6 $4y^2 - 9y + 2$

7 $5x^2 + 3x - 2$

8 $3x^2 - 13x + 12$

9 $6x^2 + 7x + 2$

10 $4x^2 - x - 3$

11 $4z^2 + 7z - 2$

12 $4x^2 + 4x - 3$

13 $2x^2 - 11x + 15$

14 $12y^2 - 13y - 14$

15 $6y^2 - y - 12$

16 $30z^2 + 13z - 3$

17 $12x^2 + xy - 6y^2$

18 $28x^2 - 45x + 18$

19 $2z^2 + 13z - 24$

20 $20x^2 - 9xy - 20y^2$

21 $18y^2 - 9yz - 2z^2$

22 $48z^2 + 22z - 15$

23 $15x^2 - 11xy - 14y^2$

24 $32x^2 + 60xz + 27z^2$

25 $24x^2 - 34x + 5$

26 $36y^2 - 11y - 12$

27 $6x^2 + xy - 35y^2$

28 $6x^2 - 43xz + 72z^2$

29 $42y^2 - 5y - 2$

30 $40y^2 + 11yz - 63z^2$

31 $2x^2 - 7x - 30$

32 $63x^2 - 69x - 20$

33 $7x^2 + 31xy + 12y^2$

34 $36y^2 + 40yz - 21z^2$

35 $9y^2 - 27y + 20$

36 $24x^2 + 58xy + 33y^2$

37 $48z^2 - 74z + 21$

38 $32x^2 - 4x - 21$

39 $20x^2 + xy - 30y^2$

40 $54y^2 - 93yz + 40z^2$

3.6 Miscellaneous Factoring

There are other polynomials that require the application of more than one of the methods described in the previous sections. In factoring such polynomials, the following steps are suggested:

1 Remove common factors if they occur.
2 Examine the remaining polynomial to see if it is factorable. If it is, perform the factoring by any method studied in the chapter.

The following examples illustrate this procedure.

EXAMPLES

Factor the following polynomials by following the steps above.

1 $x^2y - y^3$

SOLUTION

$$\begin{aligned} x^2y - y^3 &= y(x^2 - y^2) && \text{(Step 1)} \\ &= y(x - y)(x + y) && \text{(Step 2)} \end{aligned}$$

2 $x^2y - 3xy - 4y$

SOLUTION

$$\begin{aligned} x^2y - 3xy - 4y &= y(x^2 - 3x - 4) \\ &= y(x - 4)(x + 1) \end{aligned}$$

Now we briefly examine other types of factoring.

Factoring by Grouping

Some polynomials that do not contain factors common to each term may still be factorable by the method of common factors if we first group the terms of the polynomials and look for common polynomial factors in each group. For example, consider the polynomial $3xm + 3ym - 2x - 2y$. We first note that two terms contain a factor of x and the other two terms contain a factor of y, so that by grouping these terms accordingly, we obtain

$$3xm + 3ym - 2x - 2y = (3xm - 2x) + (3ym - 2y)$$

Next, factor out the common monomial x from the first group and the common monomial y from the second group, so that

$$3xm + 3ym - 2x - 2y = (3m - 2)x + (3m - 2)y$$

Factoring out the common binomial, $(3m - 2)$, we have

$$3xm + 3ym - 2x - 2y = (3m - 2)(x + y)$$

EXAMPLES

Factor the given polynomials by grouping.

1 $4y + xy + 4y^2 + x$

SOLUTION

$$\begin{aligned} 4y + xy + 4y^2 + x &= (4y + 4y^2) + (x + xy) \\ &= 4y(1 + y) + x(1 + y) \\ &= (4y + x)(1 + y) \end{aligned}$$

2 $3a + 3b - ma - mb$

SOLUTION

$$\begin{aligned} 3a + 3b - ma - mb &= (3a + 3b) + (-ma - mb) \\ &= 3(a + b) - m(a + b) \\ &= (a + b)(3 - m) \end{aligned}$$

Factoring the Sum or the Difference of Two Cubes

To factor polynomial expressions such as $A^3 + B^3$ and $A^3 - B^3$, we consider the following special products, which prove to be helpful. Let A and B represent real numbers or algebraic expressions; then

SPECIAL PRODUCT 4

$$(A + B)(A^2 - AB + B^2) = A^3 + B^3$$

To verify Special Product 4, we have

$$\begin{aligned}(A + B)(A^2 - AB + B^2) &= A(A^2 - AB + B^2) + B(A^2 - AB + B^2)\\ &= A^3 - A^2B + AB^2 + BA^2 - AB^2 + B^3\\ &= A^3 + B^3\end{aligned}$$

SPECIAL PRODUCT 5

$$(A - B)(A^2 + AB + B^2) = A^3 - B^3$$

To verify Special Product 5, we have

$$\begin{aligned}(A - B)(A^2 + AB + B^2) &= A(A^2 + AB + B^2) - B(A^2 + AB + B^2)\\ &= A^3 + A^2B + AB^2 - BA^2 - AB^2 - B^3\\ &= A^3 - B^3\end{aligned}$$

Now we reverse the process to be able to factor third-degree binomials whose terms are the cubes of some numbers or algebraic expressions. When using Special Products 4 and 5, these polynomials are written as:

1 $A^3 + B^3 = (A + B)(A^2 - AB + B^2)$
2 $A^3 - B^3 = (A - B)(A^2 + AB + B^2)$

The following examples illustrate the procedure.

EXAMPLES

Use Special Product 4 or 5 to factor the following third-degree binomials.

1 $x^3 + 8$

SOLUTION

$$x^3 + 8 = x^3 + 2^3$$

which is in the form of Special Product 4, where $A = x$ and $B = 2$, so that

$$x^3 + 2^3 = (x + 2)[x^2 - x(2) + (2)^2]$$

or

$$x^3 + 8 = (x + 2)(x^2 - 2x + 4)$$

2 $y^3 - 27$

SOLUTION

$$y^3 - 27 = y^3 - 3^3$$

which is in the form of Special Product 5, where $A = y$ and $B = 3$, so that

$$y^3 - 3^3 = (y - 3)[y^2 + y(3) + (3)^2]$$

or

$$y^3 - 27 = (y - 3)(y^2 + 3y + 9)$$

PROBLEM SET 3.6

In problems 1–10, use miscellaneous factoring and the steps suggested in this section to factor the polynomials.

1 $5x^3 - 5xy^2$

2 $28x^4y - 7x^2y^3$

3 $x^3y - 2x^2y - 3xy$

4 $3a^2x^3y - 3a^2xy^3$

5 $ax^3 - 4ax^2 + 4ax$

6 $27ab^3 - 108a^3b$

7 $4x^3 - 8x^2y - 5xy^2$

8 $6x^2y + 5xy^2 + y^3$

9 $3x^3 - 2x^2y - xy^2$

10 $3x^3 | 4x^2y + xy^2$

In problems 11–24, factor the polynomials by the method of grouping.

11 $(a + b)x + 2(a + b)$

12 $x(z + 1) + y(z + 1)$

13 $3(2x + y) - z(2x + y)$

14 $c(x + y) - 5(x + y)$

15 $mx - 4mn + mnx - 4m$

16 $cd + 7c + 2d + 14$

17 $ab - 5a - 15 + 3b$

18 $7s - 2t - 14 + st$

19 $ab - xa + a^2 - xb$

20 $3ac - ad - 2bd + 6bc$

21 $a^2x^2 - 5x^2 - 4a^2 + 20$

22 $ma^2 + mb^2 - ka^2 - kb^2$

23 $bc + 3c - 4b - 12$

24 $xz^2 + 5z^2 + x + 5 + xz + 5z$

In problems 25–40, use Special Products 4 and 5 to factor the polynomials.

25 $x^3 + y^3$

26 $x^3 + 1$

27 $8x^3 + y^3$

28 $x^3 + 8y^3$

29 $27x^3 + y^3$

30 $8x^3 + 27y^3$

31 $x^3 - 1$

32 $8x^3 - 1$

33 $27m^3 - 1$

34 $64 - a^3$

35 $125a^3 + b^3$

36 $a^3 - 125b^3$

37 $1 - 1{,}000x^3$

38 $27m^3n^3 + 64$

39 $a^3b^3 + c^3$

40 $x^3y^3z^3 + w^3$

REVIEW PROBLEM SET

In problems 1–10, factor the polynomials by removing the common monomial factors.

1 $3x^2 - 5x$

2 $4x^2 + 6x^3$

3 $4x^2y + 8xy^3$

4 $6x^3z^2 - 15x^2z$

5 $3x^3y^2 + 6x^2y^4 - 9xy^3$

6 $4a^4b^3 - 8a^3b^5 + 12a^5b^2$

7 $7m^4n^2 - 14m^3n + 21m^2n^3$

8 $10x^3z^2 - 15xz^3 + 25x^2z^4$

9 $6a^3b^4c^2 - 12a^4b^2c^3 - 18a^3b^3c^2$

10 $26x^3y^4z^7 - 39x^2y^5z^4 + 13x^5y^3z^6$

In problems 11–14, factor the expressions by removing the common binomial factors.

11 $8x(y - z) + 3(y - z)$

12 $10x^2(y + 2z) + 9(y + 2z)$

13 $7x^2(a + b) + x(a + b)$

14 $4x^2(2y + 1) - x(2y + 1)$

In problems 15–24, use Special Products 1 and 2 to find the following squares of a binomial.

15 $(x + 2)^2$

16 $(x + 7)^2$

17 $(y - 3)^2$

18 $(z - 5)^2$

19 $(2x + 1)^2$

20 $(7xy - 3)^2$

21 $(3y + 4z)^2$

22 $(8y + 5z)^2$

23 $(2xy - 5)^2$

24 $(4ab + 7c)^2$

In problems 25–36, use Special Product 3 to find the products.

25 $(x + 2)(x - 2)$

26 $(y - 7)(y + 7)$

27 $(x - 5y)(x + 5y)$

28 $(xy - 10)(xy + 10)$

29 $(3a + 8b)(3a - 8b)$

30 $(7z - 9x)(7z + 9x)$

31 $(4y + 5z)(4y - 5z)$

32 $(10y + 11z)(10y - 11z)$

33 $(11 - 3y)(11 + 3y)$

34 $(5 - 7xy)(5 + 7xy)$

35 $(7 - 4xyz)(7 + 4xyz)$

36 $(9 - 10xyz)(9 + 10xyz)$

In problems 37–48, use Special Product 1 or 2 to factor the polynomials.

37 $1 - 14xy + 49x^2y^2$

38 $x^2 + 10xy + 25y^2$

39 $100 - 60ab + 9a^2b^2$

40 $121 - 22abc + a^2b^2c^2$

41 $x^2 + 4x + 4$

42 $z^2 - 12z + 36$

43 $y^2 - 6y + 9$

44 $9a^2 + 12ac + 4c^2$

45 $4x^2 + 28xy + 49y^2$

46 $36x^2 - 84xz + 49z^2$

47 $16x^2y^2 - 24xy + 9$

48 $1 + 10y^2 + 25y^4$

In problems 49–60, use Special Product 3 to factor the polynomials.

49 $x^2 - 81$

50 $y^2 - 25$

51 $25 - 4y^2$

52 $36x^2 - 49$

53 $16a^2 - 81b^2$

54 $25y^2 - 9z^2$

55 $4x^2y^2 - 9z^2$

56 $16a^2b^2c^2 - 9$

57 $1 - 144x^2y^2$

58 $9 - 169a^2b^2$

59 $100 - 81a^2b^2c^2$

60 $81 - 16x^2y^2z^2$

In problems 61–80, factor the quadratic trinomials.

61 $x^2 + 5x + 4$

62 $z^2 - 10z + 9$

63 $x^2 + 3x - 4$

64 $x^2 + 6x - 16$

65 $y^2 - 8y + 7$

66 $x^2 - 10x - 11$

67 $x^2 - 3xy - 54y^2$

68 $y^2 + 12yz - 28z^2$

69 $a^2 + 18ab + 77b^2$

70 $a^2b^2 - 3abc - 10c^2$

71 $2x^2 + x - 3$

72 $5x^2 - 11x + 2$

73 $6x^2 + 7x - 3$

74 $6x^2 + 25x + 4$

75 $12y^2 - 11y + 2$

76 $33x^2 - 20x - 32$

77 $12x^2 + 25xy + 12y^2$

78 $10y^2 - 13yz - 30z^2$

79 $30y^2 - yz - 20z^2$

80 $48x^2 + 10xy - 3y^2$

In problems 81–86, factor the polynomials by the method of grouping.

81 $2ax - 3a + 2bx - 3b$

82 $m^2a^2 + n^2b^2 + m^2b^2 + n^2a^2$

83 $2x^3 - y^2 - x^2 + 2xy^2$

84 $a^2x^2 + 3a^2 - 3b^2 - b^2x^2$

85 $ax^2 + bx^2 + 2ax + 2bx + 3a + 3b$

86 $2m^2x + 6mx + 10x - 5y - 3my - m^2y$

In problems 87–98, use the methods described in Section 3.6 to factor the polynomials.

87 $x^3y^2 - 9xy^4$

88 $x^5 - 16xy^4$

89 $x^3y + 4x^2y^2 + 3xy^3$

90 $2ax^3 - ax^2z - 6axz^2$

91 $16x^4 - 81$

92 $16 - x^2 + 2xy - y^2$

93 $a^2 - 4ab + 4b^2 - 9$

94 $ky^5 - 7ky^3 + 6ky$

95 $x^2 - a^2 + 4xy + 2a + 4y^2 - 1$

96 $x^2 + 4y^2 - m^2 - n^2 + 2mn - 4xy$

97 $x^4 + 2x^2 - 15$

98 $2x^4 + x^2y^2 - 6y^4$

In problems 99–104, use Special Product 4 or 5 to factor the expressions.

99 $x^3 + 27$

100 $y^3 - 64$

101 $8y^3 - z^3$

102 $8x^3 + 125y^3$

103 $27 + a^3b^3$

104 $64x^3y^3 + 27z^3$

CHAPTER 4

Algebra of Fractions

4 ALGEBRA OF FRACTIONS

Introduction

In Chapters 2 and 3 we discussed operations on polynomial expressions. In this chapter we learn how to use the quotient of polynomials in various operations. We first begin with the operations of rational numbers. Later in the chapter we see that the principles governing operations with rational expressions are extensions of those for rational numbers.

4.1 Multiplication and Division of Fractions

We indicated in Section 1.3 that a rational number can be represented in the form of $\frac{a}{b}$, where a and b are integers and $b \neq 0$. In keeping with the number–numeral distinction, it may be said that a number has many equivalent "names." For example, the rational number that is midway between 1 and 2 may be "named" by the fractions

$$\frac{3}{2}, \quad \frac{18}{12}, \quad \frac{-21}{-14}, \quad \frac{150}{100}, \quad \frac{10 + 5}{6 + 4}, \quad \text{etc.}$$

Two fractions are said to be *equivalent* if they have the same value but different forms, or if they correspond to the same point on the number line. For example, $\frac{1}{2}$ and $\frac{4}{8}$ are equivalent, $\frac{4}{3}$ and $\frac{12}{9}$ are equivalent, and $-\frac{2}{3}$ and $-\frac{6}{9}$ are equivalent. It should be noted that the number line cannot serve as a formal definition of equivalent fractions, because it is difficult to tell whether two fractions such as $\frac{34}{102}$ and $\frac{26}{78}$ correspond to the same point. However, a simple test can be used to answer such questions. Compare the equivalent fractions $\frac{1}{2}$ and $\frac{4}{8}$. Notice that the "cross" products 1×8 and 2×4 are the same. Also, $-\frac{2}{3}$ and $-\frac{6}{9}$ are equivalent, since $(-2) \times 9 = 3 \times (-6)$. This pattern can be generalized by the following property.

PROPERTY 1

Two fractions $\frac{a}{b}$ and $\frac{c}{d}$, with $b \neq 0$ and $d \neq 0$, are equivalent if and only if $ad = bc$.

In this book, however, we denote two equivalent fractions by writing an equal sign between them. Accordingly, Property 1 can be expressed in symbols as follows:

$$\frac{a}{b} = \frac{c}{d} \quad \text{if and only if} \quad ad = bc$$

EXAMPLE

Indicate which pairs of fractions are equivalent.

a) $\frac{5}{7}$ and $\frac{15}{21}$ b) $\frac{-4}{11}$ and $\frac{4}{-11}$ c) $\frac{6}{7}$ and $\frac{5}{8}$

SOLUTION

a) Since $5 \times 21 = 7 \times 15$, then $\frac{5}{7} = \frac{15}{21}$.

b) Since $(-4) \times (-11) = 11 \times 4$, then $\frac{-4}{11} = \frac{4}{-11}$.

c) Since $6 \times 8 \neq 5 \times 7$, then $\frac{6}{7} \neq \frac{5}{8}$.

If the numerator and denominator of a fraction are integers and the denominator is not zero, the fraction is said to be in *lowest terms* when the numerator and the denominator have no common factor other than 1. For example, $\frac{2}{5}$, $\frac{5}{7}$, and $\frac{3}{17}$ are in lowest terms, whereas $\frac{4}{6}$ and $\frac{3}{9}$ are not, since each fraction has a common factor in its numerator and denominator. The procedure of expressing a given fraction as an equivalent fraction in lowest terms is called *reducing the fraction to lowest terms.* To reduce a fraction to lowest terms, we use the following property, called the *fundamental principle of fractions.* If $\frac{a}{b}$ is a fraction where $b \neq 0$, and if $k \neq 0$, then

$$\frac{ak}{bk} = \frac{a}{b}$$

EXAMPLES

In examples 1 and 2, reduce each fraction to an equivalent fraction in its lowest terms.

1 $\frac{35}{55}$

SOLUTION

$$\frac{35}{55} = \frac{7 \times 5}{11 \times 5} = \frac{7}{11}$$

2 $\frac{26}{91}$

SOLUTION

$$\frac{26}{91} = \frac{2 \times 13}{7 \times 13} = \frac{2}{7}$$

In examples 3 and 4, find the missing number for each of the equivalent fractions.

3 $\frac{5}{7} = \frac{?}{63}$

SOLUTION

$$\frac{5}{7} = \frac{5 \times 9}{7 \times 9} = \frac{45}{63}$$

so the unknown number is 45.

4 $\frac{13}{17} = \frac{?}{119}$

SOLUTION

$$\frac{13}{17} = \frac{13 \times 7}{17 \times 7} = \frac{91}{119}$$

so the unknown number is 91.

Now, we combine fractions using the operations of multiplication, division, addition, and subtraction.

Multiplication of Fractions

To multiply two fractions, multiply their numerators and multiply their denominators. That is, if $\frac{a}{b}$ and $\frac{c}{d}$ are any fractions, their product is given by

$$\frac{a}{b} \cdot \frac{c}{d} = \frac{ac}{bd}$$

The following examples illustrate this rule.

EXAMPLES

Find the following products and reduce them to lowest terms.

1 $\frac{4}{8} \times \frac{5}{6}$

SOLUTION

$$\frac{4}{8} \times \frac{5}{6} = \frac{4 \times 5}{8 \times 6} = \frac{20}{48} = \frac{5 \times 4}{12 \times 4} = \frac{5}{12}$$

2 $\frac{5}{11} \times \frac{-22}{65}$

SOLUTION

$$\frac{5}{11} \times \frac{-22}{65} = \frac{5 \times (-22)}{11 \times 65} = \frac{-110}{715} = \frac{55 \times (-2)}{55 \times 13} = \frac{-2}{13}$$

Division of Fractions

Recall from Section 1.7, Inverse Property 6b, that $a \cdot \frac{1}{a} = \frac{1}{a} \cdot a = 1$, $a \neq 0$. $\frac{1}{a}$ is the multiplicative inverse or the reciprocal of a. Using the rule for multiplication of fractions and the fundamental principle of fractions, we see that

$$\frac{a}{b} \cdot \frac{b}{a} = 1 \quad \text{provided that} \quad \frac{a}{b} \neq 0$$

Therefore, the *multiplicative inverse*, or the *reciprocal* of $\frac{a}{b}$, is $\frac{b}{a}$. For example, the reciprocal of $\frac{2}{3}$ is $\frac{3}{2}$ and the reciprocal of $\frac{-7}{9}$ is $\frac{9}{-7}$. In Section 1.5, we defined division of numbers a and b, with $b \neq 0$, by

$$a \div b = a \cdot \frac{1}{b}$$

This definition can be applied to the division of fractions, that is, if $\frac{a}{b}$ and $\frac{c}{d}$ are rational numbers, with $\frac{c}{d} \neq 0$, then

$$\frac{a}{b} \div \frac{c}{d} = \frac{a}{b} \cdot \frac{1}{\frac{c}{d}} = \frac{a}{b} \cdot \frac{d}{c} = \frac{ad}{bc}$$

EXAMPLES

Perform the following divisions and reduce them to lowest terms.

1 $\frac{5}{8} \div \frac{3}{4}$

SOLUTION

$$\frac{5}{8} \div \frac{3}{4} = \frac{5}{8} \times \frac{4}{3} = \frac{20}{24} = \frac{4 \times 5}{4 \times 6} = \frac{5}{6}$$

2 $\frac{51}{95} \div \frac{34}{57}$

SOLUTION

$$\frac{51}{95} \div \frac{34}{57} = \frac{51}{95} \times \frac{57}{34} = \frac{51 \times 57}{95 \times 34} = \frac{3 \times 17 \times 3 \times 19}{5 \times 19 \times 2 \times 17} = \frac{9}{10}$$

The order of operations discussed in Section 1.1 can now be applied to fractions, as the following example shows.

EXAMPLE

Perform the following operations and reduce the result to lowest terms:

$$\frac{10}{8} \times \frac{8}{5} \div \frac{7}{15} \times \frac{21}{40}$$

SOLUTION. Using the order of operations by performing multiplications and divisions from left to right, we have

$$\frac{10}{8} \times \frac{8}{5} \div \frac{7}{15} \times \frac{21}{40} = 2 \div \frac{7}{15} \times \frac{21}{40}$$

$$= 2 \times \frac{15}{7} \times \frac{21}{40}$$

$$= \frac{30}{7} \times \frac{21}{40} = \frac{9}{4}$$

PROBLEM SET 4.1

In problems 1–8, use the property $\frac{a}{b} = \frac{c}{d}$ if and only if $a \cdot d = b \cdot c$ to indicate which pairs of fractions are equivalent.

1 $\frac{2}{3}$ and $\frac{34}{51}$

2 $\frac{7}{11}$ and $\frac{-56}{-88}$

3 $\frac{60}{84}$ and $\frac{5}{7}$

4 $\frac{-2}{3}$ and $\frac{-56}{84}$

5 $\frac{-5}{-8}$ and $\frac{45}{72}$

6 $\frac{9}{11}$ and $\frac{36}{44}$

7 $\frac{10}{11}$ and $\frac{13}{22}$

8 $\frac{-9}{13}$ and $\frac{15}{-26}$

In problems 9–18, use the fundamental principle of fractions to reduce each fraction into an equivalent fraction in its lowest terms.

9 $\frac{15}{45}$

10 $\frac{28}{42}$

11 $\frac{77}{121}$

12 $\frac{-56}{96}$

13 $\frac{-65}{91}$

14 $\frac{98}{168}$

15 $\frac{250}{350}$

16 $\frac{156}{520}$

17 $\frac{65}{-39}$

18 $\frac{176}{464}$

In problems 19–24, use the fundamental principle of fractions to determine the missing number for each fraction.

19 $\frac{7}{8} = \frac{?}{96}$

20 $\frac{9}{13} = \frac{?}{91}$

21 $\frac{2}{3} = \frac{?}{81}$

22 $\frac{-9}{13} = \frac{?}{-52}$

23 $\frac{7}{12} = \frac{?}{-132}$

24 $\frac{-7}{17} = \frac{?}{85}$

In problems 25–38, find the products and reduce them to lowest terms.

25 $\frac{4}{5} \times \frac{10}{12}$

26 $\frac{11}{52} \times \frac{39}{80}$

27 $\frac{15}{14} \times \frac{7}{24}$

28 $\frac{14}{36} \times \frac{-24}{35}$

29 $\frac{-57}{34} \times \frac{51}{19}$

30 $\frac{-16}{105} \times \frac{155}{63}$

31 $\frac{9}{5} \times \frac{-5}{27}$

32 $\frac{4}{5} \times \frac{-55}{72}$

33 $\frac{-7}{5} \times \frac{15}{21}$

34 $\frac{-2}{5} \times \frac{3}{-4}$

35 $\frac{-32}{121} \times \frac{-33}{144}$

36 $\frac{-11}{9} \times \frac{-18}{55}$

37 $\frac{-21}{25} \times \frac{-15}{28}$

38 $\frac{-18}{35} \times \frac{-14}{27}$

In problems 39–52, perform the divisions and reduce them to lowest terms.

39 $\frac{3}{4} \div \frac{5}{7}$

40 $\frac{25}{24} \div \frac{15}{16}$

41 $\frac{5}{9} \div \frac{3}{4}$

42 $\frac{35}{24} \div \frac{15}{84}$

43 $\frac{15}{21} \div \frac{3}{5}$

44 $\frac{336}{55} \div \frac{168}{125}$

45 $\frac{-18}{35} \div \frac{9}{14}$

46 $\frac{5}{12} \div \frac{-15}{16}$

47 $\frac{1}{6} \div \frac{-13}{11}$

48 $\frac{-14}{15} \div \frac{-16}{9}$

49 $\frac{-5}{9} \div \frac{-25}{6}$

50 $\frac{-7}{45} \div \frac{-30}{14}$

51 $\frac{20}{21} \div \frac{-15}{14}$

52 $\frac{-16}{105} \div \frac{-93}{155}$

In problems 53–58, find the products and quotients and reduce them to lowest terms.

53 $\frac{8}{5} \times \frac{15}{14} \div \frac{24}{7}$

54 $\frac{1}{3} \times \frac{1}{2} \div \frac{2}{5} \times \frac{3}{7}$

55 $\frac{2}{5} \times \frac{3}{7} \div \frac{3}{2} \times \frac{35}{2}$

56 $\frac{1}{3} \div \frac{5}{2} \times \frac{3}{4} \div \frac{7}{5}$

57 $\frac{2}{3} \div \frac{4}{5} \times \frac{3}{7} \times \frac{14}{9}$

58 $\frac{10}{9} \times \frac{12}{15} \div \frac{3}{5} \times \frac{10}{21}$

4.2 Addition and Subtraction of Fractions

In this section we shall develop methods for adding and subtracting fractions. First, we consider adding and subtracting "like" fractions, that is, fractions with the same denominator.

Addition and Subtraction of Like Fractions

The addition of two like fractions may be accomplished by a direct application of the distributive property. For example, to add $\frac{5}{9}$ and $\frac{2}{9}$, we have

$$\frac{5}{9} + \frac{2}{9} = \left(5 \times \frac{1}{9}\right) + \left(2 \times \frac{1}{9}\right)$$

Using the distributive property, we obtain

$$\frac{5}{9} + \frac{2}{9} = (5 + 2) \times \frac{1}{9} = 7 \times \frac{1}{9} = \frac{7}{9}$$

In general, to add like fractions $\frac{a}{c}$ and $\frac{b}{c}$, we use the fact that $\frac{a}{c} = a \cdot \frac{1}{c}$ and $\frac{b}{c} = b \cdot \frac{1}{c}$, where $c \neq 0$, so that

$$\begin{aligned} \frac{a}{c} + \frac{b}{c} &= a \cdot \frac{1}{c} + b \cdot \frac{1}{c} \\ &= (a + b) \cdot \frac{1}{c} \qquad \text{(Distributive property)} \\ &= \frac{a + b}{c} \end{aligned}$$

The following examples illustrate this rule.

EXAMPLES

Perform the following additions.

1 $\frac{2}{7} + \frac{3}{7}$

SOLUTION. Since $\frac{2}{7}$ and $\frac{3}{7}$ have the same denominator, we apply the rule above, so that

$$\frac{2}{7} + \frac{3}{7} = \frac{2 + 3}{7} = \frac{5}{7}$$

2 $\frac{11}{17} + \frac{5}{17}$

SOLUTION

$$\frac{11}{17} + \frac{5}{17} = \frac{11 + 5}{17} = \frac{16}{17}$$

To subtract like fractions, we use the definition of subtraction of real numbers in terms of addition, so $a - b = a + (-b)$, where $-b$, the additive inverse of b, has the property that $b + (-b) = 0$. Recall from Section 1.6 that $-\left(\frac{c}{b}\right) = \frac{-c}{b}$, so that

$$\frac{a}{b} - \frac{c}{b} = \frac{a}{b} + \left(\frac{-c}{b}\right) = \frac{a}{b} + \frac{-c}{b} = \frac{a - c}{b}$$

This property is illustrated by the following examples.

EXAMPLES

Perform the following subtractions.

1 $\frac{8}{11} - \frac{5}{11}$

SOLUTION

$$\frac{8}{11} - \frac{5}{11} = \frac{8-5}{11} = \frac{3}{11}$$

2 $\frac{8}{9} - \frac{3}{9}$

SOLUTION

$$\frac{8}{9} - \frac{3}{9} = \frac{8-3}{9} = \frac{5}{9}$$

Addition and Subtraction of Unlike Fractions

Unlike fractions are those with different denominators. To add or subtract unlike fractions, we first find a common denominator; then we use the fundamental principle of fractions to change each fraction into an equivalent fraction which has the common denominator for its denominator. For example, to add $\frac{1}{3}$ and $\frac{1}{2}$, we have

$$\frac{1}{3} + \frac{1}{2} = \frac{1 \times 2}{3 \times 2} + \frac{1 \times 3}{2 \times 3} = \frac{2}{6} + \frac{3}{6} = \frac{2+3}{6} = \frac{5}{6}$$

In general, if $\frac{a}{b}$ and $\frac{c}{d}$ are fractions, with $b \neq 0$ and $d \neq 0$, then

$$\frac{a}{b} + \frac{c}{d} = \frac{ad+bc}{bd}$$

This result is stated in words as follows: To add unlike fractions, change them to equivalent fractions having the same denominator, and then follow the rule for adding like fractions. This is illustrated by the following examples.

EXAMPLES

Perform the following additions.

1 $\frac{2}{3} + \frac{3}{4}$

SOLUTION

$$\frac{2}{3} + \frac{3}{4} = \frac{2(4) + 3(3)}{(3)(4)} = \frac{8 + 9}{12} = \frac{17}{12}$$

2 $\frac{5}{7} + \frac{3}{5}$

SOLUTION

$$\frac{5}{7} + \frac{3}{5} = \frac{5(5) + 3(7)}{(7)(5)} = \frac{25 + 21}{35} = \frac{46}{35}$$

3 $\frac{7}{9} + \frac{3}{8}$

SOLUTION

$$\frac{7}{9} + \frac{3}{8} = \frac{7(8) + 3(9)}{(9)(8)} = \frac{56 + 27}{72} = \frac{83}{72}$$

The rule for subtracting unlike fractions can be derived from the rule for addition above and by using the definition of subtraction, $x - y = x + (-y)$. If $\frac{a}{b}$ and $\frac{c}{d}$ are fractions, with $b \neq 0$ and $d \neq 0$, then

$$\frac{a}{b} - \frac{c}{d} = \frac{a}{b} + \left(-\frac{c}{d}\right) = \frac{ad + (-c)b}{bd} = \frac{ad - cb}{bd}$$

The following examples illustrate this rule.

EXAMPLES

Perform the following subtractions.

1 $\frac{2}{5} - \frac{1}{4}$

SOLUTION

$$\frac{2}{5} - \frac{1}{4} = \frac{2(4) - 1(5)}{(5)(4)} = \frac{8 - 5}{20} = \frac{3}{20}$$

2 $\frac{2}{3} - \frac{3}{7}$

SOLUTION

$$\frac{2}{3} - \frac{3}{7} = \frac{2(7) - 3(3)}{(3)(7)} = \frac{14 - 9}{21} = \frac{5}{21}$$

3 $\frac{4}{5} - \frac{5}{13}$

SOLUTION

$$\frac{4}{5} - \frac{5}{13} = \frac{4(13) - 5(5)}{(5)(13)} = \frac{52 - 25}{65} = \frac{27}{65}$$

There are occasions when a direct application of the addition or subtraction rules is not the most convenient approach. Such an example is as follows:

$$\frac{3}{8} + \frac{5}{6} = \frac{3(6) + 8(5)}{(8)(6)} = \frac{18 + 40}{48} = \frac{58}{48} = \frac{29}{24}$$

Although this process is always correct, and it is especially useful if the denominators have no common factors, it can lead to unnecessary complications if the fractions have a common factor, such as the example above. Therefore, we shall seek a simpler way. For example, the addition above can be performed as follows:

$$\frac{3}{8} + \frac{5}{6} = \frac{3 \times 3}{8 \times 3} + \frac{5 \times 4}{6 \times 4} = \frac{9}{24} + \frac{20}{24} = \frac{9 + 20}{24} = \frac{29}{24}$$

Note that in the second solution we converted the fractions into equivalent fractions which have a common denominator by using the fundamental principle of fractions. In this case, we found the smallest positive integer that both denominators divide into, which is called the *least common denominator*, and it is abbreviated by L.C.D. To find the L.C.D. of a set of fractions, we adopt the following steps:

1 Factor each denominator into a product of prime numbers or into a product of prime numbers and -1.
2 List each prime factor with the largest exponent it has in any factored denominator; then find the product of the factors so listed.

This is illustrated by the following examples.

EXAMPLES

Express the following fractions in an equivalent form having the L.C.D. of the given set as a denominator.

1 $\frac{1}{24}$ and $\frac{1}{15}$

SOLUTION. First, we write the denominators in factored form, so that

$$24 = 2 \times 2 \times 2 \times 3 = 2^3(3)$$
$$15 = 3 \times 5$$

The various prime factors are 2, 3, and 5. Listing each of these with the largest exponent it has in any factored form, we have 2^3, 3, and 5. The L.C.D. of these fractions is $(2^3)(3)(5) = 120$. Using the fundamental principle of fractions, we have

$$\frac{1}{24} = \frac{1 \times 5}{24 \times 5} = \frac{5}{120}$$
$$\frac{1}{15} = \frac{1 \times 8}{15 \times 8} = \frac{8}{120}$$

2 $\frac{3}{14}$ and $\frac{11}{35}$

SOLUTION. The denominators are expressed in factored form as

$$14 = 2 \times 7$$
$$35 = 5 \times 7$$

The various prime factors are 2, 5, and 7. Listing each of these with the largest exponent it has in any factored form, we have 2, 5, and 7. The L.C.D. of these fractions is $2 \times 5 \times 7 = 70$. Using the fundamental principle of fractions, we have

$$\frac{3}{14} = \frac{3 \times 5}{14 \times 5} = \frac{15}{70}$$
$$\frac{11}{35} = \frac{11 \times 2}{35 \times 2} = \frac{22}{70}$$

3 $\frac{7}{12}$, $\frac{1}{10}$, and $\frac{2}{45}$

SOLUTION. Since $12 = 2 \times 2 \times 3 = 2^2(3)$, $10 = 2 \times 5$, and $45 = 5 \times 3 \times 3 = 5(3^2)$, the L.C.D. of the fractions is $(2^2) \cdot (3^2)(5) = 180$. Using the fundamental principle of fractions, we have

$$\frac{7}{12} = \frac{7 \times 15}{12 \times 15} = \frac{105}{180}$$

$$\frac{1}{10} = \frac{1 \times 18}{10 \times 18} = \frac{18}{180}$$

$$\frac{2}{45} = \frac{2 \times 4}{45 \times 4} = \frac{8}{180}$$

The use of the L.C.D. in adding or subtracting fractions usually simplifies the computation. This is illustrated as follows.

EXAMPLES

Perform the following operations and simplify.

1 $\frac{1}{24} + \frac{1}{15}$

SOLUTION. The L.C.D. of the fractions is 120 (see the preceding example 1). Thus,

$$\frac{1}{24} + \frac{1}{15} = \frac{1 \times 5}{24 \times 5} + \frac{1 \times 8}{15 \times 8} = \frac{5}{120} + \frac{8}{120}$$

$$= \frac{5 + 8}{120} = \frac{13}{120}$$

2 $\frac{11}{35} - \frac{3}{14}$

SOLUTION. The L.C.D. of the fractions is 70 (see the preceding example 2). Thus,

$$\frac{11}{35} - \frac{3}{14} = \frac{11 \times 2}{35 \times 2} - \frac{3 \times 5}{14 \times 5} = \frac{22}{70} - \frac{15}{70}$$

$$= \frac{22 - 15}{70} = \frac{7}{70} = \frac{1}{10}$$

3 $\frac{7}{12} + \frac{1}{10} - \frac{2}{45}$

SOLUTION. The L.C.D. of the fractions is 180 (see the preceding example 3). Thus,

$$\frac{7}{12}+\frac{1}{10}-\frac{2}{45}=\frac{7\times 15}{12\times 15}+\frac{1\times 18}{10\times 18}-\frac{2\times 4}{45\times 4}$$
$$=\frac{105}{180}+\frac{18}{180}-\frac{8}{180}=\frac{105+18-8}{180}$$
$$=\frac{115}{180}=\frac{23}{36}$$

In adding or subtracting numbers such as $5\frac{1}{6}$ and $3\frac{5}{8}$, it is convenient to write these numbers as single fractions. Numbers such as $5\frac{1}{6}$ and $3\frac{5}{8}$ are called *mixed numbers.* Each mixed number can be represented as the sum of an integer and a rational number. For example, the numbers above are written as

$$5\frac{1}{6}=5+\frac{1}{6} \quad \text{and} \quad 3\frac{5}{8}=3+\frac{5}{8}$$

To express these numbers as single fractions, we have

$$5\frac{1}{6}=5+\frac{1}{6}=5\left(\frac{6}{6}\right)+\frac{1}{6}=\frac{30}{6}+\frac{1}{6}=\frac{31}{6}$$
$$3\frac{5}{8}=3+\frac{5}{8}=3\left(\frac{8}{8}\right)+\frac{5}{8}=\frac{24}{8}+\frac{5}{8}=\frac{29}{8}$$

EXAMPLE

Determine the sum of these mixed fractions.

$$5\frac{1}{6}+3\frac{5}{8}$$

SOLUTION

$$5\frac{1}{6}+3\frac{5}{8}=\frac{31}{6}+\frac{29}{8}$$

The L.C.D. of the fractions is 24, so that

$$5\frac{1}{6} + 3\frac{5}{8} = \frac{31}{6} + \frac{29}{8} = \frac{31 \times 4}{6 \times 4} + \frac{29 \times 3}{8 \times 3}$$
$$= \frac{124}{24} + \frac{87}{24} = \frac{124 + 87}{24}$$
$$= \frac{211}{24}$$

PROBLEM SET 4.2

In problems 1–20, perform the indicated operations of like fractions and reduce them to lowest terms.

1 $\frac{2}{5} + \frac{1}{5}$

2 $\frac{2}{3} + \frac{5}{3}$

3 $\frac{3}{10} + \frac{6}{10}$

4 $\frac{7}{12} + \frac{3}{12}$

5 $4\frac{3}{8} + 2\frac{1}{8}$

6 $5\frac{3}{4} + 6\frac{1}{4}$

7 $2\frac{4}{8} + 3\frac{3}{8}$

8 $11\frac{5}{6} + 8\frac{2}{6}$

9 $5\frac{5}{18} + 6\frac{11}{18}$

10 $7\frac{2}{3} + 2\frac{5}{3}$

11 $\frac{5}{11} - \frac{3}{11}$

12 $\frac{7}{13} - \frac{5}{13}$

13 $\frac{7}{15} - \frac{3}{15}$

14 $\frac{7}{9} - \frac{2}{9}$

15 $\frac{21}{25} - \frac{16}{25}$

16 $\frac{9}{49} - \frac{2}{49}$

17 $\frac{43}{48} - \frac{7}{48}$

18 $6\frac{5}{14} - 2\frac{3}{14}$

19 $11\frac{5}{6} - 8\frac{2}{6}$

20 $11\frac{36}{84} - 5\frac{23}{84}$

In problems 21–50, perform the indicated operations of unlike fractions and reduce them to lowest terms.

21 $\frac{1}{2} + \frac{3}{8}$

22 $\frac{3}{4} + \frac{1}{12}$

23 $\frac{2}{3} + \frac{1}{18}$

24 $\frac{5}{8} + \frac{1}{16}$

25 $\frac{3}{15} + \frac{1}{3}$

26 $\frac{1}{24} + \frac{5}{6}$

27 $\frac{7}{8} + \frac{1}{4}$

28 $\frac{5}{6} + \frac{7}{24}$

29 $\frac{3}{8} + \frac{5}{32}$

30 $\frac{6}{7} + \frac{8}{21}$

31 $4\frac{1}{4} + 3\frac{3}{8}$

32 $5\frac{4}{7} + 3\frac{5}{14}$

33 $\frac{1}{3} - \frac{1}{6}$

34 $\frac{7}{8} - \frac{1}{2}$

35 $\frac{14}{15} - \frac{2}{5}$

36 $\frac{4}{5} - \frac{11}{20}$

37 $\frac{5}{3} - \frac{5}{6}$

38 $\frac{3}{5} - \frac{1}{10}$

39 $3\frac{35}{63} - 2\frac{5}{7}$

40 $7\frac{11}{13} - 2\frac{5}{26}$

41 $\frac{5}{8} + \frac{5}{12}$

42 $\frac{5}{6} + \frac{5}{8}$

43 $\frac{13}{12} + \frac{5}{8}$

44 $\frac{7}{12} + \frac{4}{9}$

45 $\frac{5}{6} - \frac{5}{9}$

46 $\frac{11}{12} - \frac{5}{16}$

47 $11\frac{5}{9} - 2\frac{2}{5}$

48 $14\frac{2}{5} - 2\frac{3}{11}$

49 $\frac{13}{9} - \frac{14}{15}$

50 $\frac{5}{9} - \frac{7}{12}$

In problems 51–60, perform the indicated operations and reduce them to lowest terms.

51 $\frac{19}{6} + \frac{1}{3} + \frac{3}{5}$

52 $\frac{9}{10} + \frac{4}{5} + \frac{1}{15}$

53 $\frac{13}{12} - \frac{5}{8} + \frac{3}{4}$

54 $\frac{7}{12} - \frac{3}{2} + \frac{4}{9}$

55 $\frac{7}{8} + \frac{5}{3} - \frac{7}{4}$

56 $\frac{8}{9} - \frac{5}{12} - \frac{1}{3}$

57 $\frac{5}{6} - \frac{1}{18} + \frac{5}{12}$

58 $\frac{4}{7} + \frac{3}{14} - \frac{2}{21}$

59 $\frac{3}{5} \div \frac{2}{3} \times \frac{7}{3} + \frac{1}{5} \div \frac{2}{7} \times \frac{1}{4}$

60 $\left(\frac{3}{4} + \frac{5}{6} - \frac{2}{3}\right) \div \frac{5}{12}$

4.3 Rational Expressions

In Section 4.1 we dealt with fractions represented by $\frac{a}{b}$, where the numerator a and the denominator b are integers and $b \neq 0$. If the numerator and the denominator of a fraction are polynomial expressions, the fraction is called a *rational expression.* That is, a rational expression can be expressed in the form $\frac{P}{Q}$, where P and Q are polynomials and $Q \neq 0$. Such rational expressions are commonly referred to as algebraic fractions or simply fractions. Throughout the book, we shall use the term "fraction" to mean either an algebraic fraction or rational expression, and we shall use them interchangeably. Examples of rational expressions are $\frac{3}{x}$, $\frac{3x}{x+1}$, and $\frac{y^2-1}{4y^3+7}$.

To find specific values for rational expressions, we make use of the substitution property. For example, consider the rational expression $\frac{x+2}{x-1}$:

$$\text{if } x = 2 \qquad \text{we have } \frac{2+2}{2-1} = \frac{4}{1} = 4$$

$$\text{if } x = -1 \qquad \text{we have } \frac{-1+2}{-1-2} = \frac{1}{-3} = -\frac{1}{3}$$

However, if $x = 1$, we obtain $\frac{1+2}{1-1} = \frac{3}{0}$, which does not exist, since 0 does not have a multiplicative inverse. Thus, a rational expression represents a real number for any specific value substituted for the variable except for those which cause the denominator to take on the value zero. In these cases we say that the rational expression is *not defined.* For instance, if we substitute $x = 2$ in the rational expression $\frac{(x-2)^2}{(x-1)(x-2)}$, we obtain $\frac{0}{0}$; hence, the expression is not defined for $x = 2$. Also, if we substitute $x = 1$ in the expression, we have $\frac{1}{0}$. Thus, the expression is also not defined for $x = 1$.

EXAMPLE

Find the numerical value of the rational expression $\frac{2x+5}{x+7}$ when

a) $x = 2$ b) $x = -7$

SOLUTION

a) To find the numerical value of the rational expression for $x = 2$, we have

$$\frac{2(2) + 5}{2 + 7} = \frac{4 + 5}{2 + 7} = \frac{9}{9} = 1$$

b) Substituting -7 for x in the rational expression, we see that the denominator, $x + 7$, becomes 0, since $-7 + 7 = 0$. Therefore, the expression is not defined for $x = -7$.

To avoid constant repetition, let us agree that we cannot assign values to any variable that will make any denominator of a fraction zero.

Since much of our work in algebra involves the substitution property, we need a method for determining "equivalent" rational expressions. We say that two rational expressions (or fractions) are *equivalent* if they give equal real numbers for each assignment of values to their variables, except when such an assignment gives an undefined value to either (or both) expressions. As in Section 4.1, we shall denote the equivalence of two rational expressions by writing an equal sign between them. Accordingly, we can determine equivalent rational expressions by extending Property 1 in Section 4.1 to include rational expressions.

PROPERTY

If P, Q, R, and S are polynomials with Q and $S \neq 0$, the fractions $\frac{P}{Q}$ and $\frac{R}{S}$ are equivalent if and only if $PS = QR$. In symbols, we have

$$\frac{P}{Q} = \frac{R}{S} \quad \text{if and only if} \quad PS = QR$$

EXAMPLE

Indicate which pairs of rational expressions are equivalent.

a) $\frac{3x}{6x^2}$ and $\frac{1}{2x}$

b) $\frac{x^2}{xy}$ and $\frac{x}{y}$

c) $\frac{6x}{3(x + 3)}$ and $\frac{2x}{x + 3}$

d) $\frac{5}{x}$ and $\frac{7}{x}$

SOLUTION

a) Since $(3x)(2x) = (6x^2)(1)$, then $\dfrac{3x}{6x^2} = \dfrac{1}{2x}$.

b) Since $(x^2)(y) = (xy)(x)$, then $\dfrac{x^2}{xy} = \dfrac{x}{y}$.

c) Since $(6x)(x + 3) = 3(x + 3)(2x)$, then $\dfrac{6x}{3(x + 3)} = \dfrac{2x}{x + 3}$.

d) Since $(5)(x) \neq (x)(7)$, then $\dfrac{5}{x} \neq \dfrac{7}{x}$.

The notion of reducing fractions to lowest terms that was discussed in Section 4.1 can be extended to rational expressions. For instance, the expressions $\dfrac{5}{x}$, $\dfrac{x}{x + 1}$, and $\dfrac{7x}{3x^2 + 2}$ are in lowest terms, whereas $\dfrac{5x}{15x^2}$ is not in lowest terms, since $5x$ is a common factor in both the numerator and the denominator of the fraction. As in Section 4.1, we shall call the procedure of expressing a given fraction as an equivalent fraction in lowest terms, reducing the fraction to lowest terms. The concept of equivalent fractions can be used to verify that rational expressions can be reduced to lowest terms in the same way that rational numbers are reduced. For example,

$$\frac{x^2}{x^3} = \frac{x^2(1)}{x^2(x)} = \frac{1}{x} \qquad \text{since } x^2 \cdot x = x^3 \cdot 1$$

$$\frac{28x^3y}{21xy^2} = \frac{4x^2(7xy)}{3y(7xy)} = \frac{4x^2}{3y} \qquad \text{since } 28x^3y \cdot 3y = 21xy^2 \cdot 4x^2$$

Here, the fundamental principle of fractions is extended to include rational expressions. That is: If P and Q are polynomials with $Q \neq 0$ and if K represents any rational expression, $K \neq 0$, then

$$\frac{PK}{QK} = \frac{P}{Q}$$

This is further illustrated by the following examples.

EXAMPLES

Reduce each of the following fractions to lowest terms.

1 $\dfrac{3a^2}{ba}$

SOLUTION

$$\frac{3a^2}{ba} = \frac{3a(a)}{b(a)} = \frac{3a}{b}$$

2 $\dfrac{12x^2y}{9xy^2}$

SOLUTION

$$\frac{12x^2y}{9xy^2} = \frac{4x(3xy)}{3y(3xy)} = \frac{4x}{3y}$$

3 $\dfrac{12(x+1)}{15(x+1)^2}$

SOLUTION

$$\frac{12(x+1)}{15(x+1)^2} = \frac{4[3(x+1)]}{5(x+1)[3(x+1)]} = \frac{4}{5(x+1)}$$

4 $\dfrac{4x^2-1}{2x^2+x}$

SOLUTION

$$\frac{4x^2-1}{2x^2+x} = \frac{(2x-1)(2x+1)}{x(2x+1)} = \frac{2x-1}{x}$$

5 $\dfrac{x^2-4}{x^2+x-6}$

SOLUTION

$$\frac{x^2-4}{x^2+x-6} = \frac{(x-2)(x+2)}{(x-2)(x+3)} = \frac{x+2}{x+3}$$

6 $\dfrac{x^2+x-12}{x^2+4x-21}$

SOLUTION

$$\frac{x^2 + x - 12}{x^2 + 4x - 21} = \frac{(x + 4)(x - 3)}{(x + 7)(x - 3)} = \frac{x + 4}{x + 7}$$

7 $\dfrac{a^3 - ab^2}{a^3 - 2a^2b + ab^2}$

SOLUTION

$$\frac{a^3 - ab^2}{a^3 - 2a^2b + ab^2} = \frac{a(a^2 - b^2)}{a(a^2 - 2ab + b^2)} = \frac{a(a - b)(a + b)}{a(a - b)(a - b)} = \frac{a + b}{a - b}$$

The fundamental principle of fractions can also be used to change a fraction into an equivalent fraction by multiplying or dividing both the numerator and the denominator of the fraction by the same nonzero expression. This is illustrated by the following examples.

EXAMPLES

In examples 1–4, find the missing expression for each of the following equivalent fractions.

1 $\dfrac{5x}{y} = \dfrac{?}{yx}$

SOLUTION

$$\frac{5x}{y} = \frac{(5x)(x)}{y(x)} = \frac{5x^2}{yx}$$

so that the unknown expression is $5x^2$.

2 $\dfrac{b}{4a} = \dfrac{?}{24a^3b^2}$

SOLUTION

$$\frac{b}{4a} = \frac{b(6a^2b^2)}{4a(6a^2b^2)} = \frac{6a^2b^3}{24a^3b^2}$$

so that the unknown expression is $6a^2b^3$.

3 $\dfrac{x-2}{x+1} = \dfrac{?}{x^2-1}$

SOLUTION

$$\frac{x-2}{x+1} = \frac{(x-2)(x-1)}{(x+1)(x-1)} = \frac{x^2-3x+2}{x^2-1}$$

so that the unknown expression is $x^2 - 3x + 2$.

4 $\dfrac{y}{2y-1} = \dfrac{?}{4y^2-1}$

SOLUTION

$$\frac{y}{2y-1} = \frac{y(2y+1)}{(2y-1)(2y+1)} = \frac{2y^2+y}{4y^2-1}$$

so that the unknown expression is $2y^2 + y$.

5 Use the fundamental principle of fractions to show that each of the following statements is true.

a) $\dfrac{-a}{b} = \dfrac{a}{-b}$ b) $\dfrac{a}{b} = \dfrac{-a}{-b}$

SOLUTION

a) $\dfrac{-a}{b} = \dfrac{(-a)(-1)}{b(-1)} = \dfrac{a}{-b}$

b) $\dfrac{a}{b} = \dfrac{(a)(-1)}{(b)(-1)} = \dfrac{-a}{-b}$

6 Write each of the following fractions with a denominator $x - y$.

a) $\dfrac{1}{y-x}$ b) $\dfrac{-xy}{y-x}$ c) $\dfrac{a-b}{y-x}$ d) $-\dfrac{1}{y-x}$

SOLUTION

a) $\dfrac{1}{y-x} = \dfrac{(-1)1}{(-1)(y-x)} = \dfrac{-1}{x-y}$

b) $\dfrac{-xy}{y-x} = \dfrac{(-1)(-xy)}{(-1)(y-x)} = \dfrac{xy}{x-y}$

c) $\dfrac{a-b}{y-x} = \dfrac{(-1)(a-b)}{(-1)(y-x)} = \dfrac{b-a}{x-y}$

d) $\dfrac{1}{y-x} = \dfrac{1}{(-1)(y-x)} = \dfrac{1}{x-y}$

PROBLEM SET 4.3

In problems 1–6, find the numerical value of each rational expression for the indicated value of x.

1 $\dfrac{5x - 3}{6x}$; $x = 2$

2 $\dfrac{5}{x^2 - x}$; $x = 4$

3 $\dfrac{x}{x - 5}$; $x = 5$

4 $\dfrac{7x}{x^2 - 3x}$; $x = 3$

5 $\dfrac{x - 2}{3x + 6}$; $x = 2$

6 $\dfrac{x^2 - 1}{x - 1}$; $x = 1$

In problems 7–18, use the property $\dfrac{P}{Q} = \dfrac{R}{S}$ if and only if $PS = QR$ to indicate which pairs of fractions are equivalent.

7 $\dfrac{5}{a}, \dfrac{15a}{3a^2}$

8 $\dfrac{3b}{17}, \dfrac{15b^2}{85b}$

9 $\dfrac{-2}{9x}, \dfrac{2}{-9x}$

10 $\dfrac{-3a}{-37b}, \dfrac{6a}{58b}$

11 $\dfrac{x^6}{x^8}, \dfrac{1}{x^3}$

12 $\dfrac{5ab^2}{9a^2b}, \dfrac{2b}{3a}$

13 $\dfrac{x + 1}{-x}, \dfrac{-x - 1}{x}$

14 $\dfrac{5(x - 1)^2}{10(x - 1)}, \dfrac{x - 1}{2}$

15 $\frac{ax + ay}{bx + by}, \frac{a}{b}$

16 $\frac{x^2 - 9}{x - 3}, x + 3$

17 $\frac{x + y}{1}, \frac{ax + y}{a}$

18 $\frac{x + 1}{x - 1}, \frac{x + 2}{x + 3}$

In problems 19–50, use the fundamental principle of fractions to reduce each fraction into an equivalent fraction in its lowest terms.

19 $\frac{7x}{7y}$

20 $\frac{ay^2}{cy}$

21 $\frac{6x}{9y^3}$

22 $\frac{2\pi r}{6\pi r^3}$

23 $\frac{6abx}{9a^2x^2}$

24 $\frac{10xy^2}{15x^2y}$

25 $\frac{2a^3b^2x^3}{6a^2b^2x}$

26 $\frac{10x^4yz^2}{16xy^2z^2}$

27 $\frac{5x + 5y}{9x + 9y}$

28 $\frac{12a + 24b}{3a + 6b}$

29 $\frac{(x + 1)(x - 2)}{(x - 2)^2}$

30 $\frac{8(x - 7)^3}{4(x + 1)(x - 7)^2}$

31 $\dfrac{2x^2 + x}{2x^2 - x}$

32 $\dfrac{2x - y}{y - 2x}$

33 $\dfrac{5y^2 - 5y}{15y - 15}$

34 $\dfrac{4x^2 - 9}{2x - 3}$

35 $\dfrac{25x^2 - 1}{5x + 1}$

36 $\dfrac{13y^2 - 26y^3}{5 - 10y}$

37 $\dfrac{8x^2 + 2x}{4x + 1}$

38 $\dfrac{4x^2 - x}{16x^2 - 1}$

39 $\dfrac{6x + 4xy}{9 - 4y^2}$

40 $\dfrac{x^2 - 6x + 9}{3x - 9}$

41 $\dfrac{x^2 - 8x + 16}{5x - 20}$

42 $\dfrac{a^2x - b^2x}{cax - cbx}$

43 $\dfrac{x^2 - 5x}{x^2 - 7x + 10}$

44 $\dfrac{x^2 - 9}{x^2 + 5x + 6}$

45 $\dfrac{x^2 + 2x - 3}{x^2 + 5x + 6}$

46 $\dfrac{x^2 + x - 12}{x^2 + 3x - 4}$

47 $\dfrac{x^2 - 5x - 24}{x^2 - 64}$

48 $\dfrac{6a^2 + 11a + 4}{3a^2 + 7a + 4}$

49 $\dfrac{2m^2 - 10mn + 12n^2}{4m^2 - 36n^2}$

50 $\dfrac{12a^2 - 42ab + 36b^2}{12a^2 - 36ab + 27b^2}$

In problems 51–60, use the fundamental principle of fractions to determine the missing expression for each fraction.

51 $\dfrac{ax}{ay} = \dfrac{?}{y}$

52 $\dfrac{cb}{ay} = \dfrac{?}{y}$

53 $\dfrac{ax}{a^2x^3} = \dfrac{?}{ax^4}$

54 $\dfrac{5a^2x}{10ax^2} = \dfrac{?}{20a^2x^4}$

55 $\dfrac{abx}{abxy} = \dfrac{?}{ay^2bx}$

56 $\dfrac{8x^3y^3}{12xy^5} = \dfrac{?}{24x^3y^7}$

57 $\dfrac{3(a - b)}{a + b} = \dfrac{?}{(a + b)^2}$

58 $\dfrac{x^2 - y^2}{(x - y)^2} = \dfrac{?}{x - y}$

59 $\dfrac{x + 3}{x^2 - 27} = \dfrac{x^2 + 3x}{?}$

60 $\dfrac{a(x^2 - 1)}{x} = \dfrac{?}{x(x^2 + 1)}$

In problems 61–66, write each fraction as an equivalent fraction with a denominator $y - x$.

61 $\dfrac{-5}{x - y}$

62 $\dfrac{6}{-(y - x)}$

63 $-\dfrac{-xy}{x-y}$

64 $\dfrac{a+b}{x-y}$

65 $-\dfrac{x+y}{(x-y)(y+x)}$

66 $\dfrac{-(3x-3y)}{(x-y)^2}$

4.4 Multiplication and Division of Rational Expressions

In this section we extend the multiplication and division of fractions that was discussed in Section 4.1 to include multiplication and division of rational expressions.

Multiplication of Rational Expressions

The rule for multiplying rational expressions is similar to that for fractions. That is, if $\dfrac{P}{Q}$ and $\dfrac{R}{S}$ are rational expressions, then

$$\frac{P}{Q}\cdot\frac{R}{S}=\frac{P\cdot R}{Q\cdot S}$$

The following is the statement of the rule in words: To multiply two rational expressions, we obtain a rational expression whose numerator is the product of the numerators and whose denominator is the product of the denominators. The following examples illustrate this rule.

EXAMPLES

Find the following products and reduce them to lowest terms.

1 $\dfrac{9x^2}{16}\cdot\dfrac{4}{3x}$

SOLUTION

$$\frac{9x^2}{16}\cdot\frac{4}{3x}=\frac{(9x^2)(4)}{(16)(3x)}=\frac{36x^2}{48x}=\frac{3x(12x)}{4(12x)}=\frac{3x}{4}$$

2 $\dfrac{3y}{4x^2}\cdot\dfrac{6x^3y}{9y^2}$

SOLUTION

$$\frac{3y}{4x^2}\cdot\frac{6x^3y}{9y^2}=\frac{(3y)(6x^3y)}{(4x^2)(9y^2)}=\frac{18x^3y^2}{36x^2y^2}=\frac{x(18x^2y^2)}{2(18x^2y^2)}=\frac{x}{2}$$

3 $\dfrac{3x+12}{8}\cdot\dfrac{24x}{9x+36}$

SOLUTION

$$\frac{3x+12}{8}\cdot\frac{24x}{9x+36}=\frac{(3x+12)(24x)}{8(9x+36)}=\frac{3(x+4)(24x)}{(8)(9)(x+4)}$$
$$=\frac{(72)(x)(x+4)}{(72)(x+4)}=x$$

4 $\dfrac{5-3x}{7}\cdot\dfrac{21}{3x-5}$

SOLUTION

$$\frac{5-3x}{7}\cdot\frac{21}{3x-5}=\frac{(5-3x)(21)}{7(3x-5)}=\frac{(-21)(3x-5)}{7(3x-5)}$$
$$=\frac{(-3)(7)(3x-5)}{7(3x-5)}=-3$$

5 $\dfrac{x^2-1}{3x^5}\cdot\dfrac{12x^4}{x+1}$

SOLUTION

$$\frac{x^2-1}{3x^5}\cdot\frac{12x^4}{x+1}=\frac{(x^2-1)(12x^4)}{(3x^5)(x+1)}=\frac{(x-1)(x+1)(12x^4)}{(3x^5)(x+1)}$$
$$=\frac{(3x^4)(x+1)(x-1)(4)}{(3x^4)(x+1)(x)}=\frac{4(x-1)}{x}=\frac{4x-4}{x}$$

6 $\dfrac{4x+12}{2x-10}\cdot\dfrac{x^2-x-20}{x^2-9}$

SOLUTION

$$\begin{aligned}\frac{4x+12}{2x-10}\cdot\frac{x^2-x-20}{x^2-9} &= \frac{(4x+12)(x^2-x-20)}{(2x-10)(x^2-9)}\\ &= \frac{4(x+3)(x-5)(x+4)}{2(x-5)(x+3)(x-3)}\\ &= \frac{2(x+4)}{x-3} = \frac{2x+8}{x-3}\end{aligned}$$

7 $\dfrac{x^2+14x-15}{x^2+4x-5}\cdot\dfrac{x^2+6x-27}{x^2+12x-45}$

SOLUTION

$$\begin{aligned}\frac{x^2+14x-15}{x^2+4x-5}\cdot\frac{x^2+6x-27}{x^2+12x-45} &= \frac{(x^2+14x-15)(x^2+6x-27)}{(x^2+4x-5)(x^2+12x-45)}\\ &= \frac{(x+15)(x-1)(x+9)(x-3)}{(x+5)(x-1)(x+15)(x-3)}\\ &= \frac{x+9}{x+5}\end{aligned}$$

Division of Rational Expressions

The rule for dividing two rational expressions is similar to that for dividing fractions. We first recall from Section 4.1 that the multiplicative inverse or the reciprocal of $\frac{a}{b}$ is $\frac{b}{a}$. This property can be extended to rational expressions. That is, if $\frac{R}{S}$ is a rational expression with $\frac{R}{S} \neq 0$, then $\frac{R}{S}\cdot\frac{S}{R} = 1$, so that the multiplicative inverse or the reciprocal of $\frac{R}{S}$ is $\frac{S}{R}$.

Now, the rule for dividing rational expressions will be stated as follows: If $\frac{P}{Q}$ and $\frac{R}{S}$ are rational expressions, with $\frac{R}{S} \neq 0$, then

$$\frac{P}{Q} \div \frac{R}{S} = \frac{P}{Q}\cdot\frac{1}{R/S} = \frac{P}{Q}\cdot\frac{S}{R} = \frac{PS}{QR}$$

This rule is stated in words as follows: To divide one fraction by another, multiply the first fraction by the reciprocal of the second fraction.

EXAMPLES

Find the following quotients and reduce them to lowest terms.

1 $\dfrac{16a^2}{5b^3} \div \dfrac{8a}{25b^3}$

SOLUTION

$$\frac{16a^2}{5b^3} \div \frac{8a}{25b^3} = \frac{16a^2}{5b^3} \cdot \frac{25b^3}{8a} = \frac{(16a^2)(25b^3)}{(5b^3)(8a)}$$
$$= \frac{(2)(5)(a)(40ab^3)}{40ab^3} = 10a$$

2 $\dfrac{x^2}{2y} \div \dfrac{x^4}{4y^3}$

SOLUTION

$$\frac{x^2}{2y} \div \frac{x^4}{4y^3} = \frac{x^2}{2y} \cdot \frac{4y^3}{x^4} = \frac{(x^2)(4y^3)}{(2y)(x^4)}$$
$$= \frac{(2y^2)(2x^2y)}{x^2(2x^2y)} = \frac{2y^2}{x^2}$$

3 $\dfrac{x+1}{18} \div \dfrac{2x+2}{3}$

SOLUTION

$$\frac{x+1}{18} \div \frac{2x+2}{3} = \frac{x+1}{18} \cdot \frac{3}{2x+2} = \frac{(x+1)(3)}{(18)(2x+2)}$$
$$= \frac{(x+1)(3)}{(18)(2)(x+1)} = \frac{3(x+1)}{(12)(3)(x+1)} = \frac{1}{12}$$

4 $\dfrac{x+2}{y} \div \dfrac{(x+2)^2}{y^2}$

SOLUTION

$$\frac{x+2}{y} \div \frac{(x+2)^2}{y^2} = \frac{x+2}{y} \cdot \frac{y^2}{(x+2)^2} = \frac{(x+2)y^2}{y(x+2)^2}$$
$$= \frac{(y)[y(x+2)]}{(x+2)[y(x+2)]} = \frac{y}{x+2}$$

5 $\dfrac{x^2 - 4}{x^2 + 4x} \div \dfrac{x^2 - 2x}{x + 4}$

SOLUTION

$$\frac{x^2 - 4}{x^2 + 4x} \div \frac{x^2 - 2x}{x + 4} = \frac{x^2 - 4}{x^2 + 4x} \cdot \frac{x + 4}{x^2 - 2x} = \frac{(x^2 - 4)(x + 4)}{(x^2 + 4x)(x^2 - 2x)}$$

$$= \frac{(x - 2)(x + 2)(x + 4)}{x(x + 4)x(x - 2)} = \frac{x + 2}{x^2}$$

6 $\dfrac{x^2 - x - 6}{x^2 + x - 12} \div \dfrac{x^2 + 2x - 3}{x^2 + 3x - 4}$

SOLUTION

$$\frac{x^2 - x - 6}{x^2 + x - 12} \div \frac{x^2 + 2x - 3}{x^2 + 3x - 4} = \frac{x^2 - x - 6}{x^2 + x - 12} \cdot \frac{x^2 + 3x - 4}{x^2 + 2x - 3}$$

$$= \frac{(x^2 - x - 6)(x^2 + 3x - 4)}{(x^2 + x - 12)(x^2 + 2x - 3)}$$

$$= \frac{(x - 3)(x + 2)(x + 4)(x - 1)}{(x - 3)(x + 4)(x + 3)(x - 1)}$$

$$= \frac{x + 2}{x + 3}$$

7 $\dfrac{x^2 - x - 12}{x^2 - 9} \cdot \dfrac{x^2 - 2x - 3}{x^2 - 2x - 8} \div \dfrac{x^2 + x}{x + 2}$

SOLUTION

$$\frac{x^2 - x - 12}{x^2 - 9} \cdot \frac{x^2 - 2x - 3}{x^2 - 2x - 8} \div \frac{x^2 + x}{x + 2}$$

$$= \frac{x^2 - x - 12}{x^2 - 9} \cdot \frac{x^2 - 2x - 3}{x^2 - 2x - 8} \cdot \frac{x + 2}{x^2 + x}$$

$$= \frac{(x^2 - x - 12)(x^2 - 2x - 3)(x + 2)}{(x^2 - 9)(x^2 - 2x - 8)(x^2 + x)}$$

$$= \frac{(x - 4)(x + 3)(x - 3)(x + 1)(x + 2)}{(x - 3)(x + 3)(x - 4)(x + 2)(x)(x + 1)}$$

$$= \frac{1}{x}$$

PROBLEM SET 4.4

In problems 1–32, find the products and reduce them to lowest terms.

1 $\frac{7a}{3b} \cdot \frac{4b}{14a^2}$

2 $\frac{4x^2}{3y} \cdot \frac{12y^2}{8x^3}$

3 $\frac{7x^2y}{3a^2b} \cdot \frac{6a}{14x^2y}$

4 $\frac{3x^4}{4a^2y^2} \cdot \frac{8a^6y^2}{6x^6}$

5 $\frac{5x^2y}{3x^2y} \cdot \frac{6x}{10x^2} \cdot \frac{11x}{8y^4}$

6 $\frac{15xyz}{16a^2} \cdot \frac{12xa^2}{25x^2y^2z} \cdot \frac{5}{x^3}$

7 $\frac{12}{x} \cdot \frac{3x^2 - 5x}{4} \cdot \frac{3}{16x^2}$

8 $\frac{6x}{34} \cdot \frac{17x - 51}{12x} \cdot \frac{-49}{3x}$

9 $\frac{y + 4}{5y} \cdot \frac{15y^2}{(y + 4)^2}$

10 $\frac{11 - 3x}{7} \cdot \frac{28}{3x - 11}$

11 $\frac{x + y}{13} \cdot \frac{39}{(x + y)^2}$

12 $\frac{17(x - 3)}{y^2} \cdot \frac{4y^2}{34(x - 3)}$

13 $\frac{5x + 30}{14} \cdot \frac{28x}{7x + 42}$

14 $\frac{21}{16 - 8y} \cdot \frac{2 - y}{28}$

15 $\frac{9 - x^2}{x + 3} \cdot \frac{x}{3 - x}$

16 $\frac{x^2 - 16}{39} \cdot \frac{3(x + 4)}{x - 4}$

17 $\frac{x + 4}{x^2} \cdot \frac{x^4}{x^2 - 16}$

18 $\dfrac{x^2 - 36}{x + 5} \cdot \dfrac{x^2 - 25}{x + 6}$

19 $\dfrac{3x + 12}{3x - 15} \cdot \dfrac{x^2 - 25}{x^2 + 4x}$

20 $\dfrac{x^2 + xy}{xy} \cdot \dfrac{y}{x^2 - y^2}$

21 $\dfrac{x^2 - 6x}{x - 6} \cdot \dfrac{x + 3}{x}$

22 $\dfrac{18 - 2y^2}{x^3 - x} \cdot \dfrac{x - 1}{2y + 6}$

23 $\dfrac{a^2 - b^2}{(a + b)^2} \cdot \dfrac{3a + 3b}{6a}$

24 $\dfrac{6(x + 2)}{3(x - 1)^2} \cdot \dfrac{12(x - 1)}{(x + 2)^2}$

25 $\dfrac{81 - m^2}{m + 9} \cdot \dfrac{m}{9 - m}$

26 $\dfrac{x^2 - 6x + 5}{x - 1} \cdot \dfrac{x - 1}{x - 5}$

27 $\dfrac{a^2 + 8a + 16}{a^2 - 9} \cdot \dfrac{a - 3}{a + 4}$

28 $\dfrac{m^2 - 2m - 24}{m^2 - m - 30} \cdot \dfrac{(m + 5)^2}{m^2 - 16}$

29 $\dfrac{x^2 + 2x - 8}{6x^2 + 6x - 72} \cdot \dfrac{4x^2 - 36}{x^2 + 3x - 10}$

30 $\dfrac{4x^2 + 6xy - 4y^2}{36x^2 - 9y^2} \cdot \dfrac{6x + 3y}{4x + 8y}$

31 $\dfrac{3a^2 + ab - 2b^2}{3a^2 - ab - 2b^2} \cdot \dfrac{9a^2 + 3ab - 2b^2}{9a^2 - 3ab - 2b^2}$

32 $\dfrac{2m^2 - 3mn - 2n^2}{36m^2 - 9n^2} \cdot \dfrac{6m^2 + 15mn - 9n^2}{m^2 + mn - 6n^2}$

In problems 33–58, find the quotients and reduce them to lowest terms.

33 $\dfrac{14x^2}{10b^2} \div \dfrac{21x^2}{15b^2}$

34 $\dfrac{5a}{12yz^2} \div \dfrac{15a^2}{18y^2z^2}$

35 $\dfrac{34a}{4x^2y} \div \dfrac{17a^3}{2xy}$

36 $\dfrac{25x^2}{(3yz)^2} \div \dfrac{100x^3}{18yz^3}$

37 $\dfrac{21a^2}{40b^3} \div \dfrac{35a^4}{36b^3}$

38 $\dfrac{28x^2y^2}{15a} \div \dfrac{3xy^4}{75a^2}$

39 $\dfrac{10x^2}{(x+y)^2} \div \dfrac{5x}{x+y}$

40 $\dfrac{8a^2}{a^2-16} \div \dfrac{4a}{a-4}$

41 $\dfrac{c^2-b^2}{x^2-y^2} \div \dfrac{c+b}{x-y}$

42 $\dfrac{5}{2x+3y} \div \dfrac{10}{4x^2-9y^2}$

43 $\dfrac{6x+9y}{2x-3y} \div \dfrac{2x+3y}{2x-3y}$

44 $\dfrac{3ab+ac}{3b^2-bc} \div \dfrac{3b^2+bc}{6b-2c}$

45 $\dfrac{6x^2-54}{x^4} \div \dfrac{x+3}{x^2}$

46 $\dfrac{4-x^2}{2+x} \div \dfrac{x^2-3x+2}{x-1}$

47 $\dfrac{x^2-9}{x^2-x-2} \div \dfrac{x-3}{x^2+x-6}$

48 $\dfrac{x^2-x-2}{x^2-x-6} \div \dfrac{x^2-2x}{2x+x^2}$

49 $\dfrac{9x^2-1}{8x+6} \div \dfrac{3x^2-4x+1}{4x^2-x-3}$

50 $\dfrac{b^2-2b+1}{4b^2-1} \div \dfrac{4b-4}{6b+3}$

51 $\dfrac{6x^2+3x-9}{4x^2+12x+9} \div \dfrac{3x^2-27}{x^2+6x+9}$

52 $\dfrac{2x^2 - 9x - 18}{4x^2 - 27x + 18} \div \dfrac{4x^2 + 3x - 1}{3x^2 + 5x + 2}$

53 $\dfrac{18x - 27}{2 - 4x - 6x^2} \div \dfrac{6x^2 - 9x}{4 - 36x^2}$

54 $\dfrac{12x^2 - 14x - 6}{-15x^2 + x + 2} \div \dfrac{16x^2 - 36}{10x^2 + 11x - 6}$

55 $\dfrac{x^3y - xy^3}{x^3 - 2x^2y + xy^2} \cdot \dfrac{1}{x + y} \div \dfrac{2x^2y - 2xy^2}{x^2 - y^2}$

56 $\dfrac{x^3 + x^2 - 12x}{x^2 + 5x + 6} \cdot \dfrac{x^2 - 4}{x^2 - 9} \div \dfrac{x^3 - 16x}{x^2 - 3x - 10}$

57 $\dfrac{x^2 - 5x + 4}{x^2 + 5x + 6} \cdot \dfrac{x^2 - 3x - 18}{x^2 - 6x + 5} \div \dfrac{x^2 + 3x - 28}{x^2 - x - 20}$

58 $\dfrac{x^2 - 8x + 7}{x^2 - 4} \cdot \dfrac{x^2 + 4x + 4}{x^2 - 6x - 7} \div \dfrac{x^2 - x - 6}{x^2 - 5x + 6}$

4.5 Addition and Subtraction of Rational Expressions

Fractions are either rational numbers or rational expressions. Hence, we can extend the rules in Section 4.2 to adding and subtracting rational expressions. First, we consider adding and subtracting "like" rational expressions (fractions), that is, rational expressions (fractions) with the same denominator.

Addition and Subtraction of Like Rational Expressions

Recall that $\dfrac{a}{b} = a \cdot \dfrac{1}{b}$ for real numbers a and b, $b \neq 0$. This is also true for polynomials; that is, if P and Q are polynomials, then $\dfrac{P}{Q} = P \cdot \dfrac{1}{Q}$, where $Q \neq 0$. We can apply this definition and the distributive property of real numbers to add like fractions as follows:

$$\frac{P}{Q} + \frac{R}{Q} = P \cdot \frac{1}{Q} + R \cdot \frac{1}{Q} = (P + R) \cdot \frac{1}{Q} = \frac{P + R}{Q}$$

This result can be stated as follows: To add like rational expressions (fractions), add the numerators to find the numerator of the sum, and retain the common denominator as the denominator of the sum.

EXAMPLES

Determine each of the following sums and reduce to lowest terms.

1 $\frac{5}{x} + \frac{8}{x}$

SOLUTION

$$\frac{5}{x} + \frac{8}{x} = \frac{5 + 8}{x} = \frac{13}{x}$$

2 $\frac{x + 1}{4} + \frac{3(x + 1)}{4}$

SOLUTION

$$\frac{x + 1}{4} + \frac{3(x + 1)}{4} = \frac{x + 1 + 3(x + 1)}{4} = \frac{x + 1 + 3x + 3}{4}$$

$$= \frac{4x + 4}{4}$$

$$= \frac{4(x + 1)}{4} = x + 1$$

3 $\frac{x + y}{5} + \frac{x - y}{5}$

SOLUTION

$$\frac{x + y}{5} + \frac{x - y}{5} = \frac{x + y + x - y}{5} = \frac{2x}{5}$$

4 $\frac{x^2}{x + 3} + \frac{3x}{x + 3}$

SOLUTION

$$\frac{x^2}{x + 3} + \frac{3x}{x + 3} = \frac{x^2 + 3x}{x + 3} = \frac{x(x + 3)}{x + 3} = x$$

5 $\frac{x^2}{x - 2} + \frac{-4}{x - 2}$

SOLUTION

$$\frac{x^2}{x-2} + \frac{-4}{x-2} = \frac{x^2-4}{x-2} = \frac{(x-2)(x+2)}{x-2} = x+2$$

In Section 4.2 we indicated that the definition of subtraction, that is, $x - y = x + (-y)$, can be used to subtract fractions. This definition can be extended to rational expressions. The additive inverse of $\frac{P}{Q}$ is denoted by $-\frac{P}{Q}$, so that $\frac{P}{Q} + \left(-\frac{P}{Q}\right) = 0$. We can follow a similar procedure to that of Section 4.2 for subtracting two like rational expressions: If $\frac{P}{Q}$ and $\frac{R}{Q}$ are rational expressions with $Q \neq 0$, then

$$\frac{P}{Q} - \frac{R}{Q} = \frac{P}{Q} + \left(-\frac{R}{Q}\right) = \frac{P-R}{Q}$$

This is further illustrated by the following examples.

EXAMPLES

Perform the following subtractions and reduce to lowest terms.

1 $\frac{13}{x} - \frac{7}{x}$

SOLUTION

$$\frac{13}{x} - \frac{7}{x} = \frac{13-7}{x} = \frac{6}{x}$$

2 $\frac{7}{9y} - \frac{2}{9y}$

SOLUTION

$$\frac{7}{9y} - \frac{2}{9y} = \frac{7-2}{9y} = \frac{5}{9y}$$

3 $\frac{3x}{x-4} - \frac{12}{x-4}$

SOLUTION

$$\frac{3x}{x-4} - \frac{12}{x-4} = \frac{3x-12}{x-4} = \frac{3(x-4)}{x-4} = 3$$

4 $\dfrac{x^2}{x-7} - \dfrac{49}{x-7}$

SOLUTION

$$\frac{x^2}{x-7} - \frac{49}{x-7} = \frac{x^2-49}{x-7} = \frac{(x-7)(x+7)}{x-7} = x+7$$

Addition and Subtraction of Unlike Fractions

Now, we consider the sum and difference of *unlike fractions* (fractions with different denominators). The procedure of adding and subtracting unlike fractions can be extended to rational expressions. For example, to add the two rational expressions $\frac{x}{5}$ and $\frac{x}{8}$, we can write

$$\frac{x}{5} + \frac{x}{8} = \frac{8x}{40} + \frac{5x}{40}$$

so that

$$\frac{x}{5} + \frac{x}{8} = \frac{8x+5x}{40} = \frac{13x}{40}$$

This example can be generalized by the following rule: If $\frac{P}{Q}$ and $\frac{R}{S}$ are rational expressions, with $Q \neq 0$, $S \neq 0$, then

$$\frac{P}{Q} + \frac{R}{S} = \frac{PS}{QS} + \frac{RQ}{SQ} = \frac{PS+RQ}{QS}$$

This rule is stated in words as follows: To add unlike fractions, change them to equivalent fractions having the same denominator, then follow the rule for adding like fractions.

EXAMPLES

Perform the following additions.

1 $\dfrac{x}{5} + \dfrac{x}{7}$

SOLUTION

$$\frac{x}{5} + \frac{x}{7} = \frac{7x + 5x}{5(7)} = \frac{12x}{35}$$

2 $\dfrac{3}{8y} + \dfrac{2}{9y}$

SOLUTION

$$\frac{3}{8y} + \frac{2}{9y} = \frac{3(9y) + 2(8y)}{(8y)(9y)} = \frac{27y + 16y}{72y^2} = \frac{43y}{72y^2} = \frac{43}{72y}$$

3 $\dfrac{x + 2}{x} + \dfrac{x}{x + 2}$

SOLUTION

$$\frac{x + 2}{x} + \frac{x}{x + 2} = \frac{(x + 2)(x + 2) + x(x)}{x(x + 2)} = \frac{x^2 + 4x + 4 + x^2}{x^2 + 2x}$$

$$= \frac{2x^2 + 4x + 4}{x^2 + 2x}$$

The rule for subtracting unlike rational expressions is an extension of the rule for subtracting fractions in Section 4.2: If $\dfrac{P}{Q}$ and $\dfrac{R}{S}$ with $Q \neq 0$ and $S \neq 0$ are rational expressions, then

$$\frac{P}{Q} - \frac{R}{S} = \frac{P}{Q} + \left(-\frac{R}{S}\right) = \frac{PS - QR}{QS}$$

EXAMPLES

Perform the following subtractions.

1 $\dfrac{9x}{7} - \dfrac{3x}{5}$

SOLUTION

$$\frac{9x}{7} - \frac{3x}{5} = \frac{5(9x) - 7(3x)}{7(5)} = \frac{45x - 21x}{35}$$

$$= \frac{24x}{35}$$

2 $\dfrac{8}{x+1} - \dfrac{3}{x-1}$

SOLUTION

$$\frac{8}{x+1} - \frac{3}{x-1} = \frac{8(x-1) - 3(x+1)}{(x+1)(x-1)} = \frac{8x - 8 - 3x - 3}{(x-1)(x+1)}$$

$$= \frac{5x - 11}{(x-1)(x+1)} = \frac{5x - 11}{x^2 - 1}$$

Suppose that we consider the addition of the two rational expressions $\dfrac{7}{10x}$ and $\dfrac{8}{15x}$. One way is to use the above approach, so

$$\frac{7}{10x} + \frac{8}{15x} = \frac{7(15x) + 8(10x)}{150x^2} = \frac{105x + 80x}{150x^2} = \frac{185x}{150x^2} = \frac{37}{30x}$$

This approach can prove to be inconvenient. However, we could use a procedure parallel to that in Section 4.2 by finding the L.C.D. of $\dfrac{7}{10x}$ and $\dfrac{8}{15x}$. That is, we find a polynomial of lowest degree, yielding a polynomial quotient upon division by each of the given polynomials. In this case, the L.C.D. of $10x$ and $15x$ is $30x$. Thus,

$$\frac{7}{10x} + \frac{8}{15x} = \frac{21}{30x} + \frac{16}{30x} = \frac{21 + 16}{30x} = \frac{37}{30x}$$

In the first solution, the factor "$5x$" which was included in the denominator was subsequently removed. The fact that this is an "unnecessary" inclusion is shown by the results. Notice that in the second solution we converted the fractions to equivalent fractions with a common denominator containing fewer factors than the common denominator of the first solution.

To find the L.C.D. of a set of rational expressions, proceed in the same manner as finding the L.C.D. of a set of numbers in Section 4.2.

1 Factor each denominator into a product of prime factors or into a product of prime factors and -1.
2 List each prime factor with the largest exponent it has in any factored denominator; then find the product of the factors so listed.

This procedure is further illustrated by the following examples.

EXAMPLES

Express the following fractions in equivalent form having the L.C.D. of the given set as denominators.

1 $\frac{5}{18x}$ and $\frac{7}{12xy}$

SOLUTION. First, we write the denominators in factored form, so that

$$\begin{aligned} 18x &= 2(3)(3)x = 2(3^2)x \\ 12xy &= 2(2)(3)xy = 2^2(3)xy \end{aligned}$$

The various prime factors are 2, 3, x, and y. Listing each of these with the largest exponent it has in any factored form, we have 2^2, 3^2, x, and y. The L.C.D. of these fractions is $(2^2)(3^2)xy = 36xy$. Using the fundamental principle of fractions, we have

$$\frac{5}{18x} = \frac{5(2y)}{18x(2y)} = \frac{10y}{36xy}$$

$$\frac{7}{12xy} = \frac{7(3)}{12xy(3)} = \frac{21}{36xy}$$

2 $\frac{3x}{8yz}$, $\frac{5}{12xz}$, and $\frac{7z}{6xy}$

SOLUTION. The denominators are written in factored form as

$$\begin{aligned} 8yz &= 2(2)(2)yz = 2^3yz \\ 12xz &= 2(2)(3)xz = (2^2)(3)xz \\ 6xy &= 2(3)xy \end{aligned}$$

Listing each of these with the largest exponent it has in any factored form, we have 2^3, 3, x, y, and z. The L.C.D. of these fractions is $(2^3)(3)xyz = 24xyz$. Using the fundamental principle of fractions, we have

$$\frac{3x}{8yz} = \frac{3x(3x)}{8yz(3x)} = \frac{9x^2}{24xyz}$$

$$\frac{5}{12xz} = \frac{5(2y)}{12xz(2y)} = \frac{10y}{24xyz}$$

$$\frac{7z}{6xy} = \frac{7z(4z)}{6xy(4z)} = \frac{28z^2}{24xyz}$$

3 $\dfrac{3x}{x^2 - 4}$ and $\dfrac{5}{x^2 + 2x}$

SOLUTION. The denominators are written in factored form as

$$\begin{aligned} x^2 - 4 &= (x - 2)(x + 2) \\ x^2 + 2x &= x(x + 2) \end{aligned}$$

The L.C.D. of these fractions is $(x - 2)(x + 2)x = x(x^2 - 4)$. Using the fundamental principle of fractions, we have

$$\frac{3x}{x^2 - 4} = \frac{3x}{(x - 2)(x + 2)} = \frac{3x(x)}{(x - 2)(x + 2)(x)} = \frac{3x^2}{(x - 2)(x + 2)(x)}$$

$$\frac{5}{x^2 + 2x} = \frac{5}{x(x + 2)} = \frac{5(x - 2)}{x(x + 2)(x - 2)}$$

4 $\dfrac{x}{x^2 - 1}$ and $\dfrac{2}{x^2 + 2x + 1}$

SOLUTION. The denominators are written in factored form as

$$\begin{aligned} x^2 - 1 &= (x - 1)(x + 1) \\ x^2 + 2x + 1 &= (x + 1)^2 \end{aligned}$$

The L.C.D. of these expressions is $(x - 1)(x + 1)^2 = (x^2 - 1)(x + 1)$. Using the fundamental principle of fractions, we have

$$\frac{x}{x^2 - 1} = \frac{x(x + 1)}{(x^2 - 1)(x + 1)} = \frac{x^2 + x}{(x^2 - 1)(x + 1)}$$

$$\frac{2}{x^2 + 2x + 1} = \frac{2}{(x + 1)^2} = \frac{2(x - 1)}{(x + 1)^2(x - 1)} = \frac{2(x - 1)}{(x^2 - 1)(x + 1)}$$

The use of the L.C.D. in addition and subtraction of rational expressions usually simplifies the computation, as illustrated below.

EXAMPLES

Perform the following operations and simplify.

1 $\dfrac{x}{9} + \dfrac{5x}{6}$

SOLUTION. $9 = 3 \times 3 = 3^2$ and $6 = 3 \times 2$, so the L.C.D. of the given fractions is $2(3^2) = 18$. Thus,

$$\frac{x}{9} + \frac{5x}{6} = \frac{x(2)}{9(2)} + \frac{5x(3)}{6(3)} = \frac{2x}{18} + \frac{15x}{18}$$
$$= \frac{2x + 15x}{18} = \frac{17x}{18}$$

2 $\dfrac{2x-1}{15} + \dfrac{x+3}{25}$

SOLUTION. $15 = 3 \times 5$ and $25 = 5 \times 5 = 5^2$, so the L.C.D. of the given fraction is $3(5^2) = 75$. Thus,

$$\frac{2x-1}{15} + \frac{x+3}{25} = \frac{5(2x-1)}{75} + \frac{3(x+3)}{75}$$
$$= \frac{5(2x-1) + 3(x+3)}{75} = \frac{10x - 5 + 3x + 9}{75}$$
$$= \frac{13x + 4}{75}$$

3 $\dfrac{1}{xy} - \dfrac{x}{y}$

SOLUTION. The L.C.D. of the given fractions is xy. Therefore,

$$\frac{1}{xy} - \frac{x}{y} = \frac{1}{xy} - \frac{xx}{xy} = \frac{1 - x^2}{xy}$$

4 $\dfrac{3}{x^2 - x} - \dfrac{2}{x^2 - 1}$

SOLUTION. $x^2 - x = x(x - 1)$ and $x^2 - 1 = (x - 1)(x + 1)$, so the L.C.D. is $x(x - 1)(x + 1)$. Therefore,

$$\frac{3}{x^2 - x} - \frac{2}{x^2 - 1} = \frac{3}{x(x-1)} - \frac{2}{(x-1)(x+1)}$$
$$= \frac{3(x+1)}{x(x-1)(x+1)} - \frac{2x}{x(x-1)(x+1)}$$
$$= \frac{3x + 3 - 2x}{x(x-1)(x+1)} = \frac{x+3}{x(x-1)(x+1)}$$

5 $\dfrac{x}{x+1} - \dfrac{x}{x-1} + \dfrac{2}{x^2-1}$

SOLUTION. The denominators are $x + 1$, $x - 1$, and $x^2 - 1 = (x-1)(x+1)$. The L.C.D. is $(x-1)(x+1) = x^2 - 1$. Then,

$$\begin{aligned}
\frac{x}{x+1} - \frac{x}{x-1} + \frac{2}{x^2-1} &= \frac{x}{x+1} - \frac{x}{x-1} + \frac{2}{(x-1)(x+1)} \\
&= \frac{x(x-1)}{(x+1)(x-1)} - \frac{x(x+1)}{(x-1)(x+1)} \\
&\quad + \frac{2}{(x-1)(x+1)} \\
&= \frac{x(x-1) - x(x+1) + 2}{(x+1)(x-1)} \\
&= \frac{x^2 - x - x^2 - x + 2}{(x+1)(x-1)} = \frac{-2x+2}{(x+1)(x-1)} \\
&= \frac{-2(x-1)}{(x+1)(x-1)} = \frac{-2}{x+1}
\end{aligned}$$

6 $\dfrac{3}{x^2-4} + \dfrac{5}{x^2-x-6}$

SOLUTION. The denominators are $x^2 - 4 = (x-2)(x+2)$ and $x^2 - x - 6 = (x-3)(x+2)$. The L.C.D. is $(x-2)(x+2)(x-3)$. Then,

$$\begin{aligned}
\frac{3}{x^2-4} + \frac{5}{x^2-x-6} &= \frac{3}{(x-2)(x+2)} + \frac{5}{(x+2)(x-3)} \\
&= \frac{3(x-3)}{(x-2)(x+2)(x-3)} \\
&\quad + \frac{5(x-2)}{(x+2)(x-3)(x-2)} \\
&= \frac{3(x-3) + 5(x-2)}{(x-2)(x+2)(x-3)} \\
&= \frac{3x - 9 + 5x - 10}{(x-2)(x+2)(x-3)} \\
&= \frac{8x-19}{(x-2)(x+2)(x-3)}
\end{aligned}$$

PROBLEM SET 4.5

In problems 1–20, perform the indicated operations of the like fractions and reduce them to lowest terms.

1 $\frac{7}{2x} + \frac{13}{2x}$

2 $\frac{x}{4} + \frac{3x}{4}$

3 $\frac{4}{9x} + \frac{5}{9x}$

4 $\frac{5}{3m} + \frac{4}{3m}$

5 $\frac{7}{5a} + \frac{13}{5a}$

6 $\frac{a-2}{6} + \frac{3a-1}{6}$

7 $\frac{5}{7a^2b} + \frac{2}{7a^2b}$

8 $\frac{a+b}{a^2} + \frac{-b}{a^2}$

9 $\frac{3x}{x-7} + \frac{-21}{x-7}$

10 $\frac{4}{x^2} - \frac{4-x^2}{x^2}$

11 $\frac{7}{12x} - \frac{3}{12x}$

12 $\frac{17}{13x} - \frac{4}{13x}$

13 $\frac{a^2}{a-b} - \frac{ab}{a-b}$

14 $\frac{4}{4m-4} - \frac{4m}{4m-4}$

15 $\frac{x-9}{7} - \frac{x-2}{7}$

16 $\frac{5c-15}{20} - \frac{15-35c}{20}$

17 $\frac{8}{6x} + \frac{3}{6x} - \frac{5}{6x}$

18 $\frac{5}{x} + \frac{7}{x} + \frac{x - 12}{x}$

19 $\frac{19(x + 2)}{9} - \frac{x + 2}{9}$

20 $\frac{x + x^2}{9x} - \frac{x^2 - 3x}{9x}$

In problems 21–36, perform the indicated operations of the unlike fractions and reduce them to lowest terms.

21 $\frac{x^2}{3} + \frac{x^2}{2}$

22 $\frac{15x}{9} + \frac{10x}{13}$

23 $\frac{x^2}{5} - \frac{y^2}{6}$

24 $\frac{m}{7} + \frac{5m}{6}$

25 $\frac{7a^2}{11} - \frac{3a^2}{4}$

26 $\frac{x}{y} - \frac{2y}{x}$

27 $\frac{2a + b}{5} + \frac{3a - 4b}{a}$

28 $\frac{4x - 5}{x} + \frac{5 - 3x}{3}$

29 $\frac{2x - 3}{z} + \frac{x + 2}{y}$

30 $\frac{x}{x + 1} - \frac{y}{x}$

31 $\frac{7}{y - 2} - \frac{x}{y}$

32 $\frac{1}{x + 7} - \frac{3}{x}$

33 $\dfrac{9}{x-5} + \dfrac{6}{y-2}$

34 $\dfrac{7}{x-1} - \dfrac{5}{x+1}$

35 $\dfrac{5}{x-6} + \dfrac{3}{x-4}$

36 $\dfrac{1}{x+y} - \dfrac{1}{x-y}$

In problems 37–56, express the fractions in equivalent form having the L.C.D. of the given set as denominators.

37 $\dfrac{3}{a}, \dfrac{4}{b}$

38 $\dfrac{2}{3a}, \dfrac{3}{5a^2}$

39 $\dfrac{5}{ab^2}, \dfrac{3}{a^3b}$

40 $\dfrac{a+b}{3a^3b^2}, \dfrac{5}{4a^4b}$

41 $\dfrac{x+2}{4}, \dfrac{x-5}{6}$

42 $\dfrac{3x+2}{14}, \dfrac{2x-5}{21}$

43 $\dfrac{a-2}{6}, \dfrac{5-2a}{15}$

44 $\dfrac{2}{x}, \dfrac{3}{x^3}, \dfrac{5}{x^4}$

45 $\dfrac{5}{a^2}, \dfrac{4}{ab}, \dfrac{3}{b^2}$

46 $\dfrac{1}{3a^3b^2}, \dfrac{2}{5a^4b^3}$

47 $\dfrac{2}{x^2-9}, \dfrac{4}{x+3}$

48 $\dfrac{4}{3(x-1)^2}, \dfrac{2}{x^2-1}$

49 $\frac{2}{x-1}, \frac{2x}{x^2-1}$

50 $\frac{3}{4x-8}, \frac{5}{x^2-2x}$

51 $\frac{6}{x^2-36}, \frac{3}{x^2-7x+6}$

52 $\frac{3}{x-2}, \frac{4}{x^2-3x+2}$

53 $\frac{4x}{3x+6}, \frac{5x^2}{4x+8}$

54 $\frac{x-2}{x^2+x-12}, \frac{x-1}{x^2-7x+12}$

55 $\frac{2x-1}{24x^2-72x+54}, \frac{x-2}{24x^2-20x-24}$

56 $\frac{2x-3}{2x-5}, \frac{25x^2}{4x^2-25}$

In problems 57–82, perform the operations and reduce the resulting fractions to lowest terms.

57 $\frac{x}{4} + \frac{x}{2}$

58 $\frac{5x}{12} + \frac{x}{6}$

59 $\frac{y}{15} - \frac{y}{5}$

60 $\frac{t}{21} - \frac{t}{7}$

61 $\frac{m}{26} - \frac{m}{13}$

62 $\frac{x+1}{9} + \frac{x}{3}$

63 $\frac{3x-2y}{72} - \frac{2x-3y}{48}$

64 $\frac{3x-5}{9} - \frac{2x+7}{6}$

65 $\dfrac{2x+1}{24x} - \dfrac{3x^2+2x}{36x^2}$

66 $\dfrac{3}{ab^2} - \dfrac{5}{a^2b} + \dfrac{2}{ab}$

67 $\dfrac{5x+7y}{21} - \dfrac{4y+1}{12}$

68 $\dfrac{x+2}{4} - \dfrac{x-5}{5}$

69 $\dfrac{3}{a^2} - \dfrac{2-a}{ab} + \dfrac{b-5}{b^2}$

70 $\dfrac{5}{x^2-9} + \dfrac{4}{x+3}$

71 $\dfrac{2x-y}{x^2} - \dfrac{5x+2y}{x}$

72 $\dfrac{3}{x^2-4} + \dfrac{2}{x-2}$

73 $\dfrac{4x+3}{x^2-x-2} - \dfrac{x+3}{x-2}$

74 $\dfrac{2x+5}{x^2-9} - \dfrac{2}{x-3}$

75 $\dfrac{x-3}{x^2+x-6} + \dfrac{2x+1}{x-2}$

76 $\dfrac{4m+1}{2m^2+m-6} + \dfrac{4m-5}{2m^2-7m+6}$

77 $\dfrac{x}{x+2y} - \dfrac{8y^2}{x^2-4y^2}$

78 $\dfrac{1-2a}{3a^2-13a+4} - \dfrac{2+a}{2a^2-7a-4}$

79 $\dfrac{7}{2x^2-5x-3} + \dfrac{6}{x^2-9}$

80 $\dfrac{6a}{(a-2)^2} - \dfrac{3a}{2a-4}$

81 $\dfrac{3}{x^2-4} + \dfrac{2}{x-2} - \dfrac{5}{2x}$

82 $\dfrac{x+6}{x^2+8x+15} - \dfrac{x-3}{x+3} + \dfrac{3x}{x+5}$

4.6 Complex Fractions

We have considered fractions of the form $\frac{P}{Q}$, $Q \neq 0$, where P and Q are polynomial expressions. In this section we shall consider fractions in which the numerator or denominator, or both, contain fractions. Fractions of this type are called *complex fractions*. For example, the fractions

$$\frac{\frac{5}{7}}{3}, \quad \frac{\frac{16}{x}}{\frac{11}{x}}, \quad \frac{5+\frac{3}{x}}{9}, \quad \frac{9}{\frac{x}{3-x}}, \quad \frac{\frac{5}{13}}{7+\frac{2}{x}}, \quad \text{and} \quad \frac{x+\frac{3}{x}}{\frac{6}{2+x}-\frac{x}{2-x}}$$

are complex fractions.

To simplify a complex fraction is to express it as a single fraction in lowest terms. This can be accomplished by using the following approaches:

1 Express the fraction as a quotient of simplified fractions by performing additions and subtractions in the numerator and the denominator of the complex fraction; then perform the division indicated in the resulting fraction.

2 Multiply the numerator and the denominator of the complex fraction by the least common denominator (L.C.D.) of all the fractions they contain, by making use of the fundamental principle of fractions.

EXAMPLES

1 $\frac{\frac{3}{4}}{\frac{3}{5}}$

SOLUTION

First approach

$$\frac{\frac{3}{4}}{\frac{3}{5}} = \frac{3}{4} \div \frac{3}{5} = \frac{3}{4} \times \frac{5}{3} = \frac{5}{4}$$

Second approach

The L.C.D. of the fractions $\frac{3}{4}$ and $\frac{3}{5}$ is 20, so that

$$\frac{\frac{3}{4}}{\frac{3}{5}} = \frac{\frac{3}{4} \times 20}{\frac{3}{5} \times 20} = \frac{3 \times 5}{3 \times 4} = \frac{5}{4}$$

2 $\frac{\frac{3}{4}+\frac{1}{3}}{\frac{1}{3}+\frac{1}{6}}$

SOLUTION

First approach

$$\frac{\frac{3}{4}+\frac{1}{3}}{\frac{1}{3}+\frac{1}{6}} = \frac{\frac{3(3)+4(1)}{12}}{\frac{1(2)+1}{6}} = \frac{\frac{9+4}{12}}{\frac{2+1}{6}} = \frac{\frac{13}{12}}{\frac{3}{6}} = \frac{13}{12} \div \frac{3}{6} = \frac{13}{12} \times \frac{6}{3} = \frac{13}{6}$$

Second approach

The L.C.D. of the fractions is 12, so that

$$\frac{\frac{3}{4}+\frac{1}{3}}{\frac{1}{3}+\frac{1}{6}} = \frac{(\frac{3}{4}+\frac{1}{3})12}{(\frac{1}{3}+\frac{1}{6})12} = \frac{9+4}{4+2} = \frac{13}{6}$$

3 $\dfrac{\frac{27x^3}{7}}{\frac{9x^5}{21}}$

SOLUTION

First approach

$$\frac{\frac{27x^3}{7}}{\frac{9x^5}{21}} = \frac{27x^3}{7} \div \frac{9x^5}{21} = \frac{27x^3}{7} \cdot \frac{21}{9x^5} = \frac{(27x^3)(21)}{7(9x^5)} = \frac{9}{x^2}$$

Second approach

The L.C.D. of the fractions is 21, so that

$$\frac{\frac{27x^3}{7}}{\frac{9x^5}{21}} = \frac{\frac{27x^3}{7}(21)}{\frac{9x^5}{21}(21)} = \frac{81x^3}{9x^5} = \frac{9}{x^2}$$

4 $\dfrac{x - \frac{1}{x}}{1 + \frac{1}{x}}$

SOLUTION

First approach

$$\frac{x - \frac{1}{x}}{1 + \frac{1}{x}} = \frac{\frac{x^2 - 1}{x}}{\frac{x + 1}{x}}$$

$$= \frac{x^2 - 1}{x} \div \frac{x + 1}{x}$$

$$= \frac{x^2 - 1}{x} \cdot \frac{x}{x + 1}$$

$$= \frac{(x - 1)(x + 1)x}{x(x + 1)}$$

$$= x - 1$$

Second approach

The L.C.D. of $x - \frac{1}{x}$ and $1 + \frac{1}{x}$ is x, so that

$$\frac{x - \frac{1}{x}}{1 + \frac{1}{x}} = \frac{\left(x - \frac{1}{x}\right) \cdot x}{\left(1 + \frac{1}{x}\right) \cdot x} = \frac{x^2 - 1}{x + 1}$$

$$= \frac{(x - 1)(x + 1)}{x + 1}$$

$$= x - 1$$

5 $$\frac{\frac{1}{x + 1} - 1}{1 - \frac{1}{x}}$$

SOLUTION

First approach

$$\frac{\frac{1}{x + 1} - 1}{1 - \frac{1}{x}} = \frac{\frac{1 - x - 1}{x + 1}}{\frac{x - 1}{x}}$$

$$= \frac{\frac{-x}{x + 1}}{\frac{x - 1}{x}}$$

$$= \frac{-x}{x + 1} \div \frac{x - 1}{x}$$

$$= \frac{-x}{x + 1} \cdot \frac{x}{x - 1}$$

$$= \frac{(-x)(x)}{(x + 1)(x - 1)}$$

$$= \frac{-x^2}{x^2 - 1}$$

$$= \frac{x^2}{1 - x^2}$$

Second approach

The L.C.D. of $\frac{1}{x + 1} - 1$ and $1 - \frac{1}{x}$ is $x(x + 1)$, so that

$$\frac{\frac{1}{x + 1} - 1}{1 - \frac{1}{x}} = \frac{\left(\frac{1}{x + 1} - 1\right) x(x + 1)}{\left(1 - \frac{1}{x}\right) x(x + 1)}$$

$$= \frac{x - x(x + 1)}{x(x + 1) - (x + 1)}$$

$$= \frac{x - x^2 - x}{x^2 + x - x - 1}$$

$$= \frac{-x^2}{x^2 - 1} = \frac{x^2}{1 - x^2}$$

6 $\dfrac{\dfrac{x^2 - 9}{x^2 + x}}{\dfrac{x - 3}{x^2 - 1}}$

SOLUTION

First approach

$$\frac{\dfrac{x^2 - 9}{x^2 + x}}{\dfrac{x - 3}{x^2 - 1}} = \frac{x^2 - 9}{x^2 + x} \div \frac{x - 3}{x^2 - 1}$$

$$= \frac{x^2 - 9}{x^2 + x} \cdot \frac{x^2 - 1}{x - 3}$$

$$= \frac{(x - 3)(x + 3)(x - 1)(x + 1)}{x(x + 1)(x - 3)}$$

$$= \frac{(x + 3)(x - 1)}{x}$$

$$= \frac{x^2 + 2x - 3}{x}$$

Second approach

The L.C.D. of $\dfrac{x^2 - 9}{x^2 + x}$ and $\dfrac{x - 3}{x^2 - 1}$ is $x(x - 1)(x + 1)$.

$$\frac{\dfrac{x^2 - 9}{x^2 + x}}{\dfrac{x - 3}{x^2 - 1}} = \frac{\dfrac{x^2 - 9}{x^2 + x}x(x - 1)(x + 1)}{\dfrac{x - 3}{x^2 - 1}x(x - 1)(x + 1)}$$

$$= \frac{(x^2 - 9)(x - 1)}{(x - 3)x}$$

$$= \frac{(x - 3)(x + 3)(x - 1)}{(x - 3)x}$$

$$= \frac{(x + 3)(x - 1)}{x}$$

$$= \frac{x^2 + 2x - 3}{x}$$

PROBLEM SET 4.6

In problems 1–36, express each complex fraction as a single fraction in lowest terms.

1 $\dfrac{\frac{7}{11}}{\frac{5}{44}}$

2 $\dfrac{\frac{113}{24}}{\frac{12}{9}}$

3 $\dfrac{\frac{4}{9}}{\frac{5}{21}}$

4 $\dfrac{\frac{354}{55}}{\frac{168}{125}}$

5 $\dfrac{5}{4-\frac{1}{2}}$

6 $\dfrac{\frac{2}{3}+\frac{1}{4}}{\frac{2}{3}+\frac{1}{6}}$

7 $\dfrac{\frac{7}{8}+\frac{5}{12}}{\frac{3}{4}+\frac{2}{3}}$

8 $\dfrac{\frac{11}{4}+\frac{3}{2}}{\frac{13}{4}-\frac{5}{2}}$

9 $\dfrac{-\frac{9}{5}+2}{-\frac{9}{5}+1}$

10 $\dfrac{1}{\frac{5}{3}-1}$

11 $\dfrac{2}{2-\frac{1}{3}}$

12 $\dfrac{-\frac{18}{5}+1}{12}$

13 $\dfrac{\dfrac{9ab^2}{4xy^2}}{\dfrac{b^2}{2xy}}$

14 $\dfrac{\dfrac{14x^2}{10b^2}}{\dfrac{21x^2}{15b^2}}$

15 $\dfrac{\dfrac{5x}{12yz^2}}{\dfrac{15a^2}{18y^2z^2}}$

16 $\dfrac{\dfrac{34a}{4x^2y}}{\dfrac{17a^3}{2xy}}$

17 $\dfrac{\dfrac{1}{x}}{\dfrac{2}{x} + 1}$

18 $\dfrac{\dfrac{1}{x + 5}}{1 - \dfrac{1}{x + 5}}$

19 $\dfrac{\dfrac{10x^2}{(x + y)^2}}{\dfrac{5x}{x + y}}$

20 $\dfrac{\dfrac{(d - b)(a + b)}{x^2 - y^2}}{\dfrac{d - b}{x + y}}$

21 $\dfrac{1 - \dfrac{x^2}{y^2}}{x + y}$

22 $\dfrac{\dfrac{1}{x} - \dfrac{1}{y}}{x^2 - y^2}$

23 $\dfrac{x - \dfrac{y}{7}}{\dfrac{x}{7} - y}$

24 $\dfrac{\dfrac{x}{y} - 1}{x^2 - y^2}$

25 $\dfrac{x + y}{\dfrac{1}{x} + \dfrac{1}{y}}$

26 $\dfrac{2 + \dfrac{2}{x - 1}}{\dfrac{2}{x - 1}}$

27 $\dfrac{\dfrac{1}{x + 3}}{\dfrac{1 + x}{x + 3}}$

28 $\dfrac{\dfrac{1}{x + 1}}{1 + \dfrac{1}{x + 1}}$

29 $\dfrac{9 - \dfrac{1}{x^2}}{3 + \dfrac{1}{x}}$

30 $\dfrac{3 + \dfrac{5x + 1}{x^2 - 1}}{1 + \dfrac{1}{x + 1}}$

31 $\dfrac{\dfrac{x^2 - y^2}{x + y}}{\dfrac{x}{y} + 1}$

32 $\dfrac{\dfrac{x^2 - 4}{x^2 - 4x + 3}}{\dfrac{3x + 6}{x^2 - 9}}$

33 $\dfrac{\dfrac{2}{x + 1}}{1 - \dfrac{3}{x + 1}}$

34 $\dfrac{\dfrac{3}{1-x} - \dfrac{x}{1-x^2}}{\dfrac{1}{1+x}}$

35 $\dfrac{1 - \dfrac{x}{x-y}}{1 + \dfrac{x}{x-y}}$

36 $\dfrac{1 - x + \dfrac{2x^2}{1+x}}{1 - \dfrac{1}{1+x}}$

REVIEW PROBLEM SET

In problems 1–6, indicate which pairs of fractions are equivalent.

1 $\dfrac{6}{7}$ and $\dfrac{48}{56}$

2 $\dfrac{9}{13}$ and $\dfrac{99}{143}$

3 $\dfrac{-13}{17}$ and $\dfrac{91}{-119}$

4 $\dfrac{-13}{-19}$ and $\dfrac{117}{171}$

5 $\dfrac{5}{9}$ and $\dfrac{6}{10}$

6 $\dfrac{-189}{21}$ and -8

In problems 7–16, use the fundamental principle of fractions to reduce each fraction to an equivalent fraction in its lowest terms.

7 $\dfrac{15}{27}$

8 $\dfrac{39}{78}$

9 $\dfrac{-75}{15}$

10 $\frac{17}{-153}$

11 $\frac{38}{-342}$

12 $\frac{-412}{-721}$

13 $\frac{46}{115}$

14 $\frac{-18}{33}$

15 $\frac{-312}{-630}$

16 $\frac{210}{288}$

In problems 17–26, find the products and reduce them to lowest terms.

17 $\frac{8}{13} \times \frac{39}{48}$

18 $\frac{-2}{13} \times \frac{13}{-4}$

19 $\frac{49}{30} \times \frac{10}{14}$

20 $\frac{5}{26} \times \frac{13}{125}$

21 $\frac{-40}{11} \times \frac{121}{160}$

22 $\frac{13}{15} \times \frac{-5}{39}$

23 $\frac{-4}{9} \times \frac{-21}{16}$

24 $\frac{-16}{35} \times -\frac{15}{4}$

25 $\frac{105}{144} \times \frac{84}{147}$

26 $\frac{77}{143} \times \frac{52}{105}$

In problems 27–36, perform the divisions and reduce the results to lowest terms.

27 $\frac{4}{5} \div \frac{8}{15}$

28 $\frac{7}{8} \div \frac{16}{14}$

29 $\frac{14}{9} \div \frac{42}{81}$

30 $\frac{11}{19} \div \frac{22}{57}$

31 $\frac{15}{28} \div \frac{5}{14}$

32 $\frac{39}{51} \div \frac{-13}{17}$

33 $\frac{2}{15} \div \frac{-16}{35}$

34 $\frac{-105}{143} \div \frac{-21}{26}$

35 $\frac{-105}{147} \div \frac{13}{21}$

36 $\frac{-1{,}001}{68} \div \frac{143}{102}$

In problems 37–58, perform the indicated operations and reduce the results to lowest terms.

37 $\frac{5}{33} + \frac{7}{33}$

38 $\frac{8}{9} + \frac{15}{9}$

39 $\frac{3}{14} + \frac{4}{35}$

40 $\frac{5}{28} + \frac{3}{20}$

41 $3\frac{3}{4} + 4\frac{7}{12}$

42 $\frac{5}{6} + \frac{5}{8}$

43 $2\frac{11}{15} + 3\frac{5}{18}$

44 $\frac{7}{26} + \frac{8}{39}$

45 $\frac{7}{12} + \frac{11}{24}$

46 $7\frac{5}{6} + 2\frac{11}{12}$

47 $\frac{3}{8} + \frac{4}{24}$

48 $\frac{2}{15} + \frac{3}{45}$

49 $\frac{29}{33} - \frac{17}{22}$

50 $5\frac{7}{8} - 3\frac{5}{32}$

51 $\frac{31}{35} - \frac{19}{21}$

52 $\frac{13}{33} - \frac{15}{77}$

53 $\frac{67}{91} - \frac{5}{26}$

54 $\frac{27}{8} - \frac{3}{40}$

55 $\frac{16}{21} - \frac{15}{56}$

56 $7\frac{2}{6} - 2\frac{5}{21}$

57 $\frac{3}{77} - \frac{13}{1{,}001}$

58 $\frac{8}{37} - \frac{5}{222}$

In problems 59–62, perform the indicated operations and reduce the results to lowest terms.

59 $\frac{4}{33} - \frac{7}{6} + \frac{5}{11} - \frac{4}{3}$

60 $\frac{2}{5} \times \frac{1}{4} - \left(\frac{3}{5} - \frac{4}{13}\right) \div \frac{2}{3} + \frac{7}{10}$

61 $\left(\frac{5}{6} - \frac{11}{12}\right) \div \frac{1}{36} - \frac{23}{51} \div \frac{4}{17} - \frac{4}{3}$

62 $\frac{24}{51} + \frac{105}{17}\left[\left(\frac{3}{7} + \frac{1}{35}\right) \div \frac{32}{5}\right]$

In problems 63–68, use the property $\frac{P}{Q} = \frac{R}{S}$ if and only if $PS = QR$ to indicate which pairs of fractions are equivalent.

63 $\frac{10x^2}{12xy}, \frac{5x}{6y}$

64 $\frac{2p^2qr}{6p^3q}, \frac{r}{3p}$

65 $\frac{rs^3t^2}{rst^3}, \frac{s^2}{t}$

66 $\frac{4m^2n^3}{8mn^4}, \frac{m}{2n}$

67 $\frac{5(x-y)^2}{15(x-y)}, \frac{x-y}{3}$

68 $\frac{2x-4}{x^2-6x+8}, \frac{2}{x-4}$

In problems 69–88, use the fundamental principle of fractions to reduce each fraction into an equivalent fraction in its lowest terms.

69 $\frac{-3x}{9x}$

70 $\frac{96a^2b^2}{84a^3b}$

71 $\frac{8a^3bc^2}{12a^2b^2c}$

72 $\dfrac{3x^4y^3}{6x^3y^2}$

73 $\dfrac{4(x+1)}{y(x+1)}$

74 $\dfrac{x(x+3)}{x^2(x+3)}$

75 $\dfrac{18(x-y)^3}{27(x-y)^2}$

76 $\dfrac{6(x+y)^4}{10(x+y)^5}$

77 $\dfrac{7a^4(a-b)}{9a^3(a-b)^2}$

78 $\dfrac{8(m^2-n^2)}{18(m+n)}$

79 $\dfrac{2x+2}{x^2-1}$

80 $\dfrac{x^2y+xy^2}{2x^2y-2xy^2}$

81 $\dfrac{9x-3y}{18x-54y}$

82 $\dfrac{x^2-2xy+y^2}{x^2-y^2}$

83 $\dfrac{(3x-2y)(2a-b)}{(2x-3y)(b-2a)}$

84 $\dfrac{x^2-7x+12}{x^2+3x-18}$

85 $\dfrac{(a-b)(x+y)}{(b-a)(x-y)}$

86 $\dfrac{a^2-16}{a^2-14a+40}$

87 $\dfrac{x^2-25}{x^2+8x+15}$

88 $\dfrac{x^2-x-20}{x^2+5x+4}$

In problems 89–108, find the products and reduce them to lowest terms.

89 $\dfrac{33x}{165y} \cdot \dfrac{66y^2}{132x^3}$

90 $\dfrac{9a^2}{11} \cdot \dfrac{44a^5}{72}$

91 $\dfrac{42b^5}{26a^6} \cdot \dfrac{39a^9}{7b^4}$

92 $\dfrac{15r^4}{48s^3} \cdot \dfrac{64s^4}{75r^5}$

93 $\dfrac{15a^3}{14b^2} \cdot \dfrac{7b^4}{24a^5}$

94 $\dfrac{28a^2b^2}{54ab^3} \cdot \dfrac{18a^4b}{35ab^2}$

95 $\dfrac{2x - 4}{9} \cdot \dfrac{6}{5x - 10}$

96 $\dfrac{x^2 + x}{y} \cdot \dfrac{y^2}{x + 1}$

97 $\dfrac{(x + y)^2}{a + b} \cdot \dfrac{(a + b)^2}{x + y}$

98 $\dfrac{7a^5}{a^2 - 36} \cdot \dfrac{a^2 - 5a - 6}{21a^3}$

99 $\dfrac{a^2b^2 - 9}{x^2 - x - 2} \cdot \dfrac{x^3 - x^2 - 2x}{a^2b^2 - 6ab + 9}$

100 $\dfrac{2x^2 + 4x + 2}{x - 1} \cdot \dfrac{x - 1}{(x + 1)^2}$

101 $\dfrac{x^2 - 144}{x + 4} \cdot \dfrac{x^2 - 16}{x - 12}$

102 $\dfrac{a^2 - 4b^2}{(a + b)^2} \cdot \dfrac{a^2 - b^2}{a^2 - 4ab + 4b^2}$

103 $\dfrac{x^2 - 16}{x^2 - 4x} \cdot \dfrac{x - 4}{x + 4}$

104 $\dfrac{25x^2 - 9}{x^2 - 4} \cdot \dfrac{x - 2}{5x + 3}$

105 $\dfrac{x^2 + 5x + 6}{x^2 + x - 2} \cdot \dfrac{x^2 + 3x - 4}{x^2 + 7x + 12}$

106 $\frac{x^2 + 6x - 16}{x^2 - 9x + 14} \cdot \frac{x^2 - 8x + 7}{x^2 + 10x + 16}$

107 $\frac{3x^2 + 11x + 6}{4x^2 + 16x + 7} \cdot \frac{28 + x - 2x^2}{2 + x - 3x^2}$

108 $\frac{6x^2 + 13x - 5}{6x^2 + 13x - 15} \cdot \frac{4x^2 + 11x - 3}{2x^2 + x - 10}$

In problems 109–130, find the quotients and reduce them to lowest terms.

109 $\frac{2x}{y^2} \div \frac{x}{y}$

110 $\frac{3x}{4y^2} \div \frac{x^2}{2y}$

111 $\frac{x^2}{y} \div \frac{y}{x}$

112 $\frac{7x^2}{11} \div \frac{63x^3}{44}$

113 $\frac{32x^2}{70y^5} \div \frac{24x^3}{35y^4}$

114 $\frac{63xy^8}{34x^4y^7} \div \frac{42x^6y^2}{17x^9y^3}$

115 $\frac{45x^2y^3}{21d^2e^7} \div \frac{40x^4}{112de^5}$

116 $\frac{15a^3b^2}{19k^5m^7} \div \frac{40a^2b^3}{38km^4}$

117 $\frac{7x - 14}{x^2} \div \frac{x^2 - 4}{x^3}$

118 $\frac{6x^2 - 54}{x^4} \div \frac{x + 3}{x^3}$

119 $\frac{3c - cd}{12y^5} \div \frac{c^2d^2 - 9c^2}{2y^7}$

120 $\frac{x^2 - 3x - 4}{x + 2} \div \frac{x - 4}{x^2 - 4}$

121 $\frac{x^2 - 7x + 10}{x - 3} \div \frac{x - 2}{x^2 - 9}$

122 $\dfrac{x^2 - 4}{x^2 - 4x + 3} \div \dfrac{3x + 6}{x^2 - 9}$

123 $\dfrac{x + 7}{x + 2} \div \dfrac{x^2 - 49}{x^2 - 4}$

124 $\dfrac{a - b}{(x - y)^2} \div \dfrac{a^2 - 2ab + b^2}{x^2 - y^2}$

125 $\dfrac{(x - y)^2}{a^2 - b^2} \div \dfrac{x^2 - y^2}{(a + b)^2}$

126 $\dfrac{x^2 + 3xy + 2y^2}{x^2 - 2xy} \div \dfrac{x^2 + 4xy + 3y^2}{x^2 + xy - 6y^2}$

127 $\dfrac{(x + 2)^4(x - 2)^2}{(x^4 - 16)(x^2 - 4)} \div \dfrac{x^2 + 4x + 4}{2x^3 + 8x}$

128 $\dfrac{6 - 11x - 10x^2}{2x^2 + x - 3} \div \dfrac{5x^3 - 2x^2}{3x^2 - 5x + 2}$

129 $\dfrac{25x^2 - 10x + 1}{25x^2 + 10x + 1} \cdot \dfrac{5x^2 - 5}{25x^2 - 1} \div \dfrac{30x^2 - 6x}{40x^2 + 8x}$

130 $\dfrac{m^2n + mn^2}{m^2n - n^3} \cdot \dfrac{m^2 - 2mn + n^2}{m^3 + m^2n} \div \dfrac{4}{3m^2n - 3mn^2}$

In problems 131–160, perform the indicated operations and reduce the results to lowest terms.

131 $\dfrac{7a^2}{16} + \dfrac{3a^2}{16}$

132 $\dfrac{11}{15x} + \dfrac{7}{15x}$

133 $\dfrac{9y}{16} + \dfrac{5y}{16}$

134 $\dfrac{2}{3x} + \dfrac{7}{3x}$

135 $\dfrac{a}{x^3y} + \dfrac{b}{x^3y}$

136 $\dfrac{2}{9x} + \dfrac{4}{3x}$

137 $\dfrac{2}{3x} - \dfrac{5}{3x}$

138 $\dfrac{5}{xy} - \dfrac{3}{xy}$

139 $\dfrac{2}{3x} - \dfrac{1}{4x}$

140 $\dfrac{5}{xy} - \dfrac{6}{yz} + \dfrac{7}{xz}$

141 $\dfrac{3}{2xy} - \dfrac{4}{3yz} + \dfrac{5}{4xz}$

142 $\dfrac{3x}{4} + \dfrac{2x}{3} - \dfrac{x}{12}$

143 $\dfrac{1}{x + 3} + \dfrac{1}{x - 2}$

144 $\dfrac{1}{x - y} + \dfrac{1}{x + y}$

145 $\dfrac{7}{6x^3y^2} + \dfrac{5}{12xy} + \dfrac{1}{18x^2y^3}$

146 $\dfrac{9}{x^4} + \dfrac{6}{x^2y^2} + \dfrac{1}{y^4}$

147 $\dfrac{3}{x} + \dfrac{2}{x + y} + \dfrac{1}{x(x + y)}$

148 $\dfrac{1}{x - 3y} + \dfrac{1}{x + 3y}$

149 $\dfrac{5x}{7x + 2y} - \dfrac{6y}{14x + 4y}$

150 $\dfrac{1}{x - y} - \dfrac{x - 1}{x^2 - y^2}$

151 $\dfrac{3x - 1}{3} + \dfrac{2}{x - 1}$

152 $\dfrac{3x}{1 - x - 2x^2} + \dfrac{x}{2x^2 + 3x + 1}$

153 $\dfrac{1}{x^2 - y^2} + \dfrac{1}{(x + y)^2}$

154 $\dfrac{2}{x + 2} - \dfrac{x + 1}{x^2 + 4x + 4}$

155 $\dfrac{5}{x^2 - x - 6} - \dfrac{7}{-x^2 + 3x + 10}$

156 $\dfrac{3x}{2x + 1} - \dfrac{x + 2}{6x^2 - x - 2}$

157 $\dfrac{3}{x^2 - 7x + 12} + \dfrac{2}{x^2}$

158 $\dfrac{2}{x - 2} - \dfrac{3}{x + 3} + \dfrac{4}{x - 4}$

159 $\dfrac{2}{x^2 - 4} + \dfrac{3}{x + 2} - \dfrac{1}{2x - 4}$

160 $\dfrac{1}{(x - y)^2} + \dfrac{x}{x - y} + 1$

In problems 161–174, express each complex fraction as a single fraction in lowest terms.

161 $\dfrac{\frac{2}{3}}{\frac{5}{8}}$

162 $\dfrac{\frac{4}{11}}{\frac{6}{121}}$

163 $\dfrac{\frac{2}{3} + \frac{1}{5}}{\frac{4}{3} - \frac{2}{5}}$

164 $\dfrac{2 - \frac{2}{5}}{1 + \frac{3}{5}}$

165 $\dfrac{\dfrac{x}{y}}{\dfrac{2x}{y^3}}$

166 $\dfrac{\dfrac{xy}{z}}{\dfrac{x^2y}{z^2}}$

167 $\dfrac{\dfrac{x}{x + 1}}{\dfrac{x^2}{y^4}}$

168 $\dfrac{\dfrac{x}{y^2}}{\dfrac{x^3}{y^4}}$

169 $\dfrac{x + \dfrac{1}{x}}{x - \dfrac{1}{x}}$

170 $\dfrac{\dfrac{x}{y} - 1}{\dfrac{x}{y} + 1}$

171 $\dfrac{\dfrac{x}{y} + 1}{\dfrac{y}{x} - \dfrac{x}{y}}$

172 $\dfrac{\dfrac{1}{2x - 3} - \dfrac{1}{2x + 3}}{\dfrac{1}{4x^2 - 9}}$

173 $\dfrac{\dfrac{1}{x} - \dfrac{1}{y}}{\dfrac{x^2 - y^2}{xy}}$

174 $\dfrac{\dfrac{y}{y^2 - 1} - \dfrac{1}{y + 1}}{\dfrac{y}{y - 1} + \dfrac{1}{y + 1}}$

CHAPTER 5

First-Degree Equations in One Variable

5 FIRST-DEGREE EQUATIONS IN ONE VARIABLE

Introduction

Now that we have studied the basic properties of real numbers (Chapter 1), polynomial expressions (Chapters 2 and 3), and fractional expressions (Chapter 4), we shall turn our attention to an important problem of algebra: applying many of the techniques we have learned to solve equations.

An *equation* is a mathematical statement that expresses the relation of *equality* between two expressions representing real numbers. Hence, we shall use those properties of real numbers, and the equality relation discussed in Chapter 1, to solve equations.

We conclude this chapter with an important application of first-degree equations to solve equations that contain absolute value expressions.

5.1 First-Degree Equations

As stated above, an *equation* is a mathematical statement that expresses the relation of *equality* between two expressions representing real numbers. For example, $15 - 2 = 13$ is an equation. Likewise, if x represents a real number, then $3x + 6 = 14$ is an equation, and is read: "Three times the number represented by x added to 6 is equal to 14." In general, an equation is an abbreviated form of a statement of equality that could be written in sentence form.

Equations such as $2x - 1 = 5$, $3(x + 2) = 7 - 2(x + 1)$, and $\frac{1}{3}x + 3 = 5 - x$ are called *first-degree equations* or *linear equations*. Any value of the variable x that makes the equation a valid statement is called a *solution* of the equation. Hence, to solve an equation means to find *all* values of the variable for which the equation is a valid statement. For example, in the equation $3x + 6 = 14$, $\frac{8}{3}$ is a solution to the equation since $3(\frac{8}{3}) + 6 = 8 + 6 = 14$.

The solution or solutions of an equation form a set, and this set is called the *solution set* of the equation. The solution set of the equation $3x + 6 = 14$ is designated by $\{\frac{8}{3}\}$ since the only value of x which produces a valid statement is the number $\frac{8}{3}$. An equation may have one, more than one, or no solutions. If there are no solutions, then its solution set is designated by the empty set $\emptyset$. In this chapter we shall be concerned primarily with those equations in one variable that have one solution. These equations can also be expressed in set-builder notation. For instance, we write the equation $2x - 1 = 5$ in set-builder notation as $\{x \mid 2x - 1 = 5\}$.

EXAMPLES

1 Determine if any of the numbers 3, −1, or 5 are solutions of the equation $3x - 2 = 13$.

SOLUTION. For $x = 3$, we have on the left-hand side $3(3) - 2 = 9 - 2 = 7$. Since $7 \neq 13$, 3 is not a solution.

For $x = -1$, $3(-1) - 2 = -3 - 2 = -5$; but since $-5 \neq 13$, $x = -1$ is not a solution.

For $x = 5$, $3(5) - 2 = 15 - 2 = 13$. Therefore, $x = 5$ is a solution.

2 Determine if any of the numbers −5, 3, or $\frac{-20}{7}$ are solutions of the equation $6x + 2x = -40 - 6x$.

SOLUTION. For $x = -5$, we have

$$\begin{aligned} 6(-5) + 2(-5) &\stackrel{?}{=} -40 - 6(-5) \\ -30 - 10 &\stackrel{?}{=} -40 + 30 \\ -40 &\neq -10 \end{aligned}$$

For $x = 3$, we have

$$\begin{aligned} 6(3) + 2(3) &\stackrel{?}{=} -40 - 6(3) \\ 18 + 6 &\stackrel{?}{=} -40 - 18 \\ 24 &\neq -58 \end{aligned}$$

For $x = \frac{-20}{7}$, we have

$$6\left(\frac{-20}{7}\right) + 2\left(\frac{-20}{7}\right) = -40 - 6\left(\frac{-20}{7}\right)$$

$$\frac{-120}{7} - \frac{40}{7} = -40 + \frac{120}{7}$$

$$\frac{-160}{7} = \frac{-280}{7} + \frac{120}{7}$$

$$\frac{-160}{7} = \frac{-160}{7}$$

Therefore, the numbers -5 and 3 are not solutions, while $\frac{-20}{7}$ is a solution of the given equation.

Equations can be classified into two general types: conditional equations and identical equations. *Conditional equations* are those equations that are true only for certain values of the variable. For example, if x is a real number and $x + 1 = 3$, we have placed a condition on the variable x; that is, $x = 2$. Other examples of conditional equations are

$$7x + 3 = 16$$

and

$$x^2 - 3x + 2 = 0$$

Identities are those equations that are true for all possible values of the variable. For example, the equation $x^2 - 1 = (x - 1)(x + 1)$ is an identity. Regardless of what value we assign to x, the resulting equation is true. We have already made use of identities when we discussed some of the basic properties of algebra. For example, the equations $ab = ba$, $a(bc) = (ab)c$, and $a(b + c) = ab + ac$ are identities.

Solution of First-Degree Equations

A solution of an equation is said to *satisfy* the equation because when substituted for the variable, it makes a true statement of the equation. For example, 2 satisfies the equation $5x + 3 = 13$, since $5(2) + 3 = 13$ is a true statement. Two equations are *equivalent* if all values, and no other, that satisfy one of those equations also satisfy the second equation. In other words, two equations that involve the same variable are equivalent if and only if they have the same solution sets. For example, consider the equations

$$5x + 2 = 7$$
$$5x = 5$$
$$x = 1$$

If 1 is substituted for x in each of these equations, the equations are satisfied. Also, no other value of x will satisfy any of the equations. Hence, these three equations are equivalent equations. To solve first-degree equations we shall make extensive use of some of the properties of real numbers discussed in previous chapters, in particular the following two properties:

1 ADDITION PROPERTY

The addition of the same number to both sides of an equation produces an equivalent equation. In symbols, if P, Q, and R are polynomials, and $P = Q$, then $P + R = Q + R$.

2 MULTIPLICATION PROPERTY

The multiplication of both sides of an equation by the same nonzero number produces an equivalent equation. In symbols, if P, Q, and R are polynomials, $R \neq 0$, and $P = Q$, then $PR = QR$.

EXAMPLE

Find the value of x for which the equation $3x - 1 = 11$ is true.

SOLUTION

$$3x - 1 = 11$$
$$3x - 1 + 1 = 11 + 1 \qquad \text{(Addition property)}$$
$$3x = 12$$
$$\tfrac{1}{3}(3x) = \tfrac{1}{3}(12) \qquad \text{(Multiplication property)}$$
$$x = 4$$

Note in the example that the process of solution produced three equivalent equations: $3x - 1 = 11$, $3x = 12$, and $x = 4$. It is this process of transforming the given equation $3x - 1 = 11$ into the equivalent equation $x = 4$ by applying the above properties that is called "solving the equation."

There is no established pattern for the sequence of equivalent equations that one might follow in solving a first-degree equation. One of the most commonly used techniques in solving such equations is to have the terms containing the variable collected on one side of the equality sign and the remaining terms collected on the other side. Dividing both sides of the equation by the coefficient of the variable, we obtain the solution.

EXAMPLES

Find the solution set of the following first-degree equations.

1 $2x + 5 = 19$

SOLUTION

$2x + 5 + (-5) = 19 + (-5)$ (Addition property)

so that

$$2x = 14$$

or

$$\tfrac{1}{2}(2x) = \tfrac{1}{2}(14) \qquad \text{(Multiplication property)}$$

Therefore, $x = 7$.

Check:

$$\begin{aligned} 2(7) + 5 &\stackrel{?}{=} 19 \\ 14 + 5 &\stackrel{?}{=} 19 \\ 19 &= 19 \end{aligned}$$

Thus, 7 is the solution; or, the solution set is $\{7\}$.

2 $3x + 2 = x - 6$

SOLUTION. We have $3x + 2 = x - 6$. Now rewrite the equation by eliminating the variable x from the right side of the equation:

$3x + 2 + (-x) = x - 6 + (-x)$ (Addition property)

or

$$2x + 2 = -6 \qquad \text{(Why?)}$$

then

$2x + 2 + (-2) = -6 + (-2)$ (Addition property)

so that

$$2x = -8$$

then

$$\tfrac{1}{2}(2x) = \tfrac{1}{2}(-8) \qquad \text{(Multiplication property)}$$

Therefore, $x = -4$.

Check:

$$3(-4) + 2 \stackrel{?}{=} -4 - 6$$
$$-12 + 2 \stackrel{?}{=} -4 - 6$$
$$-10 = -10$$

Thus, -4 is the solution. The solution set is $\{-4\}$.

3 $3(x - 1) = 5 - x$

SOLUTION

$$3(x - 1) = 5 - x$$
$$3x - 3 = 5 - x \quad \text{(Distributive property)}$$
$$3x - 3 + 3 = 5 - x + 3 \quad \text{(Addition property)}$$

or

$$3x = 8 - x$$

then

$$3x + x = 8 - x + x \quad \text{(Addition property)}$$

so that

$$4x = 8$$

or

$$\tfrac{1}{4}(4x) = \tfrac{1}{4}(8) \quad \text{(Multiplication property)}$$

Therefore, $x = 2$.

Check:

$$3(2 - 1) \stackrel{?}{=} 5 - 2$$
$$3(1) \stackrel{?}{=} 3$$
$$3 = 3$$

Thus, 2 is the solution. The solution set is $\{2\}$.

4 $5x - 4 + x + 5 = 19$

SOLUTION

$$5x - 4 + x + 5 = 19$$

or

$$6x + 1 = 19 \qquad \text{(Why?)}$$

so that

$$6x + 1 + (-1) = 19 + (-1) \qquad \text{(Why?)}$$

or

$$6x = 18$$

so that

$$\tfrac{1}{6}(6x) = \tfrac{1}{6}(18)$$

Therefore, $x = 3$.

Check:

$$\begin{aligned} 5(3) - 4 + 3 + 5 &\stackrel{?}{=} 19 \\ 15 - 4 + 8 &\stackrel{?}{=} 19 \\ 19 &= 19 \end{aligned}$$

Thus, 3 is the solution. The solution set is $\{3\}$.

5 $\{x \mid 5 - 4x = 9\}$

SOLUTION

$$\{x \mid 5 - 4x = 9\} = \{x \mid -4x = 4\} = \{x \mid x = -1\}$$

Therefore, the solution set is $\{-1\}$.

The procedure of solving first-degree equations with one variable depends on one or all of the following steps.

1 Perform all operations indicated by parentheses or symbols of grouping.
2 Combine all "similar terms" on the left and right sides of the equation.
3 Decide which side the variable should be on to make solving the equation easier (numerical coefficient to be positive for the variable).
4 Move all terms containing the variable to the desired side and all constants to the other side by using the additive inverse of each term to be moved and the addition property.
5 Again combine all "similar terms" on the left and right sides of the equation.
6 Eliminate the factor (numerical coefficient of the variable) by using the multiplicative inverse of the factor and the multiplication property.

PROBLEM SET 5.1

In problems 1–6, determine which of the given numbers is a solution of the given equation.

1 $3x + 1 = 7$; $-1, 0, 1, 2$

2 $2x - 1 = x - 1$; $0, 1, 2$

3 $2(x + 1) - 3 = 1$; $-1, 0, 1$

4 $3t - 2t + 5 = 9$; $0, 2, 4$

5 $r + 1 = 6r - 9$; $0, 1, 2$

6 $4t + 1 - 2t - 5 = 6$; $0, 5, -5$

In problems 7–46, find the solution set of the given equations.

7 $x + 1 = 3$

8 $x + 7 = 10$

9 $x - 10 = 15$

10 $x - 13 = 5$

11 $2x + 1 = x + 5$

12 $3x - 7 = 2x + 1$

13 $3x - 4 = 2x - 7$

14 $7x + 13 = 6x + 1$

15 $2y - 2 = 3y - 5$

16 $3t - 1 = 4t + 3$

17 $13t - 7 = 14t + 3$

18 $23t + 1 = 24t + 5$

19 $t + 3t = 8$

20 $5t + 1 = 6$

21 $5 + 3r = r + 3$

22 $2s - 1 = 5s + 8$

23 $7x - 5x + 5 = 13$

24 $8t + 5 = 5t + 8$

25 $5x - 2 = 2x + 10$

26 $3 - 2r + 5 = 3r - 2$

27 $2t + 3t - 4 = 21$

28 $3x + 2x - 6 - 4x = 13$

29 $6x - 8 + x = 34$

30 $2p + 7p - 10 = 17$

31 $16x - 8 + 3x = 49$

32 $8x + 12 = 40 - x + 8$

33 $21 - y + 4 = 30 - 6y$

34 $51 - 7t + 5 = 8 - 9t + 54$

35 $3x - 2x - 2 = 2x - 2$

36 $4x = 48x - 30 - 102$

37 $7x - 27 + 15x = 4x - 9$

38 $34 - 3x = 56 - 8x + 23$

39 $23p + 4 = 36p - 27 + 7 - p$

40 $3x - 2x = 24 + 2x - 5$

41 $15 - 25s = 7 - 5 + 2s$

42 $4y + 30 = 54 - 33y$

43 $5t + (t - 12) = 6 - (t + 4)$

44 $37 - (4r - 6) = -(9r - 5) + 73$

45 $7x - 3(9 - 5x) = 4x - 9x$

46 $6(x - 10) + 3(2x - 7) = -45$

In problems 47–52, write the solution set of the given equations.

47 $\{x \mid 2x + 1 = 7\}$

48 $\{t \mid 3t - 2 = 4\}$

49 $\{r \mid 5(3 - 2r) = 4(4r + 20)\}$

50 $\{p \mid 4(2p + 1) = 6p\}$

51 $\{y \mid -2(-y + 3) + y = 4y - 1\}$

52 $\{x \mid 2(-3x - 4) = 5(1 - x)\}$

5.2 Fractional Equations

Equations of the form $\frac{x+3}{4} = 6$ and $\frac{x-1}{2x} = 3$ are called *fractional equations.* The procedure employed in solving either of these equations is exactly the same. We must, however, take care to recognize the restrictions if the variable appears in the denominator of a fraction. As stated in Section 4.3, the variables in any fraction may not be assigned values that will involve a division by zero, since division by zero is undefined in the real numbers.

The technique for solving equations which contain fractions is as follows:

1 Determine the L.C.D. of the fractions.
2 Multiply both sides of the equation by the L.C.D. in order to clear the equations of fractions.
3 Solve the resulting equation by the steps previously illustrated.

EXAMPLES

1 $\frac{x}{3} = \frac{5}{9}$

SOLUTION. The L.C.D. of both sides of the equation is 9. Thus,

$$9\left(\frac{x}{3}\right) = 9\left(\frac{5}{9}\right)$$

or

$$3x = 5$$

so that

$$x = \frac{5}{3}$$

Check:

$$\frac{\frac{5}{3}}{3} = \frac{5}{3} \div 3 = \frac{5}{3} \times \frac{1}{3} = \frac{5}{9}$$

Therefore, the solution set is $\{\frac{5}{3}\}$.

2 $\dfrac{x}{4} + \dfrac{2x}{3} = 5$

SOLUTION. $\dfrac{x}{4} + \dfrac{2x}{3} = 5$. The L.C.D. is 12. We have

$$12\left(\frac{x}{4} + \frac{2x}{3}\right) = 12(5) \qquad \text{(Multiplication property)}$$

$$12\left(\frac{x}{4}\right) + 12\left(\frac{2x}{3}\right) = 12(5) \qquad \text{(Distributive property)}$$

or

$$3x + 8x = 60$$

so that

$$11x = 60 \qquad \text{or} \qquad x = \tfrac{60}{11}$$

Check:

$$\frac{\frac{60}{11}}{4} + \frac{2(\frac{60}{11})}{3} \stackrel{?}{=} 5$$

$$\tfrac{60}{44} + \tfrac{120}{33} \stackrel{?}{=} 5$$

$$\tfrac{180}{132} + \tfrac{480}{132} \stackrel{?}{=} 5$$

$$\tfrac{660}{132} \stackrel{?}{=} 5$$

$$5 = 5$$

Thus, $\frac{60}{11}$ is the solution. The solution set is $\{\frac{60}{11}\}$.

3 $\dfrac{x-2}{3} - \dfrac{x-3}{5} = \dfrac{13}{15}$

SOLUTION. $\dfrac{x-2}{3} - \dfrac{x-3}{5} = \dfrac{13}{15}$. The L.C.D. is 15. Hence,

$$15\left(\frac{x-2}{3} - \frac{x-3}{5}\right) = \tfrac{13}{15}(15) \qquad \text{(Why?)}$$

$$15\left(\frac{x-2}{3}\right) - 15\left(\frac{x-3}{5}\right) = \tfrac{13}{15}(15)$$

so that

$$5x - 10 - (3x - 9) = 13 \qquad \text{(Why?)}$$

That is,

$$5x - 10 - 3x + 9 = 13$$

or

$$2x - 1 = 13 \qquad \text{(Why?)}$$
$$2x = 14$$

or

$$x = 7$$

Check:

$$\frac{7-2}{3} - \frac{7-3}{5} \stackrel{?}{=} \frac{13}{15}$$
$$\tfrac{5}{3} - \tfrac{4}{5} \stackrel{?}{=} \tfrac{13}{15}$$
$$\tfrac{25}{15} - \tfrac{12}{15} \stackrel{?}{=} \tfrac{13}{15}$$
$$\tfrac{13}{15} = \tfrac{13}{15}$$

Hence, 7 is the solution. The solution set is $\{7\}$.

4 $\dfrac{1}{x} + \dfrac{1}{2} = \dfrac{5}{6x} + \dfrac{1}{3}$

SOLUTION. $\dfrac{1}{x} + \dfrac{1}{2} = \dfrac{5}{6x} + \dfrac{1}{3}$. Multiplying both sides by the L.C.D. $6x$, we have

$$6x\left(\frac{1}{x} + \frac{1}{2}\right) = 6x\left(\frac{5}{6x} + \frac{1}{3}\right)$$
$$6x\left(\frac{1}{x}\right) + 6x(\tfrac{1}{2}) = 6x\left(\frac{5}{6x}\right) + 6x(\tfrac{1}{3})$$

That is,

$$6 + 3x = 5 + 2x \qquad \text{or} \qquad 3x - 2x = 5 - 6$$

so that $x = -1$. Therefore, the solution set is $\{-1\}$.

5 $\left\{x \;\middle|\; \dfrac{x+4}{3x} = \dfrac{3x-8}{4x}\right\}$

SOLUTION. Multiplying both sides of the equation by the L.C.D. $12x$, we have

$$\left\{x \,\middle|\, \frac{x+4}{3x} = \frac{3x-8}{4x}\right\} = \left\{x \,\middle|\, 12x\left(\frac{x+4}{3x}\right) = 12x\left(\frac{3x-8}{4x}\right)\right\}$$
$$= \{x \mid 4(x+4) = 3(3x-8)\}$$
$$= \{x \mid 4x + 16 = 9x - 24\}$$
$$= \{x \mid -5x = -40\}$$
$$= \{x \mid x = 8\}$$

The solution set is {8}.

PROBLEM SET 5.2

In problems 1–42, find the solution set of the given equations.

1 $\frac{x}{2} = \frac{3}{4}$

2 $\frac{t}{4} = \frac{3}{2}$

3 $\frac{p}{3} = \frac{5}{6}$

4 $\frac{4t}{13} = \frac{5}{52}$

5 $\frac{3t}{2} = \frac{7}{8}$

6 $\frac{7r}{9} = \frac{5}{18}$

7 $\frac{t}{7} + 1 = \frac{1}{7}$

8 $\frac{x}{6} + \frac{2x}{3} = 5$

9 $\frac{y}{5} + \frac{y}{3} = 4$

10 $\frac{3x}{2} + \frac{x}{8} = 13$

11 $\frac{9x}{11} + \frac{4}{11} = 2$

12 $\frac{4x}{3} + \frac{x}{4} = 38$

13 $\frac{y}{5} - \frac{y}{6} = \frac{6}{5}$

14 $\frac{2x}{3} - \frac{4x}{9} = 2$

15 $\frac{x}{5} - \frac{x}{13} = 8$

16 $\frac{x}{2} - \frac{17 + x}{5} = 2x$

17 $x - 1 = \frac{2x}{5} - \frac{9}{5}$

18 $\frac{1 - x}{3} + \frac{x + 2}{8} = 3$

19 $\frac{t + 1}{7} - \frac{t}{15} = \frac{13}{35}$

20 $\frac{p + 1}{7} + \frac{3p}{5} = \frac{109}{35}$

21 $\frac{6 - t}{3} - \frac{t - 7}{5} = \frac{1}{3}$

22 $\frac{2r + 1}{3} + \frac{2r - 1}{2} = \frac{59}{6}$

23 $\frac{x}{3} - \frac{x - 2}{5} = \frac{4}{3}$

24 $\frac{x + 9}{4} = 2 + \frac{3}{14}(2x - 3)$

25 $\frac{2x}{5} - \frac{3x - 8}{12} = \frac{67}{60}$

26 $\frac{x + 1}{5} = \frac{x + 10}{2} - 6$

27 $\frac{2y - 1}{6} + \frac{3}{2} = \frac{y + 7}{9}$

28 $\frac{x - 14}{5} + 4 = \frac{x + 16}{10}$

29 $\frac{3(x - 1)}{3} = \frac{-(x + 7)}{5}$

30 $\dfrac{x+8}{4} = \dfrac{5x-2}{6}$

31 $\dfrac{1}{2x} + \dfrac{5}{8} = \dfrac{3}{x}$

32 $\dfrac{5}{x} + \dfrac{3}{8} = \dfrac{7}{16}$

33 $\dfrac{2}{3x} + \dfrac{1}{6x} = \dfrac{1}{4}$

34 $\dfrac{1}{x} + \dfrac{2}{x} = 3 - \dfrac{3}{x}$

35 $\dfrac{2}{x} + \dfrac{x-1}{3x} = \dfrac{2}{5}$

36 $3 - \dfrac{5}{2x} = -\dfrac{1}{x}$

37 $\dfrac{8}{x-3} = \dfrac{12}{x+3}$

38 $\dfrac{9}{5x-3} = \dfrac{5}{3x+7}$

39 $\left\{x \,\middle|\, \dfrac{3(x-1)}{x+4} = -\dfrac{3}{5}\right\}$

40 $\left\{x \,\middle|\, \dfrac{x}{x+1} + 2 = \dfrac{3x}{x+2}\right\}$

41 $\left\{y \,\middle|\, \dfrac{2y}{y-2} = \dfrac{8}{y-2} + 1\right\}$

42 $\left\{t \,\middle|\, \dfrac{t}{t-1} + 2 = \dfrac{3}{t-1}\right\}$

5.3 Literal Equations and Formulas

We have already seen (Section 5.1) how to solve first-degree equations in one variable in which the constants are real numbers. In this section we will solve ***literal*** **first-degree equations,** that is, **first-degree equations in which the constants are represented by letters.** The addition and multiplication properties we used to solve first-degree equations in one variable will be used again to solve formulas for specific variables, as well as to solve literal equations. For example, to solve the literal equation $ax + b = d$ for x, in which a, b, and d are constants, we proceed as follows. We have $(ax + b) + (-b) = d + (-b)$, so that $ax = d - b$. Thus, $\left(\frac{1}{a}\right)(ax) = \left(\frac{1}{a}\right)(d - b)$ or $x = \frac{1}{a}(d - b)$. This result could also have been written as $x = \frac{d}{a} - \frac{b}{a}$. Therefore, the solution set of the equation is $\left\{\frac{d}{a} - \frac{b}{a}\right\}$.

EXAMPLES

Find the solution set for each of the following equations by solving for x.

1 $3x - a = 3a$, where a is a constant.

SOLUTION

$3x - a = 3a$

so that

$3x = 3a + a$ (Addition property)

or

$3x = 4a$

so that

$x = \frac{4a}{3}$ (Multiplication property)

Therefore, the solution set is $\left\{\frac{4a}{3}\right\}$.

2 $\frac{3x}{2} - a = x + 3a$, where a is a constant.

SOLUTION

$$\frac{3x}{2} - a = x + 3a$$

so that

$$3x - 2a = 2x + 6a \qquad \text{(Multiplication property)}$$

or

$$\begin{aligned} 3x - 2x &= 6a + 2a \qquad \text{(Addition property)} \\ x &= 8a \end{aligned}$$

Therefore, the solution set is $\{8a\}$.

3 $\{x \mid 3ax - 2 = 5b\}$, where a and b are constants.

SOLUTION

$$\begin{aligned} \{x \mid 3ax - 2 = 5b\} &= \{x \mid 3ax = 5b + 2\} \\ &= \left\{x \,\middle|\, x = \frac{5b + 2}{3a}\right\} \end{aligned}$$

Therefore, the solution set is $\left\{\frac{5b + 2}{3a}\right\}$.

4 Solve the formula $F = \frac{9}{5}C + 32$ for C.

SOLUTION

$$\begin{aligned} F &= \tfrac{9}{5}C + 32 \\ F - 32 &= \tfrac{9}{5}C + 32 - 32 \qquad \text{(Addition property)} \\ F - 32 &= \tfrac{9}{5}C \\ \tfrac{5}{9}(F - 32) &= \tfrac{5}{9} \cdot \tfrac{9}{5}C \qquad \text{(Multiplication property)} \end{aligned}$$

Hence,

$$C = \tfrac{5}{9}(F - 32)$$

PROBLEM SET 5.3

In problems 1–14, find the solution set of the given equations. Assume that a, b, and c are constant real numbers.

1 $x + a = 2a$

2 $x + 3b = 4b$

3 $x - c = 3c$

4 $x - 2a = 3a$

5 $at - a = 3a$

6 $by - 2b = 3b$

7 $5cy - a = b$

8 $7at - b = c$

9 $\{t \mid 2t + 1 = a\}$

10 $\{x \mid ax - 3 = c\}$

11 $\{x \mid 3ax - 2 = b\}$

12 $\left\{x \,\middle|\, \frac{ax}{b} + \frac{1}{b} = \frac{c}{b}\right\}$

13 $\{r \mid 5br - 7 = c\}$

14 $\{s \mid 2s + a = b - c\}$

In problems 15–34, solve the following formulas for the indicated variables.

15 $A = lw$, for l

16 $A = \frac{1}{2}bh$, for h

17 $d = rt$, for r

18 $P = 2l + 2w$, for w

19 $C = 2\pi r$, for r

20 $A = \frac{1}{2}(a + b)h$, for h

21 $I = prt$, for t

22 $V = \frac{1}{3}bh$, for b

23 $V = lwh$, for h

24 $E = mc^2$, for m

25 $P = I^2R$, for R

26 $V = \pi r^2 h$, for h

27 $C = \frac{5}{9}(F - 32)$, for F

28 $V = \frac{1}{3}\pi r^2 h$, for h

29 $PV = C$, for V

30 $PV = nrt$, for t

31 $L = a + (n - 1)d$, for n

32 $s = vt + gt^2$, for v

33 $S = 2\pi r^2 + 2\pi rh$, for h

34 $\dfrac{1}{R} = \dfrac{1}{r} + \dfrac{1}{s}$, for R

5.4 Word Problems

One important application of first degree equations is in the solution of *word* problems (also called *story* problems). We shall see that the ability to solve linear equations will be useful in dealing with such problems as finding plane areas from geometry, converting temperature from Fahrenheit degrees to centigrade degrees in physics, computing interest in business problems, and solving many other everyday types of problems. Since most story problems are not solved by trial and error methods, the following steps are suggested as a guide:

1. Read the problem carefully, determining what facts are given and precisely what is to be found.
2. Draw diagrams if possible to help interpret the given information and the nature of the solution.
3. Express the given information and the solution in mathematical symbols and/or numbers.
4. Express the problem in the form of an equation.
5. Solve the equation.
6. Check the solution to see if it satisfies the equation and to see if it makes sense with regard to the story problem.

EXAMPLES

1 Find a number such that when two times the number is increased by 5, the result is 11.

SOLUTION. We first note that the solution is a number. Let x be the symbol for this unknown number. Next, identify the given statement of equality and translate it to a mathematical expression. Hence, the statement "*When two times the number is increased by 5 the result is 11*" translates to the equation "$2x + 5 = 11$." Solving this equation, we have

$$2x + 5 = 11$$
$$2x = 6$$
$$x = 3$$

Therefore, the number is 3.

Check:

$$2(3) + 5 \stackrel{?}{=} 11$$
$$6 + 5 = 11$$
$$11 = 11$$

The solution is correct.

2 Find two consecutive integers whose sum is 77.

SOLUTION. We first represent the two consecutive integers by x and $x + 1$. Next, using this representation, we translate the statement of equality "two consecutive integers whose sum is 77" to the equation

$$x + x + 1 = 77 \quad \text{or} \quad 2x + 1 = 77$$

Solving this equation, we obtain

$$2x = 76 \quad \text{or} \quad x = 38 \quad \text{and} \quad x + 1 = 39$$

Hence, the two consecutive integers are 38 and 39.

Check:

$$38 + 39 \stackrel{?}{=} 77$$
$$77 = 77$$

The solution is correct.

3 Maureen is 4 years older than Marguerite. Three years ago she was twice as old as Marguerite. How old is each now?

SOLUTION. Let x years represent Marguerite's age now and $x + 4$ years represent Maureen's age now. Then, $x - 3$ = Marguerite's age three years ago and $x + 4 - 3$ or $x + 1$ = Maureen's age three years ago. Hence, translating the statement "**Three years ago she (Maureen) was twice as old as Marguerite,**" we have the equation

$$x + 1 = 2(x - 3)$$
$$x + 1 = 2x - 6$$
$$x + 7 = 2x$$
$$x = 7 \qquad \text{and} \qquad x + 4 = 11$$

Therefore, 7 years is Marguerite's age and 11 years is Maureen's age.

4 The sum of the ages of Joe and Jammie is 35. Joe is 5 years older than twice Jammie's age. How old are they?

SOLUTION. Let x years be Jammie's age; then $2x + 5$ years is Joe's age, and $2x + 5 + x$ years is the sum of their ages. Therefore,

$$2x + 5 + x = 35$$
$$3x = 30$$
$$x = 10 \qquad \text{and} \qquad 2x + 5 = 25$$

Thus, 25 years is Joe's age, and 10 years is Jammie's age.

5 Find the length and width of a rectangle whose perimeter is 13 inches and whose length is 2 inches more than twice its width.

SOLUTION. The unknowns that we are seeking are the length and the width of the rectangle. Let x represent the width in inches. Although the length is also unknown, we know that "**length is 2 inches more than twice its width,**" which translates to the mathematical expression

$(2 + 2x)$ inches

We now have the known quantity P = perimeter = 13 inches and the unknowns

Width $= x$
Length $= 2x + 2$

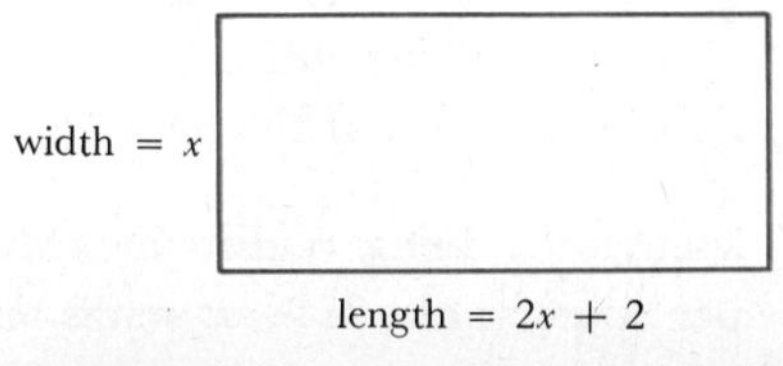

From elementary geometry (Appendix B) the **perimeter = 2(length) + 2(width)**. Therefore, we obtain the equation

$$13 = 2(2x + 2) + 2(x)$$

Now, solving the equation, we get

$$13 = 2(2x + 2) + 2(x)$$
$$13 = 4x + 4 + 2x$$
$$13 = 6x + 4$$
$$9 = 6x$$

so that

$$x = \tfrac{9}{6} = \tfrac{3}{2} \text{ inch}$$

We have found the width of the rectangle. Since the length is expressed in terms of the width, we have

$$\begin{aligned} \text{Length} &= 2x + 2 \\ &= 2(\tfrac{3}{2}) + 2 \\ &= 5 \text{ inches} \end{aligned}$$

Hence, width $= \frac{3}{2}$ inch and length $= 5$ inches.

6 When 33 is divided by a certain number, we obtain a quotient of 4 and a remainder of 5. Find the number.

SOLUTION. Let $x =$ the number. Translating the statement of equality, we obtain the equation

$$\frac{33}{x} = 4 + \frac{5}{x}$$

Solving for x, we have

$$33 = 4x + 5$$
$$4x = 28$$
$$x = 7$$

The number is 7.

7. Tom started for a town 75 miles away at the rate of 30 miles an hour. After he traveled part of the way, road construction forced him to reduce his speed to 22 miles an hour for the remainder of the trip. If the entire trip took 2 hours and 38 minutes, how far did he travel at the reduced speed?

SOLUTION. Let $x =$ distance traveled at the reduced rate (22 miles per hour). Then $75 - x =$ distance traveled at 30 miles per hour.

Using the formula that distance = rate × time, or time = distance ÷ rate, we have

$$\text{Time at 22 mph} = \frac{x}{22}$$

$$\text{Time at 30 mph} = \frac{75 - x}{30}$$

so that the total time can be expressed as

$$\frac{x}{22} + \frac{75 - x}{30}$$

The equation is

$$\frac{x}{22} + \frac{75 - x}{30} = 2 \text{ hours, } 38 \text{ minutes}$$

$$\frac{x}{22} + \frac{75 - x}{30} = \frac{79}{30} \qquad (2 \text{ hours, } 38 \text{ minutes} = \tfrac{79}{30} \text{ hours})$$

Solving this equation, we have

$$30x + 22(75 - x) = (79)(22)$$

$$30x + 1{,}650 - 22x = 1{,}738$$

$$8x = 88 \qquad \text{or} \qquad x = 11$$

Therefore, he traveled 11 miles at the reduced speed.

8 A water tank can be filled by an intake pipe in 4 hours, and can be emptied by a drainpipe in 5 hours. How long would it take to fill the tank with both pipes open?

SOLUTION. Let t be the number of hours required to fill the tank with both pipes open. The rate of flow for the intake pipe is one-fourth of a tank per hour, and the rate of flow for the drainpipe is one-fifth of a tank per hour. Hence, $(\frac{1}{4})t = \frac{t}{4}$ = the fraction of a tank that the intake pipe can fill in t hours; and $(\frac{1}{5})t = \frac{t}{5}$ = the fraction of a tank that the drainpipe can empty in t hours.

Since both pipes are open and one tank is to be filled, we have

$$\frac{t}{4} - \frac{t}{5} = 1 \qquad \text{or} \qquad 5t - 4t = 20$$

Hence

$$t = 20$$

Therefore, the time that takes to fill the tank is 20 hours.

Check:

$$\frac{t}{4} = \frac{20}{4} = 5 \text{ tankfuls}$$
$$\frac{t}{5} = \frac{20}{5} = 4 \text{ tankfuls}$$

and $5 - 4 = 1$ tank filled. Hence, $t = 20$ is the correct solution.

9 Charlotte can punch a complete set of tabulating cards in 2 hours. Darlene can punch a similar set in 3 hours. If both work together, how long will it take to complete a third set of cards?

SOLUTION. Let t be the number of hours required to punch the cards when both girls work together. Since Charlotte's rate is one-half of t and Darlene's rate is one-third of t, we have $\frac{1}{2}t = \frac{t}{2} =$ the part of the job that Charlotte does in t hours and $\frac{1}{3}t = \frac{t}{3} =$ the part of the job that Darlene does in t hours. Since there is 1 job to be done, we have the equation

$$\frac{t}{2} + \frac{t}{3} = 1$$

Solving, we obtain

$$3t + 2t = 6$$
$$5t = 6$$
$$t = \tfrac{6}{5} \text{ hours}$$

Check:

$$\frac{t}{2} = \frac{\frac{6}{5}}{2} = \frac{3}{5} \text{ set of cards}$$
$$\frac{t}{3} = \frac{\frac{6}{5}}{3} = \frac{2}{5} \text{ set of cards}$$

and

$$\tfrac{3}{5} + \tfrac{2}{5} = \tfrac{5}{5} = 1 \text{ set of cards}$$

10 A man has a certain amount of money invested at 6 percent. He has \$6,000 more than that invested at 5 percent. His total income from the two investments is at the average rate of $5\frac{1}{4}$ percent on the entire investment. How much has he invested at each rate?

SOLUTION. Let x be the amount invested at 6 percent; then $x + 6{,}000$ is the amount invested at 5 percent, and $2x + 6{,}000$ is the amount invested at $5\frac{1}{4}$ percent. Hence, the equation will be formed as follows:

$$6\% \text{ of } x + 5\% \text{ of } (x + 6{,}000) = 5\tfrac{1}{4}\% \text{ of } (2x + 6{,}000)$$

or

$$0.06x + 0.05(x + 6{,}000) = 0.0525(2x + 6{,}000)$$

Multiplying both sides of the equation by 10,000, we have

$$\begin{aligned} 600x + 500(x + 6{,}000) &= 525(2x + 6{,}000) \\ 600x + 500x + 3{,}000{,}000 &= 1{,}050x + 3{,}150{,}000 \\ 600x + 500x - 1{,}050x &= 3{,}150{,}000 - 3{,}000{,}000 \\ 50x &= 150{,}000 \\ x &= 3{,}000 \end{aligned}$$

and

$$x + 6{,}000 = 9{,}000$$

Therefore, he invested \$3,000 at 6 percent and \$9,000 at 5 percent.

11 In a collection of coins there are four times as many nickels as dimes, and there are two less quarters than dimes. If the value of the collection is \$2.25, find the number of dimes in the collection.

SOLUTION. Let x = number of dimes in the collection. Then $4x$ = number of nickels in the collection, and $x - 2$ = number of quarters. Since the total value is \$2.25, the equation is

$$\begin{aligned} 0.10x + 0.05(4x) + 0.25(x - 2) &= 2.25 \\ 10x + 5(4x) + 25(x - 2) &= 225 \\ 10x + 20x + 25x - 50 &= 225 \\ 55x - 50 &= 225 \\ 55x &= 275 \\ x &= 5 \end{aligned}$$

Hence there are 5 dimes in the collection.

PROBLEM SET 5.4

Solve the following word problems and check your solutions.

1 Three more than twice a certain number is 57. Find the number.

2 Find two numbers whose sum is 18 if one number is 8 larger than the other.

3 Find two consecutive even integers whose sum is 46.

4 Find two consecutive odd integers whose sum is 36.

5 Are there two consecutive even integers whose sum is 92?

6 Are there two consecutive odd integers whose sum is 174?

7 Find three consecutive even integers whose sum is 66.

8 Find three consecutive even integers such that the sum of the first and the last is 84.

9 Find three consecutive odd integers whose sum is 75.

10 Find three consecutive odd integers such that the sum of the last two is 92.

11 Find four consecutive even integers such that the sum of the first and the last is 62.

12 Find two consecutive even integers such that three times the first added to twice the second gives 64.

13 Find two consecutive even integers such that seven times the first exceeds five times the second by 54.

14 Find three consecutive even integers such that the first plus twice the second plus four times the third equals 174.

15 Find three consecutive odd integers such that three times the first plus four times the second exceeds five times the last by 26.

16 Find four consecutive even integers such that the sum of the first three exceeds the fourth by 84.

17 Find four consecutive odd integers such that three times the last plus five times the first exceeds twice the first plus four times the second by 52.

18 Find a number such that if one-seventh of it is added to the number, the sum is 24.

19 One-fourth of a number is 3 greater than one-sixth of it. Find the number.

20 Two-thirds of a number plus five-sixths of the same number is equal to 42. What is the number?

21 What number must be added to both terms of the fraction $\frac{7}{16}$ to make the result equal to $\frac{1}{2}$?

22 What number must be subtracted from both terms of the fraction $\frac{9}{25}$ to make the result equal to $\frac{1}{5}$?

23 What number must be added to both terms of the fraction $\frac{3}{7}$ to make the result equal to $\frac{2}{3}$?

24 A number is added to the numerator of $\frac{5}{27}$ and the same number is subtracted from the denominator. What is this number if the result is equal to $\frac{1}{3}$?

25 Subtracting 13 from a number gives the same result as taking three-fourths of the number. Find the number.

26 A number is subtracted from the numerator of the fraction $\frac{7}{34}$ and twice the same number is added to the denominator, giving a fraction equal to $\frac{1}{6}$. What is the number?

27 One-fifth of an estate is left to a sister, one-fourth to a son, one-sixth to a brother, and the remainder, which is \$57,500, to the widow. What is the value of the estate?

28 Five times a certain number is added to the numerator of $\frac{13}{17}$ and twice the same number is subtracted from the denominator, making the result equal to $3\frac{2}{3}$. Find the amount added to the numerator and the amount subtracted from the denominator.

29 Three times a certain number is subtracted from the numerator of $\frac{17}{33}$ and twice the same number is added to its denominator, thereby changing the value of the fraction to $\frac{5}{41}$. Find the unknown number.

30 A is twice as old as B, but in 8 years he will be $\frac{6}{5}$ as old as B. How old is A now?

31 John is 4 years older than his sister, and 6 years ago he was twice his sister's age then. How old is each now?

32 Gus is 10 years older than his brother, and 6 years from now he will be twice his brother's age then. How old is each now?

33 Linda's mother is three times as old as Linda and 14 years from now she will be twice as old as Linda is then. How old is each now?

34 Roy is 3 years older than his brother, and 4 years from now the sum of their ages will be 33 years. How old is each now?

35 Doreen is 5 years younger than her brother, and 3 years ago the sum of their ages was 23 years. How old is each now?

36 A sum of money consisting of nickels, dimes, and quarters amounted to \$1.90. If there were half as many quarters as nickels, and 3 more nickels than dimes, how many coins of each kind were there?

37 In a collection there are twice as many nickels as dimes, and the value is \$8.40. How many of each kind are there?

38 In the elementary class in the Sunday School the collection consisted of pennies and nickels. There were 19 coins amounting to 47 cents. How many coins of each kind were there?

39 In a collection of nickels and quarters there are four times as many nickels as quarters, and the value of the collection is $3.60. Find the number of each kind of coin.

40 In a drawer there is $12.50 in dimes and quarters. The number of dimes is 20 greater than the number of quarters. Find the numbers of each kind of coin.

41 In a collection of nickels and dimes, there are 12 more than three times as many nickels as dimes, and the value of the collection is $6.85. Find the number of nickels and the number of dimes.

42 A man withdrew $75.00 in $1 bills and $5 bills from the bank. There were 3 more $1 bills than $5 bills. How many were there of each kind?

43 A collection consisting of nickels and dimes amounted to $4.85. How many nickels and how many dimes were there if there were 70 coins?

44 Tickets for the school play sold for 10 cents and 25 cents. How many tickets of each kind were sold if they took in $73.15 for 457 tickets?

45 Bill counted the small change in the cash drawer. There were three times as many dimes as quarters and four times as many nickels as dimes. How many coins of each kind were there if there was $16.10 in all?

46 Find the dimensions of a rectangle if its width is 8 less than its length and half its perimeter is 24.

47 The length of a rectangle is 7 inches greater than its width. If its length is increased by 2 inches and its width is decreased by 3 inches, its area is decreased by 37 square inches. Find its dimensions.

48 The sides of two squares differ by 6 inches and their areas differ by 468 square inches. Find the lengths of the sides of the squares.

49 The length of a rectangle is four times its width. What are its dimensions if its perimeter is 150 inches?

50 The length of a rectangle is 17 less than three times its width, and its perimeter is 238. Find its dimensions.

51 The length and width of a square are increased by 6 feet and 8 feet respectively, thereby increasing its area by 188 square feet. Find a side of the square.

52 The width of a rectangle is 3 feet less than the length; if its width is decreased by 2 feet and its length by 4 feet, its area is decreased by 76 square feet. Find its dimensions.

53 One side of a square is increased by 2 inches and the other side decreased by 6 inches, thereby decreasing its area by 140 square inches. Find the length of a side of the square.

54 The length of a rectangle is 16 greater than the width. If the length is decreased by 4 and the width by 10, the area is decreased by 400. What are its dimensions?

55 A man has \$8,000 invested at 5 percent and \$2,000 invested at 7 percent. How much must he invest at 8 percent to make an average of 6 percent on his investment?

56 A man invests a total of \$24,000. Part of this sum is invested at 4 percent, and the remainder at 6 percent. His total yearly interest is \$1,040. Find the amount invested at 4 percent.

57 A man borrows a certain sum at $5\frac{1}{2}$ percent interest and a sum that is \$5,000 greater at 7 percent. How much does he borrow if the interest for one year is \$1,850?

58 A man borrows a certain sum at 5 percent; he borrows twice this sum at 6 percent; and three times the smaller sum at 7 percent. The total interest for one year is \$3,040. How much does he borrow at 5 percent? How much does he borrow at each of the other rates?

59 A man invests a certain sum at $3\frac{1}{2}$ percent, twice that sum at 4 percent, and twice the second sum at 5 percent. The total income from the investments is \$5,040. How much does he invest at each rate?

60 An airplane made a trip of 680 miles in 4 hours. Part of the trip was made at 150 miles per hour, and the remainder at 180 miles per hour. How many miles were traveled at each rate?

61 One pipe can fill a tank in 18 minutes and another pipe can fill it in 24 minutes. The drainpipe can empty the tank in 15 minutes. With all pipes open, how long will it take to fill the tank?

62 How many pounds of nuts at 63 cents per pound should be mixed with 20 pounds of nuts at 90 cents per pound to give a mixture worth 78 cents per pound?

5.5 Absolute Value Equations

An equation of the form $|x - a| = b$ is referred to as an *absolute value equation.* This type of equation is not a linear equation. For example, the equation $|x - 1| = 4$ is not a first-degree equation, although it contains the first-degree polynomial $x - 1$. In order to investigate the solutions of absolute value equations, we shall review the notion of absolute value of a real number that was introduced in Chapter 1. If a and b are real numbers such that a is less than or equal to b, then the distance between a and b is considered to be the nonnegative number $b - a$ (Figure 1).

Figure 1

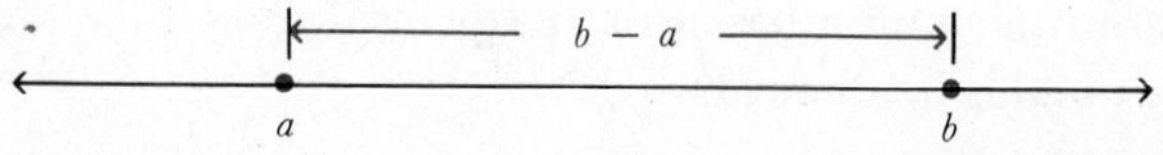

For example, the distance between -1 and 3 is given by $3 - (-1) = 4$ and the distance between -5 and -2 is equal to $(-2) - (-5) = 3$ (Figure 2).

Figure 2

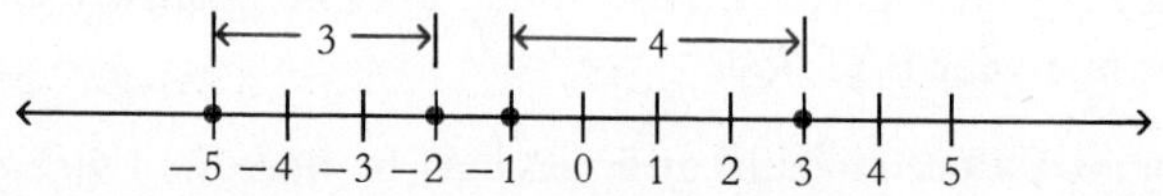

Suppose that we are interested in finding the distance between 0 and any real number x. For convenience, we will use the notation $|x|$ (which is read the *absolute value of* x) to represent the distance between x and 0. Then $|3| = 3$, $|0| = 0$, and $|-4| = 4$. This concept can be formalized in the following definition: *For x a real number, the absolute value of x, denoted by |x|, is defined by*

$$|x| = \begin{cases} x & \text{if } x \text{ is positive} \\ 0 & \text{if } x \text{ is zero} \\ -x & \text{if } x \text{ is negative} \end{cases}$$

Hence, from the definition, $|3| = 3$, because 3 is positive; $|0| = 0$; and $|-4| = -(-4) = 4$, because -4 is negative.

EXAMPLE

If $x = 4$ and $y = -7$, compute each of the following expressions.

a) $|x + 2y|$ b) $|x| + |2y|$
c) $|xy|$ d) $|x|\,|y|$
e) $|x - y|$ f) $|x| - |y|$
g) $\left|\frac{x}{y}\right|$ h) $\frac{|x|}{|y|}$

SOLUTION

a) $|x + 2y| = |4 + 2(-7)| = |4 - 14| = |-10| = 10$
b) $|x| + |2y| = |4| + |2(-7)| = |4| + |-14| = 4 + 14 = 18$
c) $|xy| = |4(-7)| = |-28| = 28$

d) $|x|\,|y| = |4|\,|-7| = (4)(7) = 28$
e) $|x - y| = |4 - (-7)| = |11| = 11$
f) $|x| - |y| = |4| - |-7| = 4 - 7 = -3$
g) $\left|\frac{x}{y}\right| = \left|\frac{4}{-7}\right| = \frac{4}{7}$
h) $\frac{|x|}{|y|} = \frac{|4|}{|-7|} = \frac{4}{7}$

We shall now use the definition of absolute value to solve absolute value equations. The procedure for solving such equations is illustrated in the following examples.

EXAMPLES

Solve each of the following absolute value equations.

1 $|x| = 3$

SOLUTION. We wish to find all numbers that are 3 units from the origin. By definition, we know that both $|3| = 3$ and $|-3| = 3$. Therefore, $x = 3$ or $x = -3$; that is, the solution set is $\{-3,3\}$. The solution can be illustrated on the real line (Figure 3).

Figure 3

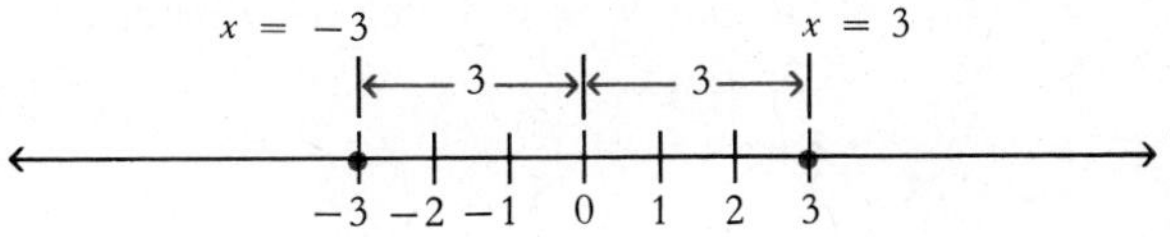

2 $|x - 2| = 5$

SOLUTION. Since $|5| = 5$ and $|-5| = 5$, then $|x - 2| = 5$ is equivalent to the two equations $x - 2 = 5$ or $x - 2 = -5$; that is, the solution set is $\{-3,7\}$ (Figure 4).

Figure 4

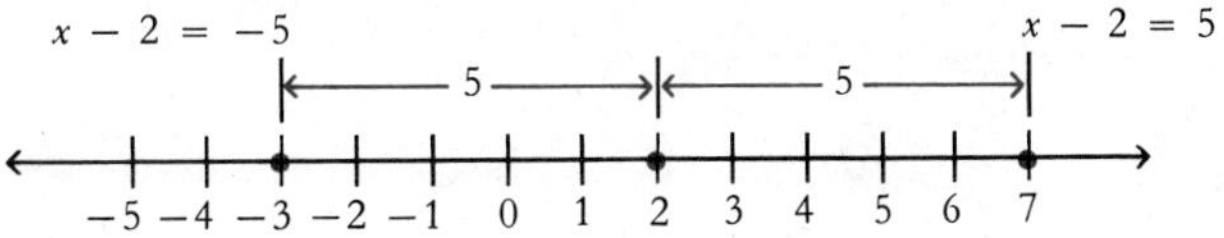

PROBLEM SET 5.5

In problems 1–10, let $x = 5$ and $y = -6$, and compute each expression.

1 $|x| + |y|$
2 $|x + y|$
3 $|x - y|$
4 $|x| - |y|$
5 $|xy|$
6 $|x|\,|y|$
7 $|y|^2$
8 $\left|\frac{x}{y}\right|$
9 $|5x| + |-3y|$
10 $5|x| - 3|y|$

In problems 11–25, find the solution set of each equation.

11 $|x| = 5$

12 $|x| = 2$

13 $|x| + 1 = 3$

14 $|x| + 3 = 7$

15 $|x| - 2 = 5$

16 $|x| - 3 = -1$

17 $|x| + 7 = 10$

18 $|3x| = 15$

19 $|2x| - 7 = 11$

20 $|5x| - 1 = 14$

21 $|x| = |-4|$

22 $|x - 1| = 2$

23 $|x + 5| = 6$

24 $|2x + 1| = 3$

25 $|-3x| = 18$

REVIEW PROBLEM SET

In problems 1–20, solve each first-degree equation for x and write the solution set.

1 $3(2 + x) + 7x = 2(x + 6)$

2 $16(2x - 4) - 76 = 4x$

3 $9(x + 2) - 15 = 6(x + 2)$

4 $6 + 2(3x - 6) = 3(x + 2)$

5 $5(x + 1) = 3(x - 2)$

6 $5x - 11 = 5(2x + 3) + 6(1 - x)$

7 $6(5 - 3x) - 23 = 4(2 - 5x)$

8 $4(3x - 5) + 8 = 3(3x - 1)$

9 $7 + \frac{3}{2}x + \frac{5}{4}x = 3x + 17$

10 $\frac{3}{2}(4x + 2) + 6 = 3x + 12$

11 $1.5x + 1.25x = 3(2 - x) + 17$

12 $7(1 - x) + 9x = 4(7 - x) - 1$

13 $8 + 4(x + 3) = 5(7 - x) + 12$

14 $12 - 6x = 6(4 - 2x) + 6$

15 $x - 7(4 + x) = 5x - 6(3 - 4x)$

16 $2(x - 1) + 3(x - 2) + 4(x - 3) = 0$

17 $3(x + 5)(x - 3) + x = 3(x + 4)(x - 2) - 5$

18 $(x + 1)(x - 2) = x^2 + 3x + 6$

19 $5(1.2 - 3.5x) = 6(3.5 - 1.2x)$

20 $4.3x + 2 = 3.5x - 2.3(x + 1) - 5$

In problems 21–30, solve each equation for x and write the solution set.

21 $2a + 3(x + b) = 5a + 6$

22 $2(x - a) = 3(2x + a + 1)$

23 $ax + b = cx + d$

24 $3xa + 2b = 6c - 3xb$

25 $(x - a)(x - b) = (x - c)(x - d)$

26 $3x - a + 5 = 5x - 3a$

27 $\frac{11x}{4} + x - a = \frac{a - x}{3}$

28 $2x + 3b = 19b + 24x$

29 $2x - 15(a + b) = -5x + 22(a - b)$

30 $17x + 3(a - 2b) = 5x + 13b$

In problems 31–45, solve each fractional equation and write the solution set.

31 $\frac{x}{3} + \frac{x}{4} = 1$

32 $\frac{x + 3}{4} = 2$

33 $\frac{x - 4}{2} = 5$

34 $\frac{x}{8} = \frac{5x}{4} - 18$

35 $\frac{x - 14}{5} + 4 = \frac{x + 16}{10}$

36 $\frac{3}{2x - 3} = \frac{1}{3}$

37 $\frac{18}{x+1} = \frac{10}{x-3}$

38 $\frac{3}{x} = \frac{6}{x+3}$

39 $\frac{2x-1}{6} + \frac{3}{2} = \frac{x+7}{6}$

40 $\frac{x+9}{4} - \frac{3(2x-3)}{14} = 2$

41 $\frac{1}{x-6} = \frac{2}{x+2}$

42 $\frac{x-3}{2} - \frac{x+1}{5} = \frac{1}{10}$

43 $\frac{2x-3}{3} - \frac{3x+4}{2} = -\frac{19}{3}$

44 $\frac{x+3}{11} + \frac{x-3}{9} = \frac{2}{9}$

45 $\frac{x+5}{5} + \frac{x-5}{2} = \frac{11}{2}$

In problems 46–59, solve each absolute value equation for x and write the solution set.

46 $|2x| = 5$

47 $|7x| = 6$

48 $|3x| + |4| = 12$

49 $|1 - 2x| = 2$

50 $|7x| - |3| = 5$

51 $|3 - x| = |-3|$

52 $|5x - 8| = 3$

53 $|3(x - 4)| = |2(-4)|$

54 $|x - 1| = 7$

55 $|x| + 2 = 3$

56 $2|x + 1| = 6$

57 $|3x - 1| = 5$

58 $|x + 1| = 0$

59 $|3x - 2| = 8$

60 In an election of the president of a faculty senate there were two candidates. The successful candidate received 5 more votes than the defeated candidate. If 243 votes were cast, how many did each candidate receive?

61 How many pounds of water must be evaporated from 50 pounds of a 3 percent salt solution so that the remaining portion will be a 5 percent solution?

62 Two airplanes left airports which are 600 miles apart and flew toward each other. One plane flew 20 miles per hour faster than the other. If they passed each other at the end of 1 hour and 12 minutes, what were their speeds?

63 A garage builder found out that it would cost $1\frac{1}{2}$ times as much to build a garage with brick as it would with aluminum siding, whereas wood siding would be $\frac{3}{4}$ the cost of aluminum siding. If it cost $1,200 to build the garage with wood siding, how much would it cost to build it with brick?

64 A student in a mathematics class having marks of 76 and 82 on two examinations, wishes to raise his average to 85. What must he make on a third examination to do this?

65 Bill invested $2,000 in a bank, part at 4 percent and the rest at 3.5 percent. At the end of one year, he received $74 in interest. How much did he invest at 4 percent?

66 The sum of two numbers is 14. Twice one of them is 1 more than twice the other. What are the numbers?

67 Three consecutive positive integers are such that their sum is 9 more than twice the smallest. What are the numbers?

CHAPTER 6

First-Degree Inequalities in One Variable

6 FIRST-DEGREE INEQUALITIES IN ONE VARIABLE

Introduction

In Chapter 5 we studied the application of the properties of the equality relation to the solution of first-degree equations. In this chapter we shall consider the solutions of first-degree inequalities such as $3x + 2$ is greater than 5, or $\dfrac{-3x+5}{2}$ is less than 3, and investigate the meaning of their relationship by considering their basic properties.

6.1 Order Relations

Recall from Chapter 1 that the real numbers are located on the real line as follows: The numbers located to the right of the origin are the positive real numbers, and they are arranged in order of increasing magnitude; and the numbers located to the left of the origin are the negative real numbers and they are arranged in order of decreasing magnitude (Figure 1). Accordingly, we agree to extend this "ordering"

Figure 1

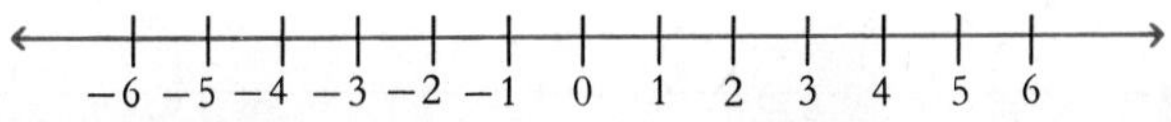

to the entire real line in the following manner: Any number whose corresponding point on the number line lies to the right of the corresponding point of a second number is said to be *greater than* (denoted by "$>$") the second number. The second number is also said to be *less than* (denoted by "$<$") the first number. For example, $7 > 3$ or $3 < 7$ and $1 > -5$ or $-5 < 1$, since the point that corresponds to the number 3 lies to the left of the point that corresponds to the number 7, and the point that corresponds to the number -5 lies to the left of the point that corresponds to the number 1 (Figure 2).

Figure 2

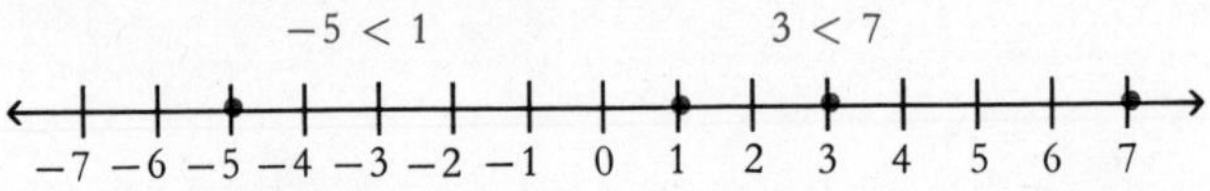

In general, if $a < b$ or $b > a$, then the point that corresponds to a lies to the left of the point that corresponds to b (Figure 3). This can be

Figure 3

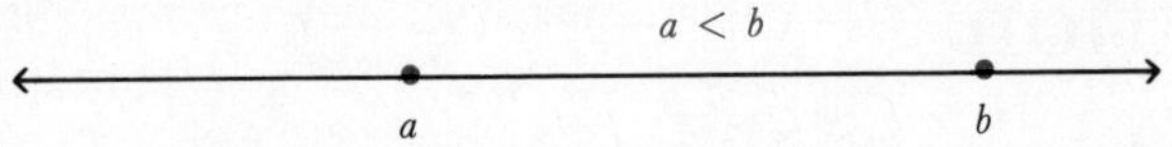

formalized by the following definition of the *order relations* "$<$" and "$>$."

DEFINITION 1

For any two real numbers a and b, *a is less than b* ($a < b$) or, equivalently, *b is greater than a* ($b > a$) if $b - a$ is a positive number.

For example, $3 < 7$ (or $7 > 3$) since the difference $7 - 3$ is 4, a positive number. Also, $-5 < -3$ (or $-3 > -5$), since $-3 - (-5) = 2$, a positive number.

Properties of Order Relations

We shall now investigate the properties of order relations. Suppose we are given two real numbers a and b. If these two numbers are located on the number line, then only one of three possible situations can occur (Figure 4). These three possible situations suggest the following property.

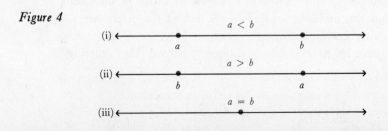

Figure 4

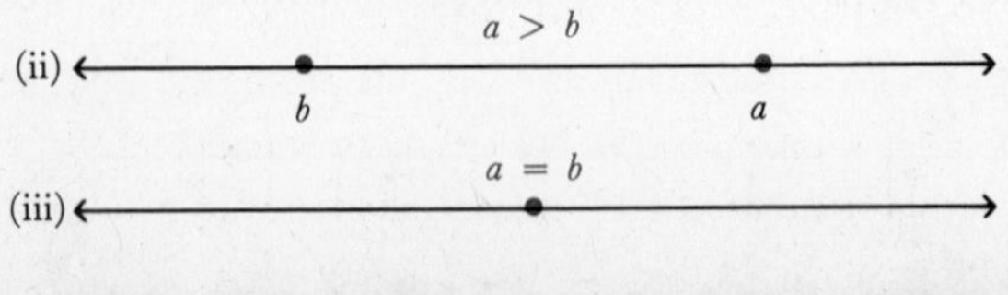

PROPERTY 1 TRICHOTOMY LAW

If a and b are real numbers, then one and only one of the following situations can hold:

$$a < b \qquad a > b \qquad \text{or} \qquad a = b$$

For example, only the first of the following statements is true.

$$-7 < 3 \qquad -7 = 3 \qquad -7 > 3$$

We can also indicate that a number is positive or negative by use of the inequality relation. If the number a is positive, then $a > 0$ (Figure 5); that is, a lies to the right of 0. Similarly, if the number b is negative

Figure 5

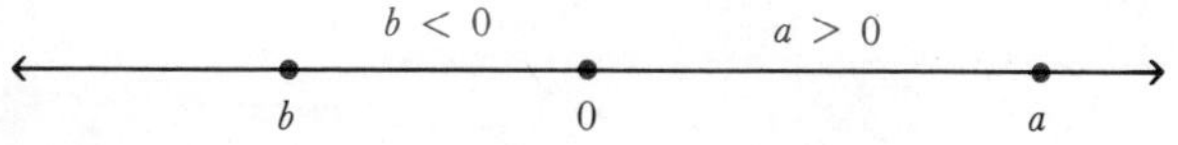

(Figure 5), then $b < 0$; that is, b lies to the left of zero.

Let us examine the following examples. Since $-3 < 1$, then -3 lies to the left of 1 and $1 < 4$ implies that 1 lies to the left of 4; therefore, -3 lies to the left of 4 and $-3 < 4$ (Figure 6*a*). Similarly, since $3 > 1$,

Figure 6

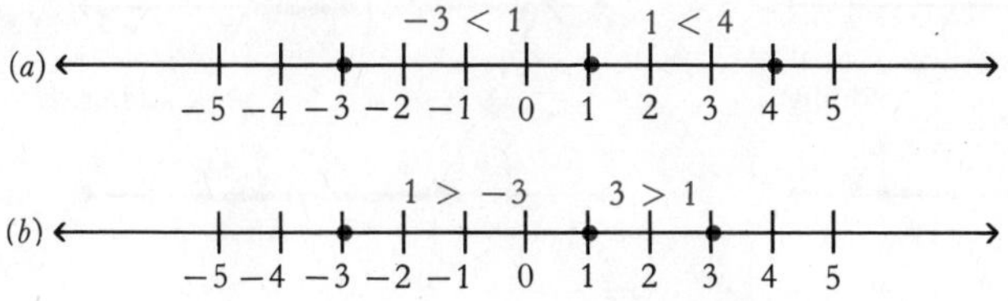

then 3 lies to the right of 1 and $1 > -3$ implies that 1 lies to the right of -3; therefore, 3 lies to the right of -3 and $3 > -3$ (Figure 6*b*). These examples can be generalized in the following property.

PROPERTY 2 TRANSITIVE LAW

For a, b, and c real numbers:

1 If $a < b$ and $b < c$, then $a < c$.
2 If $a > b$ and $b > c$, then $a > c$.

For example, since $-5 < 3$ and $3 < 7$, then $-5 < 7$; similarly, since $5 > 2$ and $2 > 1$, then $5 > 1$.

Let us now consider the effect of adding the same number to both sides of an inequality. For instance, suppose 3 is added to both sides of the inequality $2 < 4$. We have $2 + 3 < 4 + 3$ or, equivalently, $5 < 7$. The result of adding is illustrated in Figure 7. By adding 3 to both sides

Figure 7

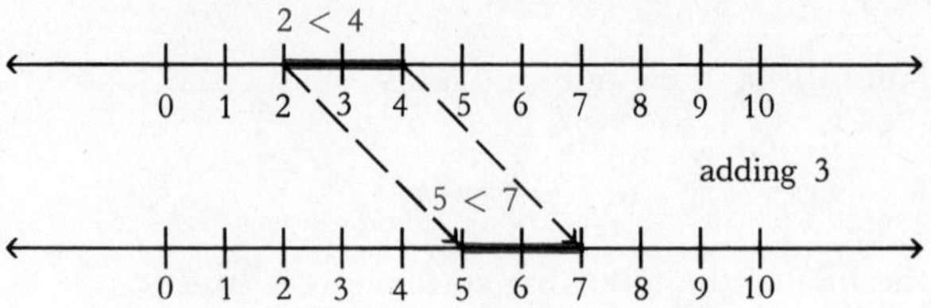

we have moved (also called translated) both points in the same direction and the same distance along the number line. The relation between the sums will be the same as between the original two numbers. This statement can be generalized as follows.

PROPERTY 3 ADDITION PROPERTY OF INEQUALITIES

For all real numbers a, b, and c:

1 If $a < b$, then $a + c < b + c$.
2 If $a > b$, then $a + c > b + c$.

Property 3 is illustrated in Figure 8, where $c > 0$.

Figure 8

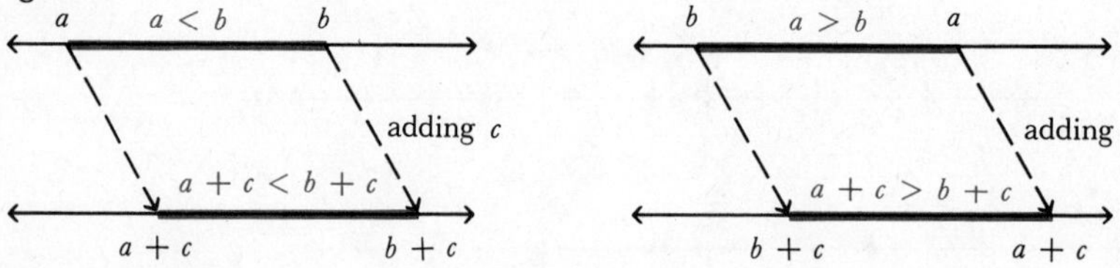

For example,

$$-4 < 3$$
$$-4 + 5 < 3 + 5$$

or

$$1 < 8$$

Similarly, for $c < 0$, we have

$$-2 > -8$$
$$-2 + (-5) > -8 + (-5)$$

or

$$-7 > -13$$

Hence, we see that the addition property of inequalities is similar to the addition property of equalities in the sense that we can add the same number to both sides of the relation and the original relation is preserved.

On the other hand, when we investigate the effect of multiplying both sides of an inequality by any real number, we find that the result of multiplication of both sides of an inequality by a number can either preserve the original relation or reverse the original relation, depending on whether we multiply by a positive number or a negative number. For example, if both sides of the inequality $3 < 5$ are multiplied by 2, the result is $6 < 10$, whereas if $3 < 5$ is multiplied on both sides by -2, we have $-6 > -10$. That is, by multiplying by a positive number, we preserve the order relation; by multiplying by a negative number, we reverse the order relation.

We shall use specific examples to illustrate the multiplication properties. Consider $-3 < 4$ and multiply both sides by 2 (Figure 9).

Figure 9

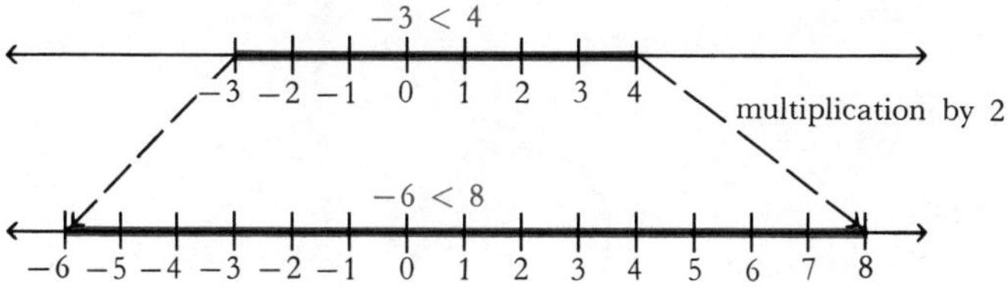

We see from the figure that multiplying both sides of the inequality $-3 < 4$ by 2 produces a new inequality $(-3)(2) < (4)(2)$. The original order relation was preserved under the operation of multiplying by a positive number.

On the other hand, if we multiply both sides of the inequality $-3 < 4$ by -2, we have $(-3)(-2) > (4)(-2)$, as shown in Figure 10.

Figure 10

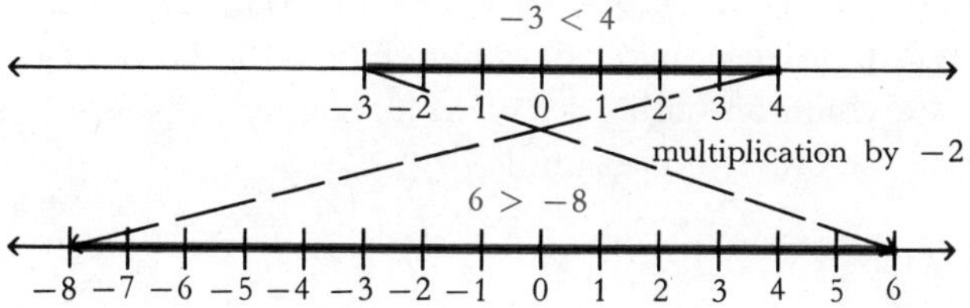

Here the multiplication by -2 reverses the order relation. In general, we have the following property.

PROPERTY 4 MULTIPLICATION PROPERTY OF INEQUALITIES

For all real numbers a, b, and c:

1 a) If $a < b$ and $c > 0$, then $ac < bc$.
  b) If $a > b$ and $c > 0$, then $ac > bc$.
2 a) If $a < b$ and $c < 0$, then $ac > bc$.
  b) If $a > b$ and $c < 0$, then $ac < bc$.

Property 4 is illustrated in Figure 11, where $c > 0$.

Figure 11

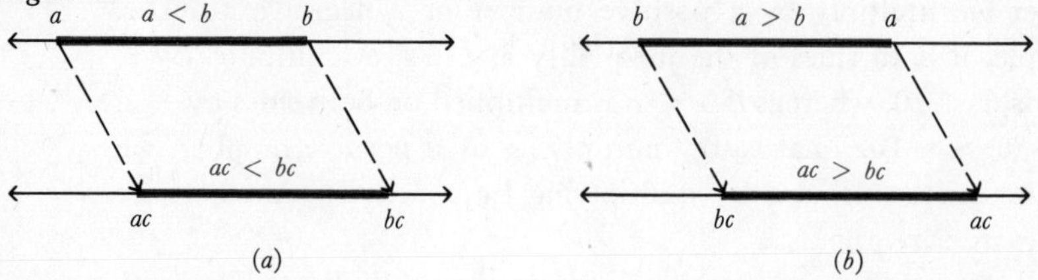

For example,

$$-3 < 5$$

Multiplying by 4, we have

$$-3(4) < 5(4) \quad \text{or} \quad -12 < 20$$

On the other hand,

$$3 > -2$$

Multiplying by -5, we have

$$3(-5) < -2(-5) \quad \text{or} \quad -15 < 10$$

It is important to realize that the above examples do not suffice as a proof of the multiplication property of inequalities; they merely serve as illustrations of it.

Since division by a nonzero real number can be expressed as a multiplication, we can conclude from Property 4 that division on both sides of an inequality by a positive number does not change the order of the inequality, whereas division on both sides of an inequality by a negative number does reverse the order. For example,

$$-2 < 3$$

Dividing by 5, we have

$$\frac{-2}{5} < \frac{3}{5}$$

Similarly, take

$$6 > -4$$

Dividing by -5, we have

$$\frac{6}{-5} < \frac{-4}{-5} \quad \text{or} \quad -\tfrac{6}{5} < \tfrac{4}{5}$$

In summary, we have the following properties of order relations:

1 If $a < b$ and c is any number, then $a + c < b + c$.
2 If $a > b$ and c is any number, then $a + c > b + c$.
3 If $a < b$ and $c > 0$, then $ac < bc$.
4 If $a > b$ and $c > 0$, then $ac > bc$.
5 If $a < b$ and $c < 0$, then $ac > bc$.
6 If $a > b$ and $c < 0$, then $ac < bc$.

We have similar properties for the equality relation (Chapter 2):

1 If $a = b$ and c is any number, then $a + c = b + c$.
2 If $a = b$ and c is any number, then $ac = bc$.

Hence we can combine the two sets of properties into one set of properties that includes both the relations of equality and inequality. (We interpret the symbol "$\leq$" to mean "less than or equal to." That is, if $a \leq b$, then either $a < b$ or $a = b$. Similarly, the symbol "$\geq$" means "greater than or equal to." That is, if $a \geq b$, then either $a > b$ or $a = b$.) Both relations cannot be true simultaneously; one or the other relation is true. Hence, we have the following results:

1 If $a \leq b$ and c is any number, then $a + c \leq b + c$.
2 If $a \geq b$ and c is any number, then $a + c \geq b + c$.
3 If $a \leq b$ and $c > 0$, then $ac \leq bc$.
4 If $a \geq b$ and $c > 0$, then $ac \geq bc$.
5 If $a \leq b$ and $c < 0$, then $ac \geq bc$.
6 If $a \geq b$ and $c < 0$, then $ac \leq bc$.

EXAMPLES

1 Illustrate each of the following sets on a number line.

a) $\{x | x < 2\}$
b) $\{x | x \geq 2\}$

SOLUTION

a) Notice that the point 2 is excluded from this set (Figure 12).

Figure 12

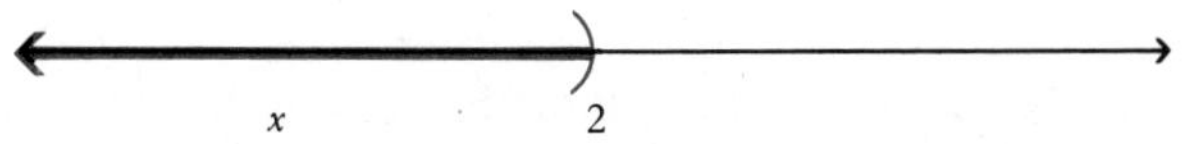

We indicate this exclusion by using the parenthesis sign, ")", to mark the number 2.

b) Here, the point 2 is included in the set (Figure 13). We indicate

Figure 13

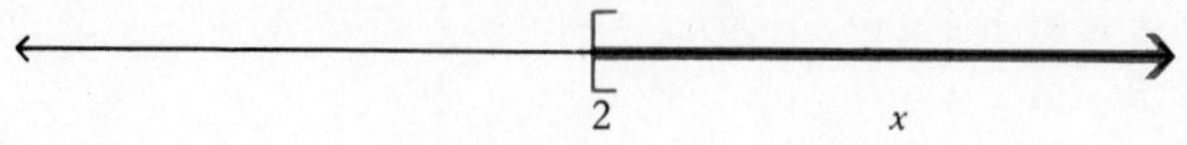

this inclusion by using the bracket sign, "[", to mark the number 2.

2 Use the symbols $<$ or $>$ to give the relationship between the first and second numbers in the following pairs.

a) 3, 18 b) $-5, -11$
c) $-6, 10$ d) $0, -5$

SOLUTION

a) $3 < 18$ or $18 > 3$, since $18 - 3 = 15 > 0$.
b) $-11 < -5$ or $-5 > -11$, since $-5 - (-11) = 6 > 0$.
c) $-6 < 10$ or $10 > -6$, since $10 - (-6) = 16 > 0$.
d) $-5 < 0$ or $0 > -5$, since $0 - (-5) = 5 > 0$.

3 Given $x < y$, what can be said about the relation between $x + 1$ and $y + 1$?

SOLUTION

$$x < y$$
$$x + 1 < y + 1 \qquad \text{(Property 1)}$$

4 Show that if $x + 2 < 5$, then $x < 3$.

SOLUTION

$$x + 2 < 5$$
$$x + 2 + (-2) < 5 + (-2) \qquad \text{(Why?)}$$
$$x < 3$$

5 If $a < b$, does it follow that $a^2 < b^2$. Illustrate your solution by an example.

SOLUTION. No. For example, if $a = -5$ and $b = 3$, we have $-5 < 3$ but $(-5)^2 = 25$ and $3^2 = 9$, so that $(-5)^2 > 3^2$.

6 If $a^2 > b^2$, does it follow that $a > b$. Illustrate your solution by an example.

SOLUTION. No. For example, if $a = -3$ and $b = 2$, we have $(-3)^2 > 2^2$, but $-3 < 2$.

PROBLEM SET 6.1

1 Show that if $a < b$, then $b - a$ is a positive number for each of the following inequalities.

a) $4 < 8$ b) $-3 < 5$
c) $-2 < -1$ d) $0 < 5$
e) $-3 < 4$ f) $-2 < 0$

2 Illustrate each of the following inequalities on the real line (see Figures 6 to 9).

a) $2 < 3$ and $2 + 3 < 3 + 3$
b) $1 < 2$ and $2 < 3$ (then $1 < 3$)
c) $-1 > -2$ and $-1 - 3 > -2 - 3$
d) $3 > 2$ and $3 - 1 > 2 - 1$
e) $3 < 4$ and $(3)(2) < (4)(2)$
f) $-2 < -1$ and $(-2)(-1) > (-1)(-1)$

3 Apply the multiplication property to the following inequalities by multiplying each side by the given number.

a) $2 < 3$, by 2 b) $2 < 5$, by -3
c) $-1 < 0$, by 2 d) $-3 < -2$, by -1
e) $x < y$, by 3 f) $x < y$, by -3

4 Indicate the order property which justifies the following statements.

a) Since $-3 < 2$, then $-3 - x < 2 - x$.
b) If $x < y$, then $-2x > -2y$.
c) If $-\frac{1}{2} < x$ and $x < y$, then $-\frac{1}{2} < y$.
d) If $x > 0$, then $-\dfrac{2}{x} < -\dfrac{1}{x}$.
e) If a is a real number, then either $a = 3$ or $a < 3$ or $a > 3$.

5 Determine if the following statements are true or false.

a) $-[2 - (4 - 5)] < 0$.
b) If $a < b$, then $\dfrac{1}{a} < \dfrac{1}{b}$ for $a, b > 0$.
c) If $5x + 7 \leq 17$, then $x \leq 2$.
d) $(-1)(-3)(-2) < -5$.
e) If $a < b$, then $-\dfrac{1}{a} < -\dfrac{1}{b}$ for $a, b > 0$.
f) If $\dfrac{1}{a} \leq \dfrac{1}{b}$, then $a \leq b$ for $a, b > 0$.

6 Give an example in each case to show that the following properties are true.

a) If $a < b$ and $b < c$, then $a < c$.
b) If $a < b$ and c is any number, then $a + c < b + c$.
c) If $a < b$ and $c < 0$, then $ac > bc$.

7 Give a numerical example to show that if $0 < a < b$, then $a^3 < b^3$.

8 If $a^3 < b^3$, does it follow that $a < b$? Use numerical examples to support your answer.

9 Give an example in each case to show that the following properties are true.

a) If $a < b$ and $c > 0$, then $\frac{a}{c} < \frac{b}{c}$.

b) If $a < b$ and $c < 0$, then $\frac{a}{c} > \frac{b}{c}$.

10 Illustrate each of the following sets on a number line.

a) $\{x|x \leq 1\}$
b) $\{x|x \geq 1\}$
c) $\{x|x < -2\} \cup \{x|x \geq 2\}$
d) $\{x|x \leq 1\} \cap \{x|x \geq -1\}$
e) $\{x|x \geq 0\} \cap \{x|x \leq 2\}$

11 Show by numerical examples that for any real number $x \neq 0$, $x^2 > 0$.

6.2 First-Degree Inequalities in One Variable

As in the case of first-degree equations, we define the *solution set* of a first-degree inequality to be the set of all numbers that make the inequality a valid statement. Also, we define *equivalent inequalities* to be inequalities that have the same solution set. Inequalities that are true for all permissible values of the variable involved such as $x^2 + 4 > 0$, where x is any real number, are called *unconditional* or *absolute* inequalities; whereas inequalities that are true only for certain values of the variables involved are called *conditional inequalities.* For example, $3x - 4 > 5$ is only true for $x > 3$. The process of solving first-degree inequalities follows the same pattern that was previously used in solving first-degree equations. We apply the properties of inequalities to the given inequality to produce a series of equivalent inequalities until an inequality of the form $x < a$ or $x > a$ is obtained. For example, let us find the solution set of the first-degree inequality $5x + 6 > 8x + 2$. By subtracting $8x$ from both sides of the equality, that is, $-8x + 5x + 6 > -8x + 8x + 2$, we have $-3x + 6 > 2$. Next, add -6 to both sides of the inequality to get $-3x + 6 - 6 > 2 - 6$ or $-3x > -4$. Dividing both sides by -3 and reversing the order of the inequality, we have $x < \frac{4}{3}$. Thus the solution set is $\{x|x < \frac{4}{3}\}$.

EXAMPLES

Solve each of the following first-degree inequalities and represent the solution sets on the real line.

1 $3x - 2 < 7$

SOLUTION

$$3x - 2 < 7$$
$$3x - 2 + 2 < 7 + 2 \qquad \text{(Addition property)}$$

We have

$$3x < 9$$
$$\tfrac{1}{3}(3x) < \tfrac{1}{3}(9) \qquad \text{(Multiplication property)}$$
$$x < 3$$

Hence, the solution set is $\{x|x < 3\}$ (Figure 1).

Figure 1

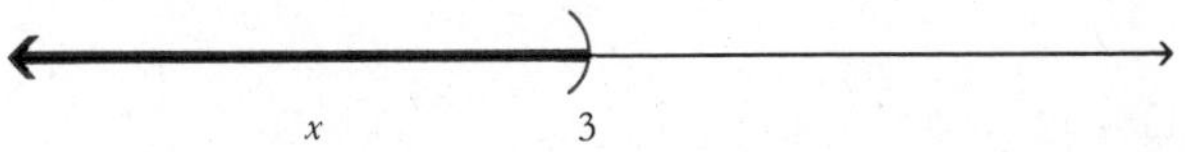

2 $x + 2 > 7x - 1$

SOLUTION. $x + 2 > 7x - 1$. Adding $-7x$ to both sides of the inequality, we get $-6x + 2 > -1$; adding -2 to both sides, we have $-6x > -3$.

Multiplying each side by $-\frac{1}{6}$ and reversing the sense of the inequality, we have $x < \frac{1}{2}$. The solution set is $\{x|x < \frac{1}{2}\}$ (Figure 2).

Figure 2

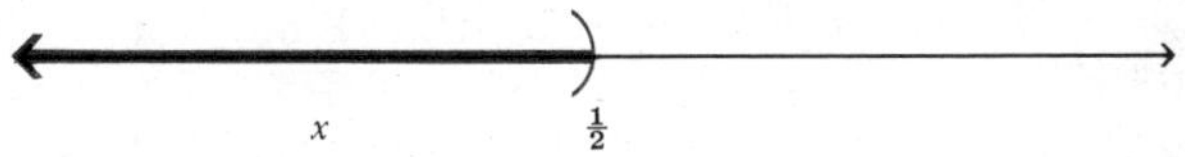

3 $3 - x \leq -4$

SOLUTION. $3 - x \leq -4$. Adding -3 to each side, we get $-x \leq -7$; multiplying both sides by -1 and reversing the sense of inequality, we get $x \geq 7$. The solution set is $\{x|x \geq 7\}$ (Figure 3).

Figure 3

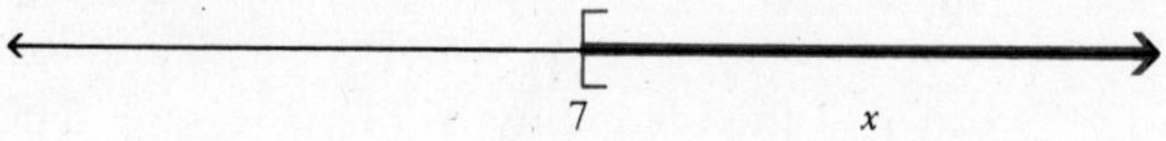

4 $\dfrac{3x + 7}{5} < 2x - 1$

SOLUTION. $\dfrac{3x + 7}{5} < 2x - 1$. Multiplying both sides by 5, we get

$3x + 7 < 10x - 5$ or $3x + 12 < 10x$ (Why?)

Adding $-3x$ to both sides, we get

$12 < 7x$

Dividing both sides by 7, we get

$\frac{12}{7} < x$ or $x > \frac{12}{7}$. The solution set is $\{x | x > \frac{12}{7}\}$ (Figure 4).

Figure 4

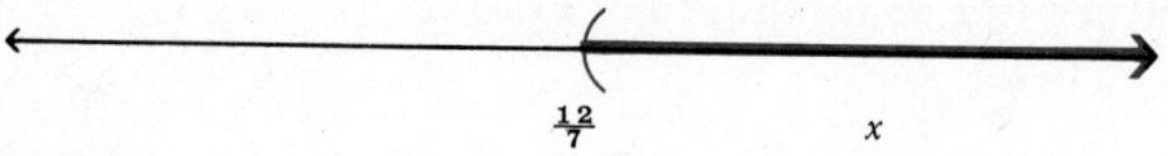

5 $\dfrac{5x - 2}{3} > \dfrac{3x - 1}{4}$

SOLUTION. $\dfrac{5x - 2}{3} > \dfrac{3x - 1}{4}$ · Multiplying both sides by 12, we get

$4(5x - 2) > 3(3x - 1)$ or $20x - 8 > 9x - 3$

Adding $-9x + 8$ to both sides, we have

$20x - 8 - 9x + 8 > 9x - 3 - 9x + 8$ or $11x > 5$

Hence

$x > \frac{5}{11}$. The solution set is $\{x | x > \frac{5}{11}\}$ (Figure 5).

Figure 5

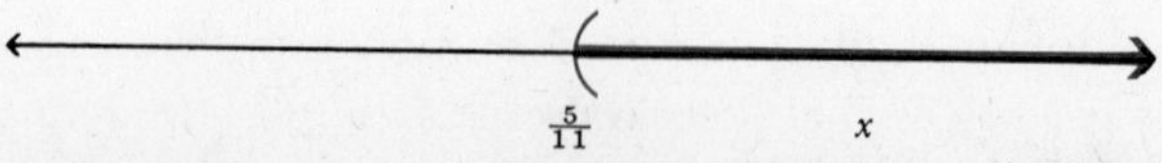

PROBLEM SET 6.2

In problems 1–44, find the solution set of the inequalities. Show the solution on the number line.

1 $x + 1 < 3$

2 $x - 1 < 2$

3 $5 \leq x - 1$

4 $t + 2 \geq 3$

5 $x + 4 \leq 0$

6 $4 > t - 3$

7 $2x - 1 < x$

8 $3x - 2 > 2x$

9 $5x - 7 \geq 4x$

10 $7x - 3 \leq 6x$

11 $5x < 5$

12 $3x > -9$

13 $4x > -12$

14 $2x \leq -6$

15 $2x - 7 > 8$

16 $5x + 6 < 13$

17 $3x - 5 > 7$

18 $3x + 1 \leq 0$

19 $5 - x < x + 3$

20 $1 - 3x \leq -4x$

21 $-3x \leq 6$

22 $-5x \geq -10$

23 $1 - 4x \geq 5$

24 $7 - 3x \leq 4$

25 $3 - 4x \leq 15$

26 $x - 7x \leq -12$

27 $\{t \mid 3t - 7 \leq 5\}$

28 $\{x \mid 5x + 1 \geq 6\}$

29 $\{x \mid 1 - 7x \geq 8\}$

30 $\{t \mid 5 - 3t \leq 13\}$

31 $3x - 2 + x \leq 5x + 6$

32 $5t - 2 + 6t > 4t - 6 + 3t$

33 $\frac{1}{2}x \leq 3$

34 $-\frac{1}{3}t \geq 1$

35 $\frac{1}{2}x - x > 2$

36 $x - 5 \leq \frac{1}{4}x$

37 $\dfrac{3x - 7}{2} \leq 13$

38 $\dfrac{1 - 5x}{2} \leq -5$

39 $\dfrac{3x - 2}{5} - 4 < 0$

40 $\dfrac{2x + 3}{3} \leq \dfrac{3x}{4}$

41 $\dfrac{2x + 1}{5} - 2 < \dfrac{3x + 1}{2}$

42 $\dfrac{3x - 7}{6} \geq 1 - \dfrac{x}{2}$

43 $\left\{x \,\middle|\, \dfrac{3 - 2x}{5} \geq \dfrac{4 - x}{2}\right\}$

44 $\left\{t \,\middle|\, \dfrac{7 + t}{3} - 1 < \dfrac{7 - t}{3} + 1\right\}$

6.3 Absolute Value Inequalities

Inequalities involving absolute value notation require more discussion than absolute value equations, although we use the same procedure as that used in finding the solution set of equations involving absolute values. For example, to find the solution set of the absolute value inequality $|x| < 3$, we draw a number line (Figure 1) and interpret

Figure 1

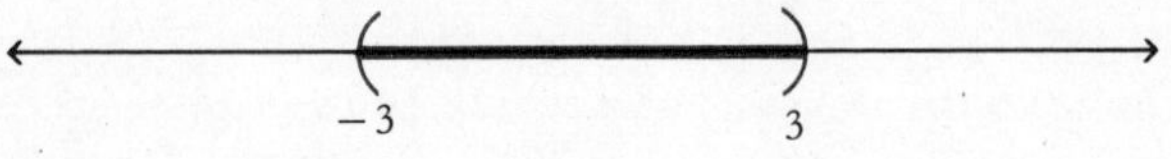

$|x| < 3$ to mean the set of all values of x that satisfy the inequality having graphs on the number line between the point labeled -3 and the point labeled 3. These points are indicated by the colored line in Figure 1. We notice then that

$$-3 < x \quad \text{and} \quad x < 3$$

Hence the solution set of $-3 < x$ is $\{x|-3 < x\}$. The solution set of $x < 3$ is $\{x|x < 3\}$. Therefore, the solution set of $-3 < x$ and $x < 3$ is $\{x|-3 < x\} \cap \{x|x < 3\}$. This set is usually written as $\{x|-3 < x < 3\}$. Now, considering the absolute value inequality $|x| > 5$, we see that $|x| > 5$ means that the set of all values that satisfy the given inequality have graphs on the number line to the right of the point labeled 5 or to the left of the point labeled -5 (Figure 2).

Figure 2

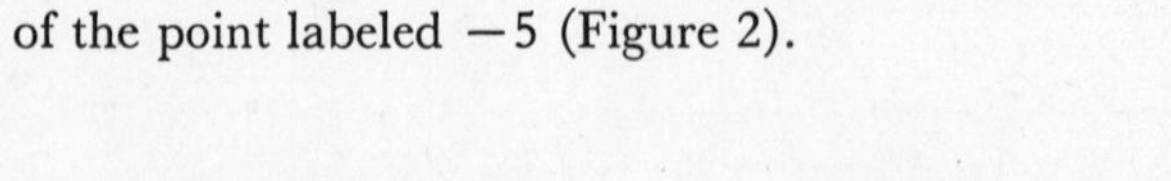

Then $\qquad x < -5 \quad \text{or} \quad x > 5$

The solution set of $x < -5$ is $\{x|x < -5\}$. The solution set of $x > 5$ is $\{x|x > 5\}$. Therefore the solution set of $x < -5$ or $x > 5$ is $\{x|x < -5\} \cup \{x|x > 5\}$. This set is written as $\{x|x < -5 \text{ or } x > 5\}$.

EXAMPLES

Solve each of the following absolute inequalities and represent the solution on the real line.

1 $|x| < 1$

SOLUTION. $|x| < 1$ means the set of all values of x that satisfy the inequality having graphs on the number line between the point labeled -1 and the point labeled 1 (Figure 3). We see then that

Figure 3

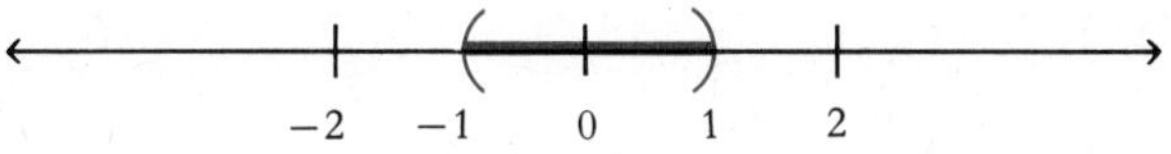

$-1 < x$ and $x < 1$. Hence the solution set is $\{x|-1 < x < 1\}$.

2 $|x| \leq 2$

SOLUTION. $|x| \leq 2$ means the set of all values of x that satisfy the inequality having graphs on the number line between (and including) the point labeled -2 and the point labeled 2 (Figure 4). We see that $-2 \leq x$ and $x \leq 2$.

Figure 4

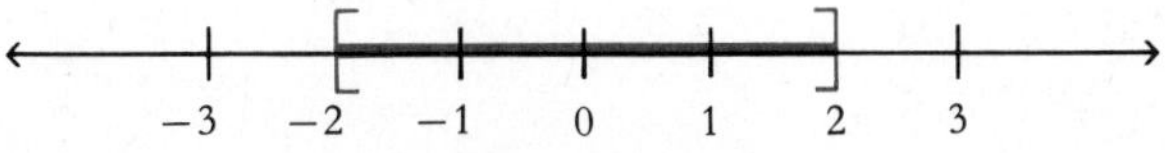

Hence the solution set is $\{x|-2 \leq x \leq 2\}$.

3 $|x| > 7$

SOLUTION. $|x| > 7$ means the set of all values that satisfy the inequality having graphs on the number line to the right of the point labeled 7 or the left of the point labeled -7 (Figure 5).

Figure 5

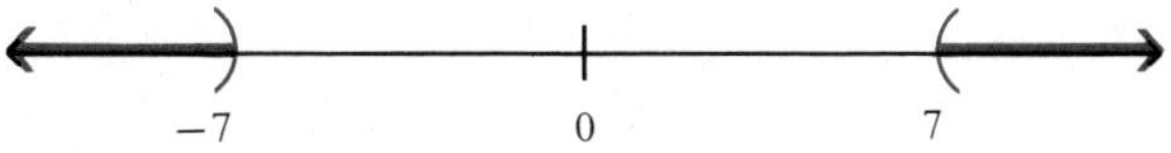

Then $x < -7$ or $x > 7$

Hence the solution set is $\{x|x < -7 \text{ or } x > 7\}$.

PROBLEM SET 6.3

In problems 1–16, solve the absolute value inequalities and represent the solutions on the real line. $\left(\textit{Note: } |ab| = |a|\,|b| \text{ and } \left|\frac{a}{b}\right| = \frac{|a|}{|b|}.\right)$

1 $|x| < 1$

2 $|x| > 2$

3 $|x| \leq 4$

4 $|2x| < 6$

5 $|3x| > 9$

6 $|-7x| \leq 28$

7 $|2x| \geq 15$

8 $|5x| > 0$

9 $|-4x| \leq 16$

10 $\left|\frac{x}{2}\right| \leq 1$

11 $|x| \geq \frac{2}{3}$

12 $\left|\frac{x}{4}\right| < 2$

13 $|5x| < 15$

14 $|-3x| \geq 6$

15 $\left|\frac{2x}{3}\right| \geq 1$

16 $\left|\frac{x}{2}\right| < \frac{5}{2}$

REVIEW PROBLEM SET

In problems 1–8, suppose that a and b are real numbers, with $a < b$. Indicate which of the following are true and which are false.

1 $a < 3b$

2 $a - 4 < b - 4$

3 $2a > -(-2)b$

4 $a + 2 < b + 3$

5 $a + c < b + c, c \in R$

6 $1/a < 1/b$

7 $a + 3 < b + 4$

8 $|a| < b$

In problems 9–14, let x, y, and z be real numbers. Which of the following are true?

9 If $0 < x < \frac{1}{2}$, then $-\frac{1}{3} < x < \frac{1}{3}$.

10 If $-\frac{1}{2} < x < \frac{1}{2}$, then $-1 < x < 1$.

11 If $x > y$, then $x - z > y - z$.

12 If $x > y$ and $z > 0$, then $x/z > y/z$.

13 If $x > y$, with $x > 0$, $y > 0$, then $x^5 > y^5$.

14 If $\frac{1}{x} \geq \frac{1}{y}$ then $x \geq y$ for $x, y > 0$.

In problems 15–40, solve each first-degree inequality. Illustrate your solution on the real line.

15 $3x + 5 < 2x + 1$

16 $3x - 5 > 2x + 7$

17 $3x - 2 < 5 - x$

18 $1 - 2x > 5 + x$

19 $1 - 3x > 1 + 2x$

20 $3x + 4 < 8x + 7$

21 $3(x - 4) < 2(4 - x)$

22 $5x + 2 > 3x + 1$

23 $3(x + 2) > 2(x + 3)$

24 $5x + 2(3x - 1) < -1 - 5(x - 3)$

25 $-3x + 7 + 5x \leq 1$

26 $3(2x + 3) - (3x + 2) > 12$

27 $2x - 3(x + 5) \leq 2$

28 $2(x + 1) + 3(x - 1) \leq 14$

29 $2(6x - 7) + 2x - 5 > 9$

30 $\{x \mid 7x - 9 - 2x \leq 15 + x - 8\}$

31 $\{x \mid 3(x + 5) + 2x + 8 \geq 31 + x\}$

32 $\{x \mid 7x + 5 + 2x \leq 7 + 3(x + 1) + 1\}$

33 $\{x \mid 2.1x + 4.6x - 5 \leq 1.7x + 30\}$

34 $\{x \mid 3x - 8 - x + 11 \leq 41 - x - 8\}$

35 $\{x \mid 9(2x + 2) - 30 \leq x + 5\}$

36 $7x - 5 + 3x \leq 5x + 10$

37 $4(3x - 5) + 8 + 3(3x - 1) \leq 2$

38 $5x - 11 \leq 5(2x + 3) + 6(1 - x)$

39 $\dfrac{4x - 9}{2} + 2x + 1 \leq 4$

40 $\dfrac{x}{3} + \dfrac{4(1 - x)}{5} \leq 1$

In problems 41–48, solve each absolute value inequality. Illustrate your solution on the real line.

41 $|x| < 7$

42 $|x| > \frac{5}{2}$

43 $|x| > 7$

44 $|x| \leq 9$

45 $|3x| < 6$

46 $|5x| \geq 10$

47 $\{x \mid |x| \leq \frac{1}{3}\}$

48 $\{x \mid |2x| < 1\}$

CHAPTER 7

Graphs in a Plane

7 GRAPHS IN A PLANE

Introduction

We saw in Chapter 1 that the real line provides us with a geometric representation of the set of real numbers R as points on a line. This geometric representation was used in investigating the order of the real numbers and illustrating the meaning of some of the operations on them. The solutions of first-degree equations and inequalities (Chapters 5 and 6, respectively) were also represented on the real line. In this chapter, we will investigate a method of representing *ordered pairs* of real numbers as points in a plane. This geometric representation will be used to display the graphs of first-degree equations and inequalities with two variables.

7.1 Cartesian Coordinate System

Consider the words "on" and "no." If we considered a word as merely a set of letters, there would be no distinction between these two words. However, since we consider a word as a set of letters with a prescribed order, "on" and "no" are different: Even though they are composed of the same letters, they have a different order.

A similar situation holds for sets. For example, the elements of the set $\{a,b\}$ do not have to be listed in any particular order. This set could be written as either $\{a,b\}$ or $\{b,a\}$; in other words, $\{a,b\} = \{b,a\}$. In contrast, (a,b) is an *ordered pair* consisting of the *first element a* and the *second element b*.

Two ordered pairs are considered to be equal when they have equal first members and equal second members. For example, $(1,2) \neq (2,1)$ even though each pair contains the same entries. Likewise, $(4,3) \neq (4,4)$ and $(7,8) \neq (-7,8)$, whereas $(9,x) = (y,8)$ if and only if $x = 8$ and $y = 9$.

EXAMPLES

1 Determine which of the following ordered pairs are equal to (6,7).

a) $(3 + 3, 4 + 3)$ b) $(5 + 1, 9 - 2)$
c) $(7 - 2, 9 - 2)$ d) $(4 + 2, 3 + 5)$

SOLUTION

a) $(3 + 3, 4 + 3) = (6,7)$ b) $(5 + 1, 9 - 2) = (6,7)$
c) $(7 - 2, 9 - 2) = (5,7) \neq (6,7)$
d) $(4 + 2, 3 + 5) = (6,8) \neq (6,7)$

2 Find the value of x in each of the following ordered pairs.

a) $(x,2) = (4,2)$ b) $(3x + 2, 3) = (5,3)$
c) $(x + 2, 3 + 5) = (5,8)$ d) $(2x, 2x - 1) = (6,5)$

SOLUTION

a) $(x,2) = (4,2)$

Hence,

$x = 4$

b) $(3x + 2, 3) = (5,3)$

Hence,

$3x + 2 = 5$
$3x = 3$ or $x = 1$

c) $(x + 2, 3 + 5) = (5,8)$

Hence,

$x + 2 = 5$
$x = 3$

d) $(2x, 2x - 1) = (6,5)$

Hence,

$2x = 6$
$x = 3$

Also,

$2x - 1 = 5$
$2(3) - 1 = 5$
$5 = 5$

It was indicated in the introduction that the set of real numbers R can be represented geometrically on a number line. Analogously, we can represent the set of all ordered pairs of real numbers as the set of points in a *plane,* using a two-dimensional indexing system called the *Cartesian coordinate system.* This system is constructed as follows.

First, two perpendicular lines L_1 and L_2 are constructed (Figure 1).

Figure 1

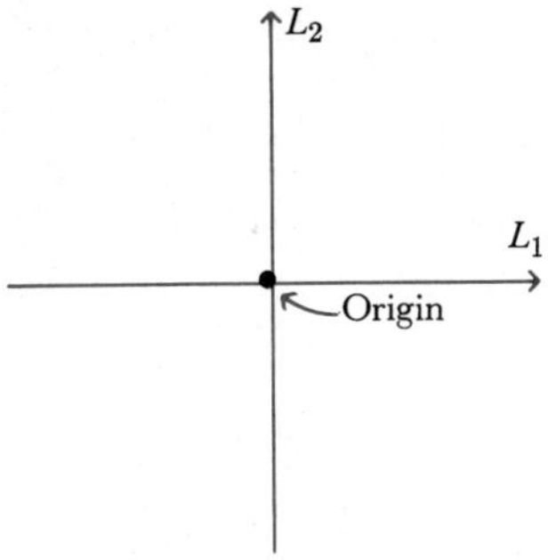

The point of intersection of the two lines is called the *origin.* Next L_1 and L_2 are scaled as real lines by using the origin as the 0 point for each of the two lines. The portion of L_2 above the origin is the positive direction and the portion below the origin is the negative direction of the number line; the portion of L_1 to the right of the origin is the positive direction and the portion to the left is the negative direction (Figure 2).

Figure 2

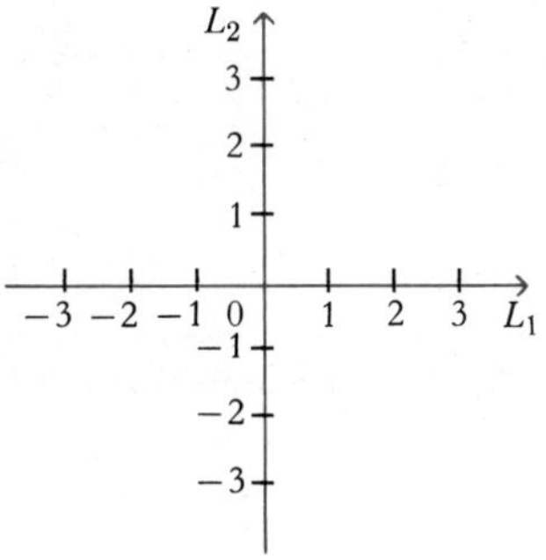

The resulting two real lines are called the *coordinate axes.* The coordinate axes L_1 and L_2 are often referred to as the *horizontal axis or the x axis* and the *vertical axis or the y axis,* respectively. Given an ordered pair of real numbers (x,y) (the first member of the pair, x, is called the *abscissa;* the second member of the pair, y, is called the *ordinate;* x and y are called the *coordinates*), we can use the coordinate system to represent (x,y) as a point on the plane as follows. The abscissa x is located on the x axis. Then a line is drawn perpendicular to the x axis at point x; the ordinate y is located on the y axis at point y and a line is drawn perpendicular to the y axis at point y. The intersection of the two lines which have just been constructed is the point in the plane used to represent the ordered pair (x,y) (Figure 3).

Figure 3

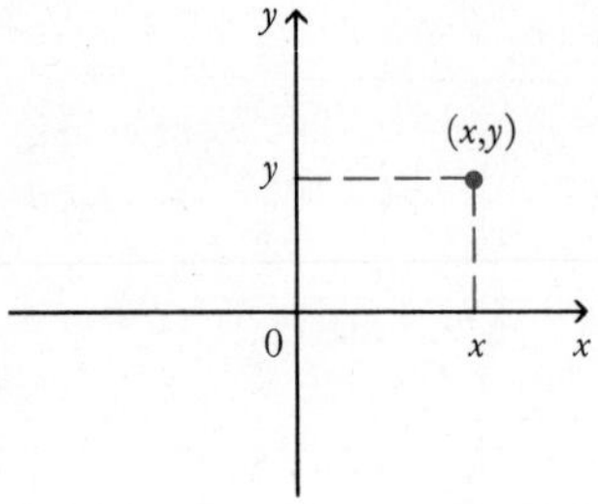

Thus, for each ordered pair (x,y) we can associate a point in the plane. Conversely, for each point in the plane, we can associate an ordered pair (x,y). For example, the ordered pair (1,1) is located by moving 1 unit to the right of 0, on the x axis, then 1 unit up from the x axis. Similarly, (7,5) is located by moving 7 units to the right of 0 and 5 units up; $(-\frac{1}{2},0)$ is located by moving $\frac{1}{2}$ unit to the left of 0 and no units up or down from the x axis (Figure 4).

Figure 4

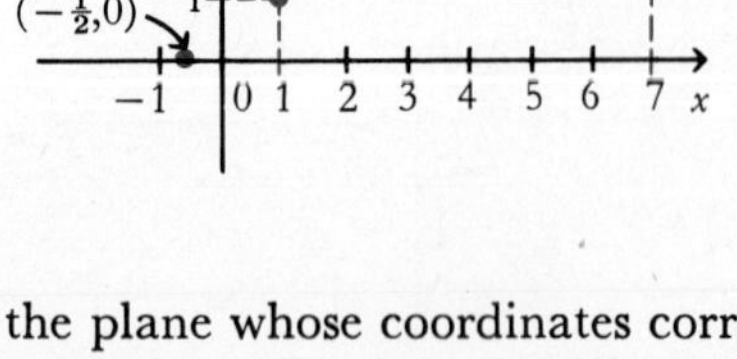

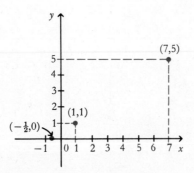

The set of all points in the plane whose coordinates correspond to the ordered pairs of a given set is called the *graph* of the set of ordered pairs. Locating these points in the plane by using the Cartesian coordinate system is called *graphing* or *plotting* the set of ordered pairs.

EXAMPLES

1 Graph the set of points whose coordinates are (2,0), (2,2), (0,2), (−2,2), (−2,0), (−2,−2), (0,−2), and (2,−2).

SOLUTION. The graph of this set of points is shown on the coordinate axis system (Figure 5).

Figure 5

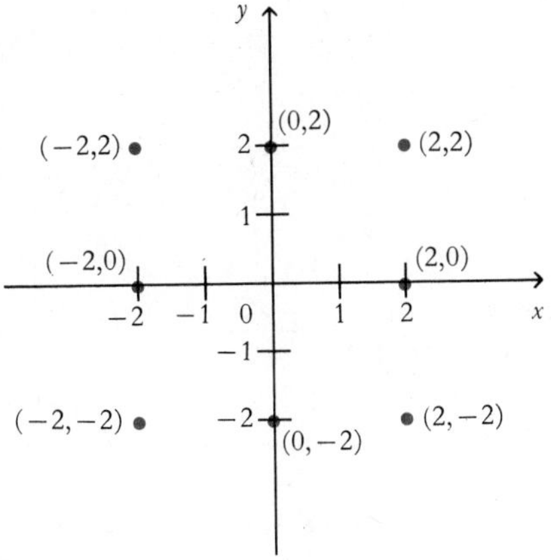

2 Graph the set of points whose coordinates are (3,3), (−2,4), (3,−1), (−3,−5), and (0,−2).

SOLUTION. The graph of this set of points is found by locating each point on the coordinate axis system (Figure 6).

Figure 6

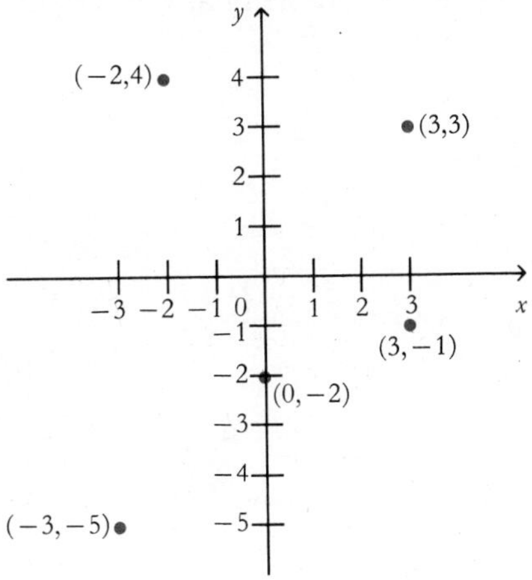

The coordinate axes divide the plane into four disjoint regions called quadrants: thus, *quadrant I* (Q_I) includes all points (x,y) such that $x > 0$ and simultaneously $y > 0$; *quadrant II* (Q_{II}) includes all points (x,y) such that $x < 0$ and simultaneously $y > 0$; *quadrant III* (Q_{III}) includes all points (x,y) such that $x < 0$ and simultaneously $y < 0$; *quadrant IV* (Q_{IV}) includes all points (x,y) such that $x > 0$ and simultaneously $y < 0$ (Figure 7).

Figure 7

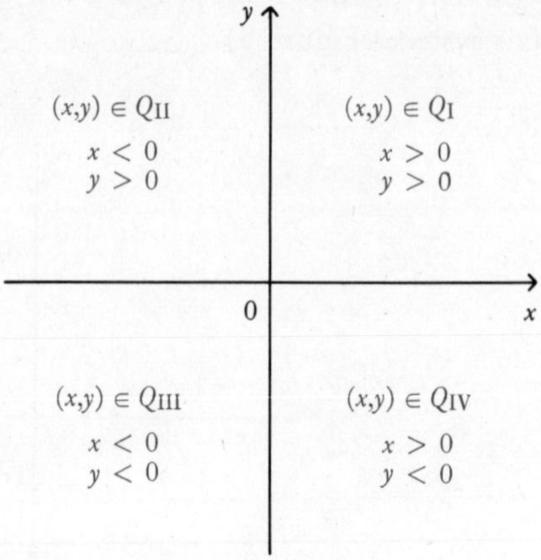

EXAMPLES

1 Plot each of the following points: $(1,-2)$, $(3,4)$, $(-2,-3)$, $(2,0)$. Indicate in which quadrant, if any, each of the points is contained.

SOLUTION. These points are plotted in Figure 8.

$(1,-2)$ lies in quadrant IV.
$(3,4)$ lies in quadrant I.
$(-2,-3)$ lies in quadrant III.
$(2,0)$ does not lie in any quadrant.

Figure 8

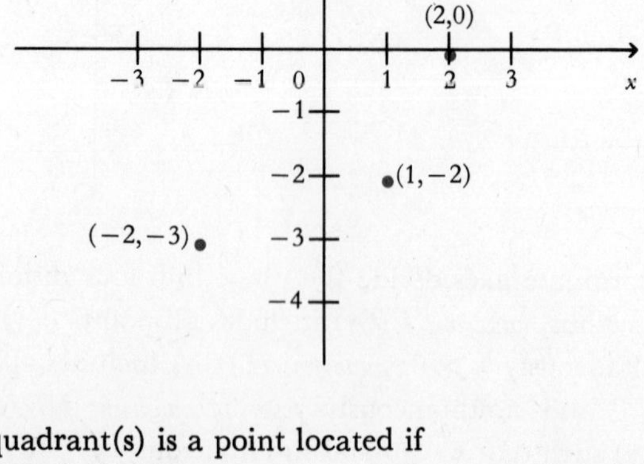

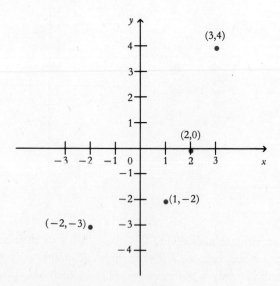

2 In what quadrant(s) is a point located if

a) Its abscissa is positive?
b) Its ordinate is negative?

c) Its abscissa is positive and its ordinate is negative?
d) Its abscissa is negative and its ordinate is negative?

SOLUTION. (Refer to Figure 9.)

Figure 9

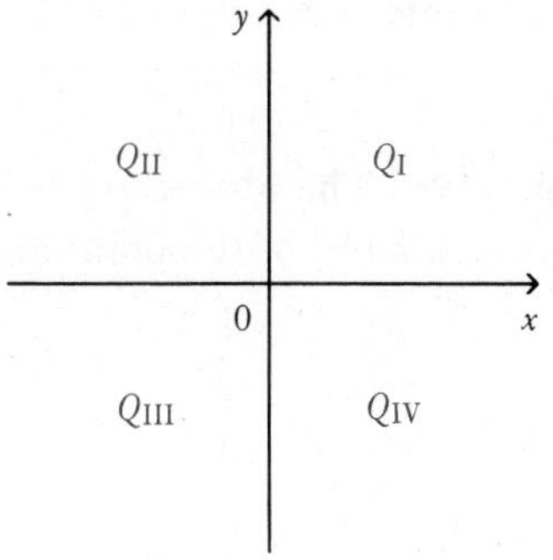

a) The abscissa (x value) is positive in quadrants I and IV.
b) The ordinate (y value) is negative in quadrants III and IV.
c) Combining the results of parts (a) and (b), these conditions are satisfied in quadrant IV.
d) Both the x value and y value are negative in quadrant III.

3 List the coordinates of each of the points that are designated by a letter in Figure 10.

Figure 10

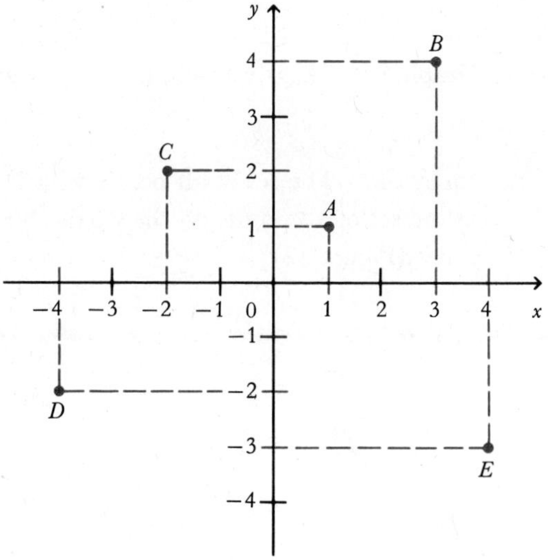

SOLUTION

A: (1,1)
B: (3,4)

C: $(-2,2)$
D: $(-4,-2)$
E: $(4,-3)$

4 What are the coordinates of a point 5 units to the left of the y axis and 3 units above the x axis?

SOLUTION. The abscissa is -5 and the ordinate is 3; therefore, the coordinates of the point are $(-5,3)$ (Figure 11).

Figure 11

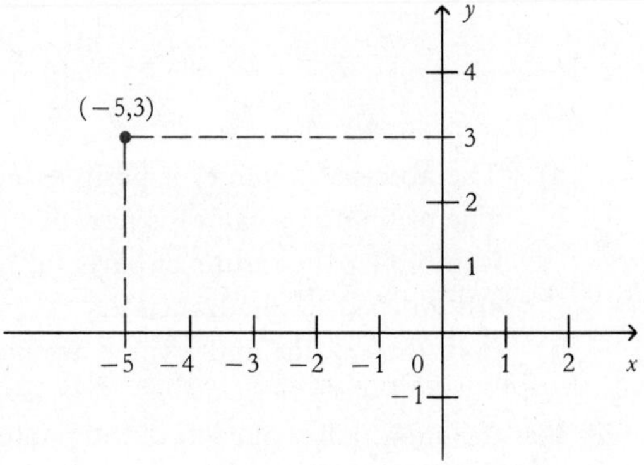

5 Graph $S = \{(x,y)|x = 0\}$, that is, the set of all points of the form $(0,y)$.

SOLUTION. The set of all points with an abscissa of 0 is the same as the set of all points on the y axis; hence, the *graph of S is the y axis* (Figure 12).

Figure 12

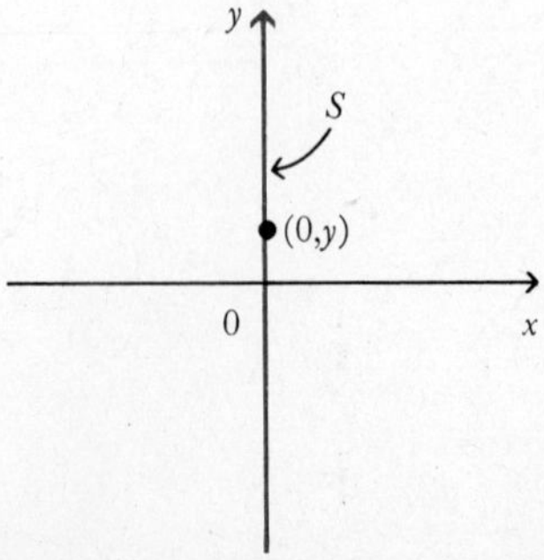

PROBLEM SET 7.1

1 Determine which of the following ordered pairs are equal to (4,5).

a) (4,7)
b) (4, 3 + 2)
c) (3 + 1, 1 + 4)
d) (2 + 2, 7 − 2)
e) (4 + 0, 8 − 3)
f) (5 + 0, 3 − 6)

2 Find the value of x in each of the following ordered pairs.

a) $(3x,2) = (6,2)$
b) $(2x - 3, 2x) = (5,8)$
c) $(x - 1, 5) = (3,5)$
d) $(3x + 7, 2x) = (-2,-6)$

3 Which of the following statements are true and which are false? Explain.

a) $(x,y) = (y,x)$
b) $\{x,y\} = \{y,x\}$
c) $(x,y) = (x,y)$
d) $\{(9y,x)\} = \{\{y,x\}\}$
e) If $(x,y) = (y,x)$, then $x = y$.

4 Plot each of the following points: (−1,2), (4,3), (3,8), (5,0), (4,−4). Indicate in which quadrant, if any, each of the points is contained.

5 Plot each of the following points and indicate what quadrant, if any, contains each of the points.

a) (3,3)
b) (−2,4)
c) $(5,\sqrt{2})$
d) (0,7)
e) (−1,−5)
f) (−3,0)
g) (−4,2)
h) (2,−4)
i) $(\frac{1}{2},-\frac{3}{2})$

6 Describe the coordinates of a point A if A

a) Lies in quadrant I
b) Lies in quadrant II
c) Lies in quadrant III
d) Lies in quadrant IV
e) Lies on the x axis between quadrant I and IV

7 Plot the points $A = (1,2)$, $B = (3,7)$, and $C = (2,-1)$ and draw the triangle ABC.

8 Plot the points $A = (-3,2)$, $B = (-3,-6)$, $C = (5,-6)$, and $D = (5,2)$. What kind of figure is $ABCD$?

9 Plot the points $A = (3,2)$, $B = (-\frac{4}{3},\frac{5}{9})$, and $C = (6,3)$. Do you think that the points A, B, and C lie on a straight line?

7.2 First-Degree Equations in Two Variables (Linear Equations)

Equations such as $y = 3x + 2$ and $2x - 3y + 5 = 0$ are examples of first-degree equations in two variables, x and y. In general, any equation that can be written in the form $Ax + By = C$, where A, B, and C are constants (both A and $B \neq$ zero) and x and y are variables, is called a *first-degree equation in two variables* or a *linear equation* in two variables. For example, while $3x + 2y = 7$ and $y = \frac{1}{3}x - 2$ are linear equations in two variables, the equations $xy = 13$, $y = 5x^2$, and $\frac{3}{x} + \frac{2}{y} = 7$ are not linear equations since they cannot be written in the form $Ax + By = C$. An ordered pair (a,b) is a *solution* of an equation in two variables x and y if the equation is true when $x = a$ and $y = b$. Let us determine which of the ordered pairs $(\frac{7}{2},1)$, $(2,2)$, and $(3,0)$ are solutions of the equation $2x - y = 6$.

For $x = \frac{7}{2}$ and $y = 1$, we have

$$2(\tfrac{7}{2}) - 1 \stackrel{?}{=} 6 \quad \text{or} \quad 6 = 6$$

Therefore $(\frac{7}{2},1)$ is a solution of the equation.

For $x = 2$ and $y = 2$, we have

$$2(2) - 2 \stackrel{?}{=} 6 \quad \text{or} \quad 2 \neq 6$$

Therefore $(2,2)$ is not a solution of the equation.

For $x = 3$ and $y = 0$, we have

$$2(3) - 0 \stackrel{?}{=} 6 \quad \text{or} \quad 6 = 6$$

Hence $(3,0)$ is a solution of the equation.

EXAMPLES

1 Determine which of the following equations are linear.

a) $5x - 2y = 3$ b) $\frac{3}{x} - \frac{5}{y} = 2$

c) $-3x + 2y = 6$ d) $3xy + 2x = 5$

SOLUTION

a) $5x - 2y = 3$ is linear since it is of the form $Ax + By = C$, where $A = 5$, $B = -2$, and $C = 3$.

b) $\frac{3}{x} - \frac{5}{y} = 2$ is not linear, since it is not of the form $Ax + By = C$.

c) $-3x + 2y = 6$ is linear since it is of the form $Ax + By = C$, where $A = -3$, $B = 2$, and $C = 6$.

d) $3xy + 2x = 5$ is not linear, since it is not of the form $Ax + By = C$.

2 Which of the given ordered pairs are solutions of the linear equation $3x + 2y = 13$? $(1,2)$, $(0,\frac{15}{2})$, $(3,2)$, $(\frac{13}{3},0)$, $(-1,8)$

SOLUTION

For (1,2) we have

$$3(1) + 2(2) \stackrel{?}{=} 13$$
$$7 \neq 13$$

Therefore (1,2) is not a solution.

For $(0,\frac{15}{2})$ we have

$$3(0) + 2(\tfrac{15}{2}) \stackrel{?}{=} 13$$
$$15 \neq 13$$

Therefore $(0,\frac{15}{2})$ is not a solution.

For (3,2) we have

$$3(3) + 2(2) \stackrel{?}{=} 13$$
$$13 = 13$$

Therefore (3,2) is a solution.

For $(\frac{13}{3},0)$ we have

$$3(\tfrac{13}{3}) + 2(0) \stackrel{?}{=} 13$$
$$13 = 13$$

Therefore $(\frac{13}{3},0)$ is a solution.

For $(-1,8)$ we have

$$3(-1) + 2(8) \stackrel{?}{=} 13$$
$$13 = 13$$

Therefore $(-1,8)$ is a solution.

3 List all the elements of the set $A = \{(x,y) \mid y = -2x + 1, x \in \{1,2,3,4\}\}$.

SOLUTION

For $x = 1, y = -2(1) + 1 = -1$.

For $x = 2, y = -2(2) + 1 = -3$.

For $x = 3, y = -2(3) + 1 = -5$.

For $x = 4, y = -2(4) + 1 = -7$.

Therefore $A = \{(1,-1), (2,-3), (3,-5), (4,-7)\}$.

4 List all the elements in the set $A = \{(x,y) | y = 2x + 1, x \in \{0,1,2,3\}\}$.

SOLUTION

For $x = 0, y = 2(0) + 1 = 1$.

For $x = 1, y = 2(1) + 1 = 3$.

For $x = 2, y = 2(2) + 1 = 5$.

For $x = 3, y = 2(3) + 1 = 7$.

Hence,

$$A = \{(0,1), (1,3), (2,5), (3,7)\}.$$

5 Compute the coordinates of five points of the set

$$A = \{(x,y) | 3x - 2y = 5\}$$

and then plot these points on the coordinate system.

SOLUTION. (Refer to Figure 1.) Solving $3x - 2y = 5$ for y, we have

$$3x = 5 + 2y$$
$$3x - 5 = 2y$$

or

$$y = \frac{3x - 5}{2}$$

Let x take on the values 1, 3, 5, 7, and 9.

For $x = 1$, $y = \dfrac{3(1) - 5}{2} = \dfrac{-2}{2} = -1$, or $(1,-1) \in A$.

For $x = 3$, $y = \dfrac{3(3) - 5}{2} = \dfrac{4}{2} = 2$, or $(3,2) \in A$.

For $x = 5$, $y = \dfrac{3(5) - 5}{2} = \dfrac{10}{2} = 5$, or $(5,5) \in A$.

For $x = 7$, $y = \dfrac{3(7) - 5}{2} = \dfrac{16}{2} = 8$, or $(7,8) \in A$.

For $x = 9$, $y = \dfrac{3(9) - 5}{2} = \dfrac{22}{2} = 11$, or $(9,11) \in A$.

Figure 1

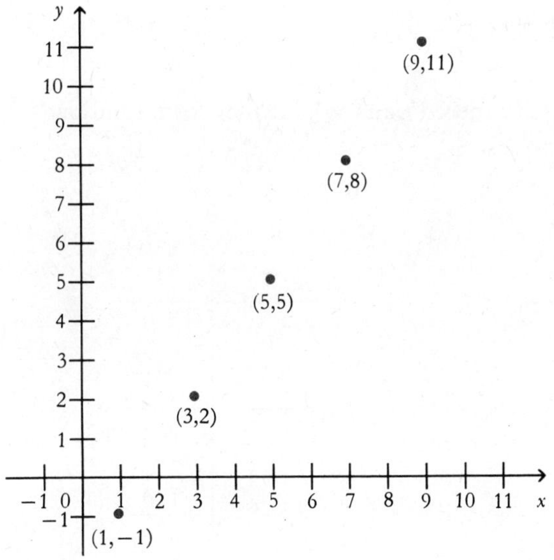

PROBLEM SET 7.2

1 Determine which of the following are linear equations.

a) $y = 5x - \frac{3}{2}$ b) $\frac{y}{5} = -2x + \frac{5}{2}$

c) $3x - y = -\frac{13}{2}$ d) $y = -\frac{1}{3}x$

e) $\frac{5}{x} - \frac{2}{y} = 3$ f) $0.03x - 0.65y = \sqrt{3}$

2 Which of the ordered pairs are solutions of the linear equation $y = 6 - 3x$?

a) $(5,-9)$ b) $(10,2)$
c) $(3,-3)$ d) $(15,-39)$
e) $(-2,12)$

3 List all the elements in each of the following sets.

a) $\{(x,y) \mid y = 5x - 2, x \in \{-1,2,3,5\}\}$
b) $\{(x,y) \mid y = -2x + 3, x \in \{-2,0,1,2,3\}\}$
c) $\{(x,y) \mid y = 4x - 7, x \in \{-1,0,1,2,3\}\}$
d) $\{(x,y) \mid y = \frac{1}{4}x + 1, x \in \{-3,-2,-1,0,1,2\}\}$
e) $\{(x,y) \mid y = -\frac{1}{3}x - 2, x \in \{-3,-2,-1,0,1,2,3\}\}$

4 Compute the coordinates of four points in the following sets, and then plot these points on a Cartesian coordinate system.

a) $A = \{(x,y) \mid y = -x + 5\}$
b) $A = \{(x,y) \mid y = 5x - 3\}$
c) $A = \{(x,y) \mid y = 2x + 1\}$

d) $A = \{(x,y) | y = -2x + 7\}$
e) $A = \{(x,y) | y = -\frac{1}{2}x - 2\}$
f) $A = \{(x,y) | y = \frac{1}{3}x + 4\}$

5 Complete the indicated ordered pairs so that they form solutions of the given linear equations.

a) $2x + y = -2$

x	y
-1	___
-2	___
___	-2
___	1
3	___
4	___

b) $4x + 3y = -7$

x	y
0	___
-1	___
___	1
___	2
3	___
4	___

c) $3x - 2y = 2$

x	y
0	___
1	___
2	___
___	3
___	-2
___	5

d) $4x + 3y = -7$

x	y
-2	___
-1	___
___	0
___	-2
___	-3

7.3 Graphs of Linear Equations

First-degree equations in two variables have solutions that are ordered pairs of real numbers. Hence, these solutions can be graphed on a Cartesian coordinate system. For example, the graph of $\{(x,y) | y = 3x$ and $x \in \{-2,-1,0,1,2\}\}$ is given in Figure 1.

Figure 1

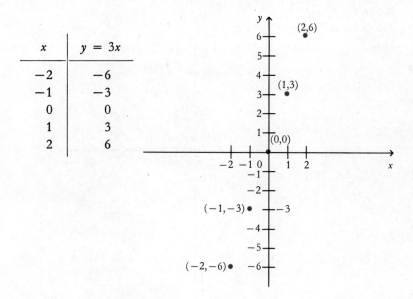

x	$y = 3x$
-2	-6
-1	-3
0	0
1	3
2	6

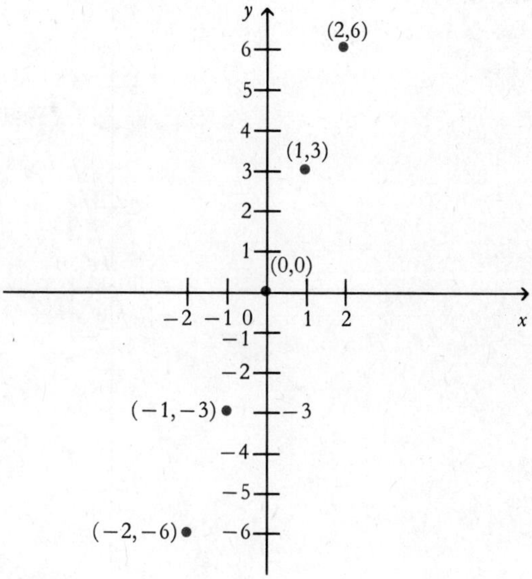

Figure 1 consists of the points $(-2,-6)$, $(-1,-3)$, $(0,0)$, $(1,3)$, and $(2,6)$, which appear to lie on a straight line. If other solutions of $y = 3x$ are graphed, they will also appear to lie on the same straight line (Figure 2).

Figure 2

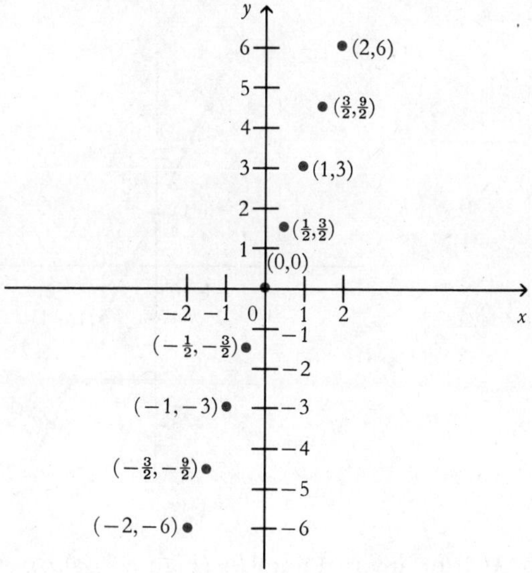

Now consider the graph of $y = 3x$ over the set of all real numbers x for which y is a real number. Each solution that is graphed will continue to lie on the same straight line; hence, by connecting those points we get the graph of $y = 3x$, which appears to be a straight line (Figure 3).

Figure 3

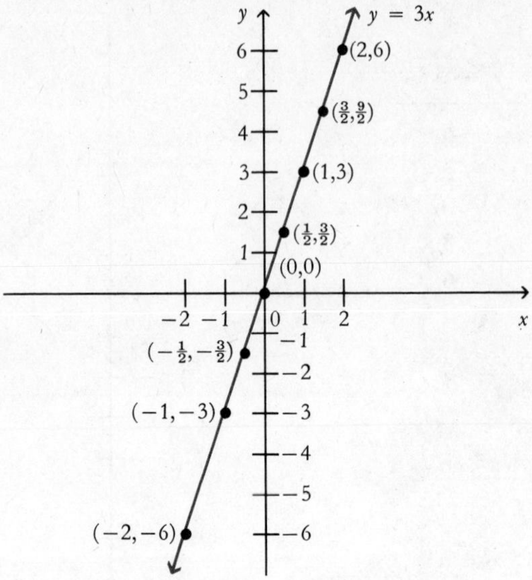

Now consider the graph of $y = -2x + 1$. Notice that if $x = 0$, then $y = -2(0) + 1 = 1$; thus the ordered pair (0,1) is on the graph of $y = -2x + 1$; if $x = \frac{1}{2}$, then $y = -2(\frac{1}{2}) + 1 = 0$, so that the ordered pair $(\frac{1}{2},0)$ is on the graph of $y = -2x + 1$. Continuing in this manner, we can form the following table:

Figure 4

x	y
0	1
$\frac{1}{2}$	0
1	−1
2	−3
−1	3

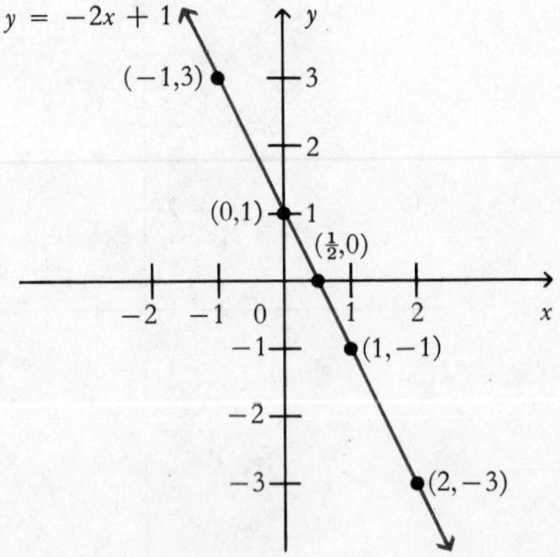

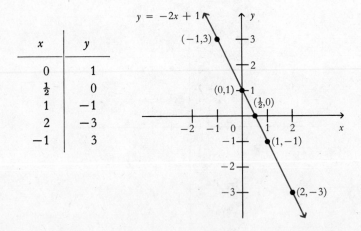

The graph of $y = -2x + 1$ for any real number (Figure 4) also appears to be a straight line. It can be shown that the coordinates of each point on the line in Figure 4 satisfy $y = -2x + 1$, and, conversely, every solution of the equation $y = -2x + 1$ corresponds to a point on the line.

In general, *the graph of any first-degree equation in two variables is a straight line, and, conversely, a straight line is the graph of a first-degree equation in two variables.* Since we know from geometry that a straight line is completely determined by two distinct points, it will be enough to locate two points in order to graph the line. For example, let us graph $y = 3x$. We know that (0,0) and (1,3) are solutions to the equation $y = 3x$; hence, if we graph P_1: (0,0) and P_2: (1,3), the graph of $y = 3x$ is the straight line determined by the points P_1 and P_2 (Figure 5).

Figure 5

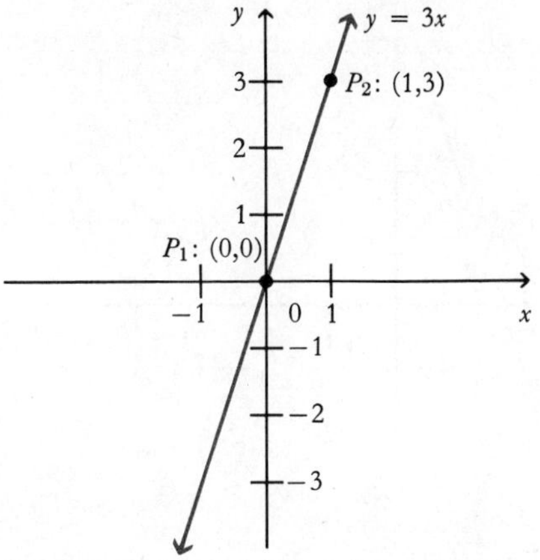

Similarly, the graph of $\{(x,y) \mid y = -2x + 1\}$ is determined by P_1: (0,1) and P_2: (1,−1) (or any other two points) (Figure 6).

Figure 6

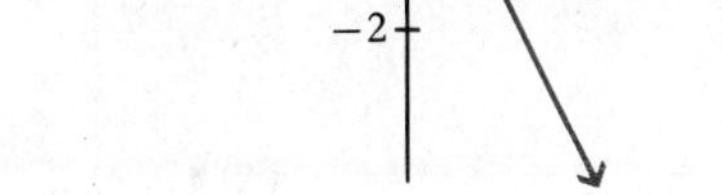

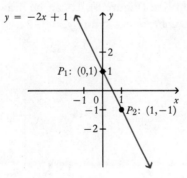

Thus, we need to find only two solutions of a linear equation such as $3x - y + 5 = 0$ to determine its graph. Two solutions that are easy to find are those with first and second components respectively zero; that is, the solutions are of the form $(0,y)$ and $(x,0)$. For example, if we set $x = 0$ in the equation $3x - y + 5 = 0$, we get $-y + 5 = 0$, so that $y = 5$. Therefore, (0,5) is a point on the graph. Letting $y = 0$, we have $3x + 5 = 0$, so that $x = -\frac{5}{3}$ and $(-\frac{5}{3},0)$ is also a point on the graph. Hence, $(-\frac{5}{3},0)$ and (0,5) are the points where the graph crosses the

x axis and y axis, respectively. For this reason, we call the number $-\frac{5}{3}$ the x *intercept* and the number 5 the y *intercept* of the graph (Figure 7).

Figure 7

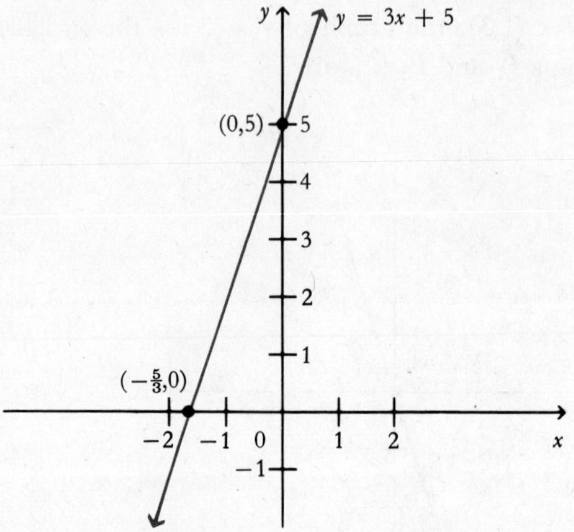

EXAMPLES

Graph each of the following linear equations. Give the x intercept and the y intercept of each graph.

1 $y = -2x$

SOLUTION. The x intercept is (0,0) and the y intercept is (0,0). The graph is a straight line (Figure 8). It can be drawn by locating any two points whose coordinates satisfy $y = -2x$.

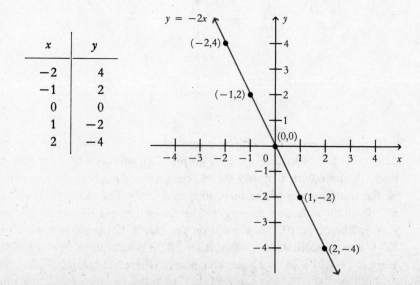

Figure 8

x	y
−2	4
−1	2
0	0
1	−2
2	−4

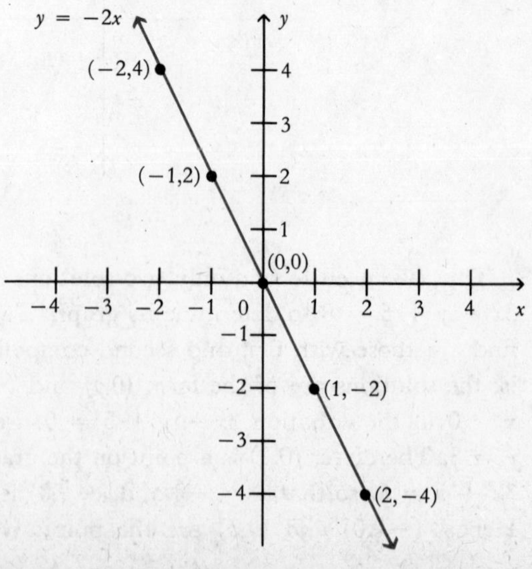

2 $y = x - 2$

SOLUTION. The x intercept is (2,0) and the y intercept is (0,−2). The graph is a straight line (Figure 9).

Figure 9

x	y
−2	−4
−1	−3
0	−2
1	−1
2	0

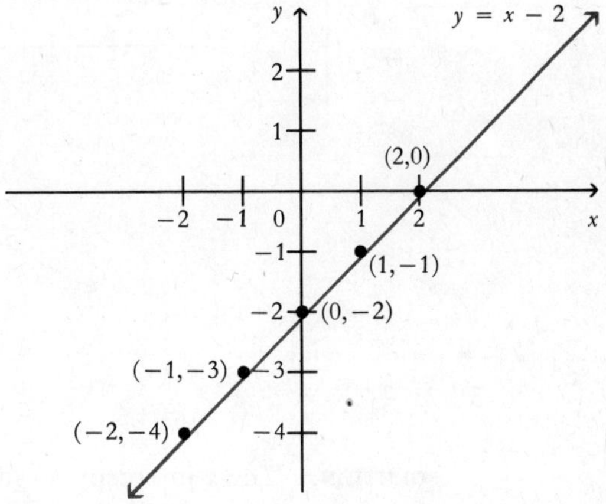

3 $y = -2x + 4$

SOLUTION. Since the x intercept is (2,0) and the y intercept is (0,4), two solutions of $y = -2x + 4$ are (0,4) and (2,0). The graph is a straight line (Figure 10).

Figure 10

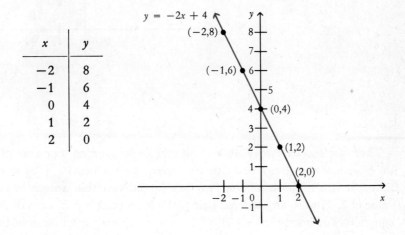

x	y
−2	8
−1	6
0	4
1	2
2	0

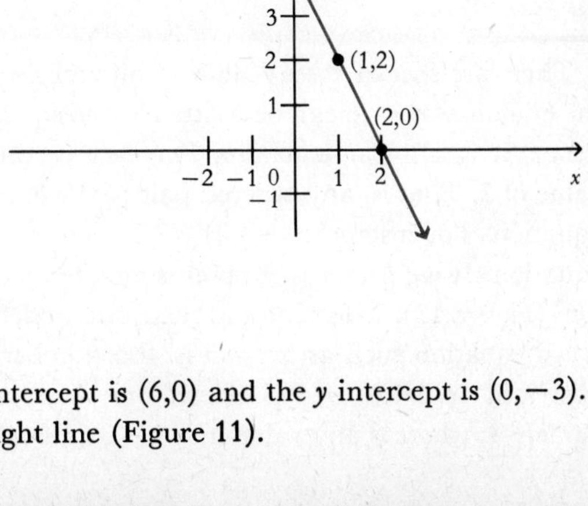

4 $x - 2y = 6$

SOLUTION. The x intercept is (6,0) and the y intercept is (0,−3). The graph is a straight line (Figure 11).

Figure 11

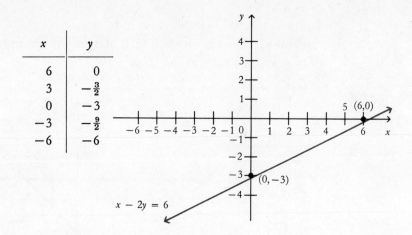

x	y
6	0
3	$-\frac{3}{2}$
0	-3
-3	$-\frac{9}{2}$
-6	-6

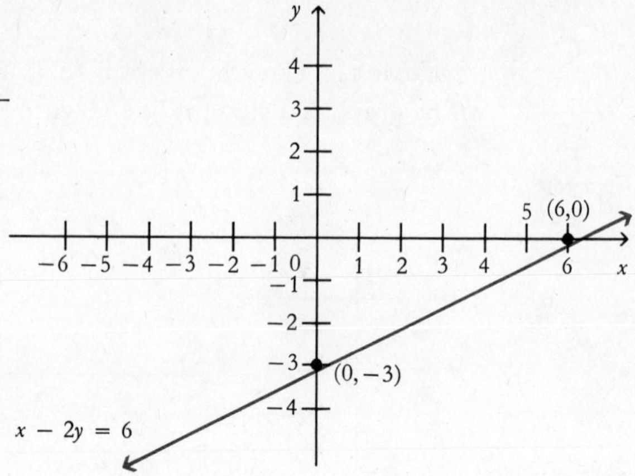

5 $3x + 2y = 5$

SOLUTION. The x intercept is $(\frac{5}{3},0)$ and the y intercept is $(0,\frac{5}{2})$. The graph is a straight line (Figure 12).

Figure 12

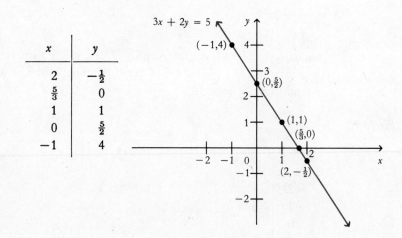

x	y
2	$-\frac{1}{2}$
$\frac{5}{3}$	0
1	1
0	$\frac{5}{2}$
-1	4

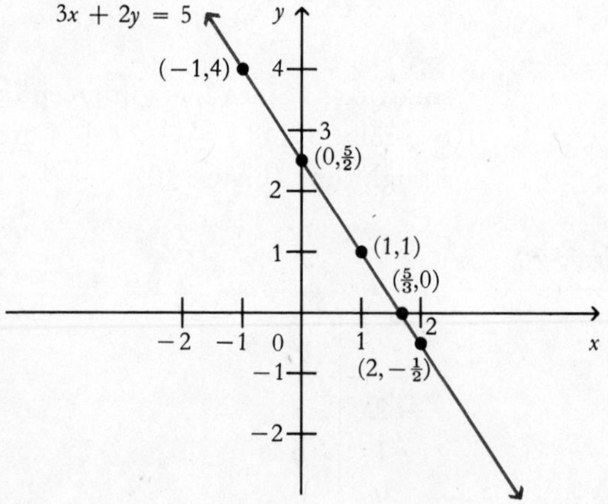

There are special cases of linear equations worth noting. For example, the equation $3y = 6$ can be written in the equivalent form $0x + 3y = 6$; hence, it is a linear equation. For each x, this equation assigns to y a value of 2. That is, any ordered pair of the form $(x,2)$ is a solution of the equation. For instance, $(-4,2)$, $(0,2)$, and $(3,2)$ are all solutions of the equation. If we graph these points and connect them, we get a straight line (Figure 13). This kind of line is often referred to as a *horizontal* line.

An equation such as $3x = 5$ is also a linear equation since it can be expressed as $3x + 0y = 5$. Here, only one value is permissible for x, namely $\frac{5}{3}$, whereas any value may be assigned to y. That is, any ordered

Figure 13

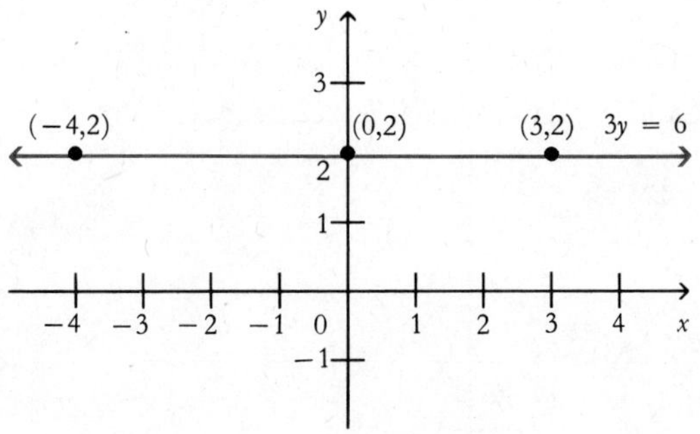

pair of the form $(\frac{5}{3},y)$ is a solution of this equation. If we choose two solutions, say $(\frac{5}{3},1)$ and $(\frac{5}{3},-1)$, we can complete the graph (Figure 14).

Figure 14

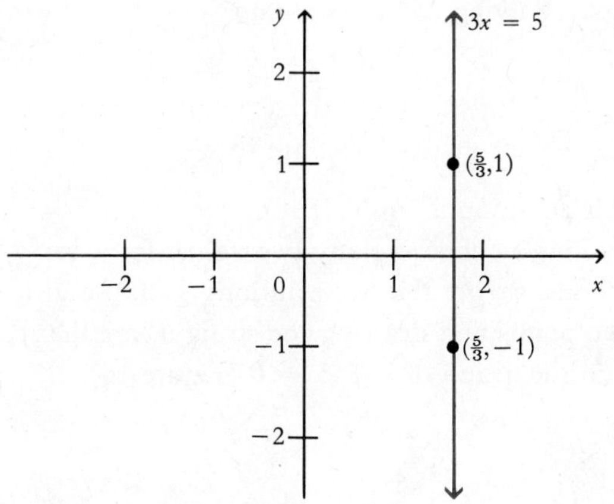

This kind of line is often referred to as a *vertical* line.

EXAMPLES

Graph each of the following linear equations.

1 $2y = 7$

SOLUTION. This equation is equivalent to $0x + 2y = 7$. Hence, by choosing any two values for x, say $x = 1$ and $x = -3$, and solving for y

$$2y = 7$$
$$y = \frac{7}{2}$$

we have the two solutions $(1,\frac{7}{2})$ and $(-3,\frac{7}{2})$. Graphing these two ordered pairs and connecting the points we get a straight line (Figure 15).

Figure 15

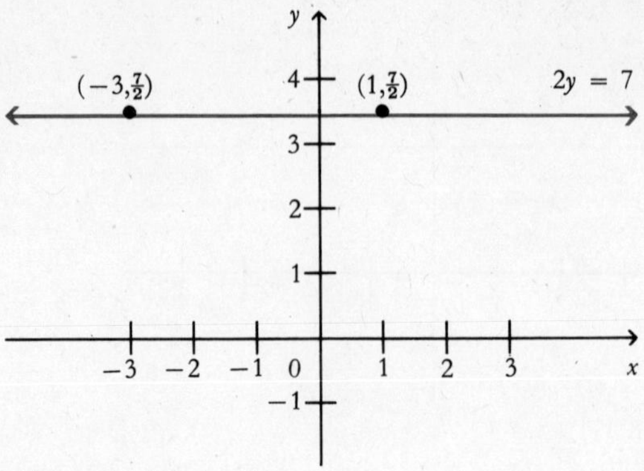

2 $x + 3 = 0$

SOLUTION. Solving for x, we obtain

$$x + 3 = 0$$
$$x = -3$$

Since this equation is equivalent to $x + 0y + 3 = 0$, we can find two solutions by choosing any two values for y, say $y = 0$ and $y = 3$. Thus, we get the two solutions $(-3,0)$ and $(-3,3)$. Graphing these two points and drawing the straight line that contains them, we obtain the graph of $x + 3 = 0$ (Figure 16).

Figure 16

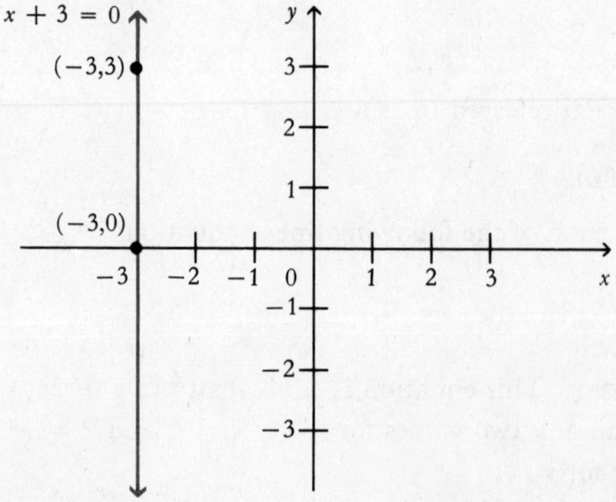

PROBLEM SET 7.3

In problems 1–12, sketch the graph of each equation.

1 $y = 2x, x \in \{1,2\}$

2 $y = 2x, x \in \{-2,-1,0,1,2\}$

3 $y = 2x, x \in \{-3,-2,-1,0,1,2,3\}$

4 $y = 2x$

5 $y = 4x + 3$

6 $y = 3x + 2$

7 $2x + 3y + 6 = 0$

8 $4x - 3y - 12 = 0$

9 $3x = 11$

10 $2y = 9$

11 $x + 5 = 0$

12 $2y + 3 = 0$

In problems 13–25, find the x intercept, the y intercept, and sketch the graph of each line.

13 $\{(x,y) | y = -3x + 5\}$

14 $3x - y + 1 = 0$

15 $2x - 3y - 9 = 0$

16 $-x + 2y - 11 = 0$

17 $y = -7x + 2$

18 $y = 3x - 5$

19 $-4x + 3y = 0$

20 $2x = 4y + 1$

21 $\frac{x}{2} + \frac{y}{-1} = 1$

22 $y - 2 = -5(x - 1)$

23 $y = 3 - 2x$

24 $x + 4y - 3 = 0$

25 $y = -\frac{3}{4}x + \frac{1}{2}$

In problems 26–29, graph each equation. Are they linear equations?

26 $y = |x|$

27 $y = |x - 1|$

28 $y = |x| + 2$

29 $y = -|x| + 2$

7.4 Graphs of Linear Inequalities

We saw in Section 7.3 that the solutions of linear equations such as $y = 3x + 6$ are ordered pairs of numbers. In this section we will see that inequalities involving two variables such as $y < 3x + 6$, $5x - 7 \geq y$, and $y \geq \frac{5x}{4} + 3$ also have ordered pairs of numbers as solutions. Linear equations, such as $y = 3x + 6$, can be viewed as describing three sets of points in a plane. One set is the set of points that lie on the line, another is the set of points that lie above the line, and the third is the set of points that lie below the line (Figure 1).

Figure 1

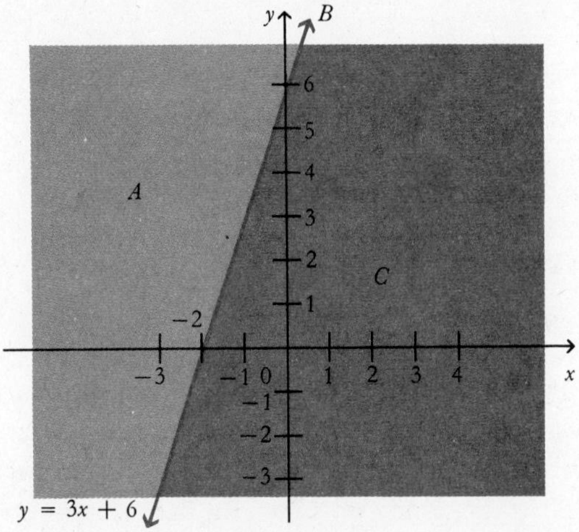

In Figure 1, A represents the set of points above the line $y = 3x + 6$, B represents the set of points on the line $y = 3x + 6$, and C represents the set of points below the line $y = 3x + 6$.

The set B can be described as $B = \{(x,y) | y = 3x + 6\}$. That is, B is the set of points whose coordinates satisfy the equation $y = 3x + 6$. Sets A and C can be described in a similar manner. In set A, all points lie above the line $y = 3x + 6$; that is, their y coordinates must be greater than the y coordinates of all points on the line having the same x coordinate (Figure 2). Here, $(-1,7) \in A$, since $7 > 3$.

Thus, the set A can be described as $A = \{(x,y) | y > 3x + 6\}$. That is, A is the set of ordered pairs which contain points such as (0,7), (1,10), and (−1,7). Similarly, a point that lies below the line $y = 3x + 6$ has a y coordinate less than the y coordinate of a point on the line having the same x coordinate (Figure 2). Thus, the set C can be described as $C = \{(x,y) | y < 3x + 6\}$. That is, C is the set of ordered pairs which contains points such as (0,5), (1,8), and (−1,−2).

Figure 2

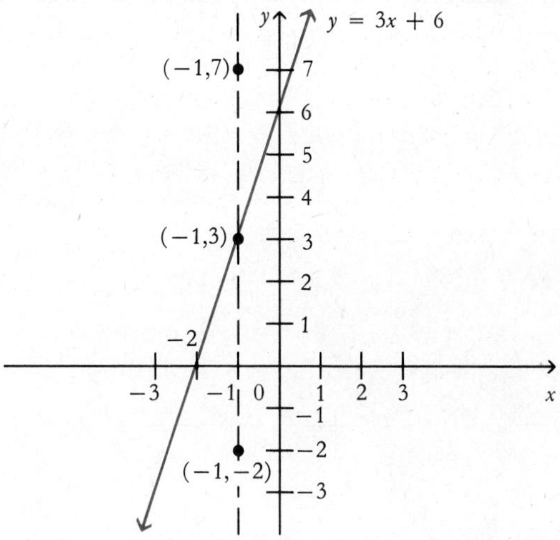

The set A of points above the line $y = 3x + 6$ is called the *graph of the inequality* $y > 3x + 6$, whereas the set C of points below the line $y = 3x + 6$ is called the *graph of the inequality* $y < 3x + 6$.

EXAMPLES

Graph each of the following sets.

1 $\{(x,y) | y < x + 2\}$

SOLUTION. The graph of this set is the graph of the linear inequality $y < x + 2$ which is the set of points that lie below the line $y = x + 2$ (Figure 3).

Figure 3

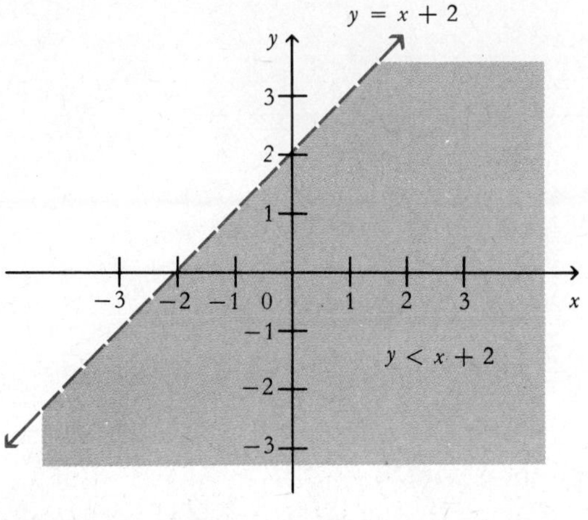

2 $\{(x,y) | y > 2x - 1\}$

SOLUTION. The graph of this set is the graph of the linear inequality $y > 2x - 1$ which is the set of points that lie above the line $y = 2x - 1$ (Figure 4).

Figure 4

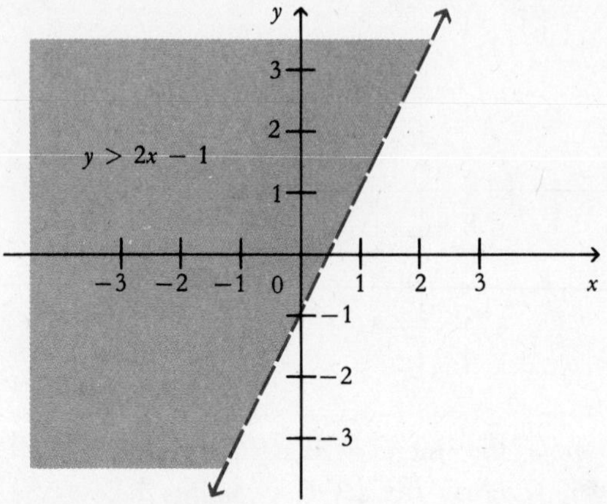

3 $\{(x,y) | y \leq 2x\}$

SOLUTION. The required graph (Figure 5) is the set of points that lie on or below the line $y = 2x$, since $y \leq 2x$ means $y < 2x$ or $y = 2x$.

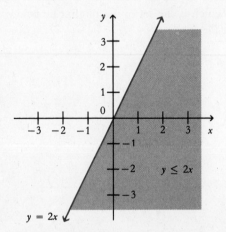

Figure 5

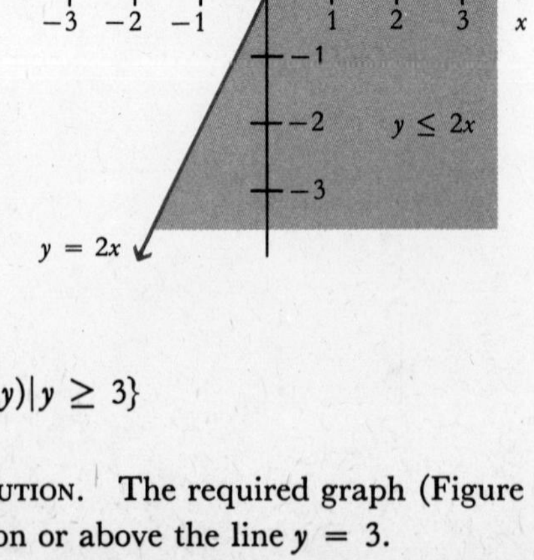

4 $\{(x,y) | y \geq 3\}$

SOLUTION. The required graph (Figure 6) is the set of points that lie on or above the line $y = 3$.

Figure 6

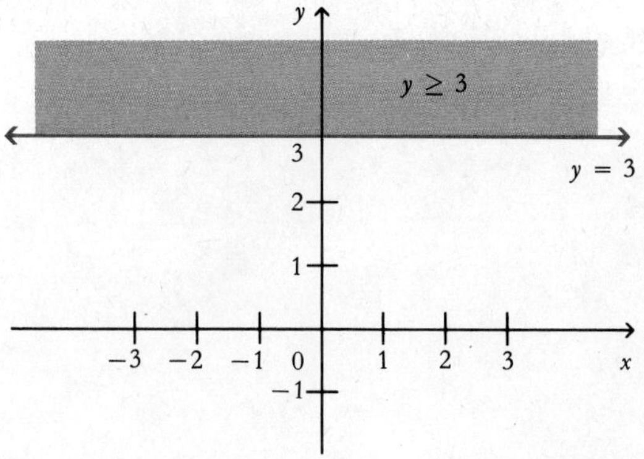

5 $\{(x,y) | y < -1\}$

SOLUTION. The required graph (Figure 7) is the set of points that lie below the line $y = -1$.

Figure 7

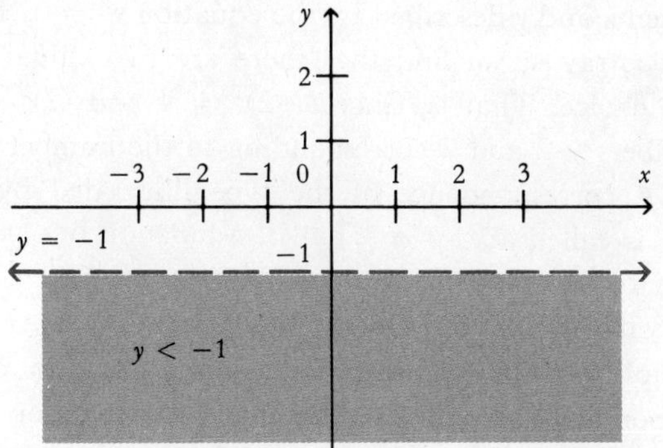

PROBLEM SET 7.4

Graph the following linear inequalities.

1 $y > 3x$

2 $y \leq -4x$

3 $y > -2x + 1$

4 $2y > 3x + 4$

5 $y + 2x < 3$

6 $3x + 2y > 4$

7 $y - 3x + 5 < 0$

8 $\{(x,y) \mid 2y + 5x - 3 \geq 0\}$

9 $\{(x,y) \mid y \geq -3\}$

10 $\{(x,y) \mid y \leq 7\}$

7.5 Functions

In this section we wish to consider a special kind of correspondence that can exist between variables. For example, consider the correspondence between the two variables x and y described by the equation $y = 2x + 1$. For each value of x that we choose, there is one and only one value of y corresponding to that choice. For example, if we let $x = 3$, then $y = 2(3) + 1 = 7$, and no other value. On the other hand, consider the correspondence between x and y described by the equation $y^2 = x$. If we choose some value of x, say 4, we find that there are two values of y corresponding to this choice. That is, since $(-2)^2 = 4$ and $(2)^2 = 4$, we have the two numbers -2 and 2 corresponding to the number 4 in this correspondence. A correspondence of the type illustrated by the equation $y = 2x + 1$ is called a *function*. That is, a function is a correspondence that assigns to each member in a certain set, called the *domain* of the function, exactly one member in a seond set, called the *range* of the function. In the example $y = 2x + 1$, since we assign a value to x first to determine a corresponding value of y, we say that y is a function of x.

Functions can also be described in terms of ordered pairs. For example, the function determined by the equation $y = 2x + 1$ can also be described as

$$\{(x,y) | y = 2x + 1\}$$

Thus, we can also describe a function as a *set of ordered pairs such that no two different ordered pairs have the same first element.* To say that "no two different ordered pairs have the same first element" is the same as saying that "there is one and only one value of the y variable for each choice of the x variable." In this case, the set of all of the first elements of the ordered pairs is called the domain of the function, and the set of all the second elements of the ordered pairs is called the range of the function.

From the above definition we conclude that all functions are sets of ordered pairs; however, not all sets of ordered pairs are functions. For example, $\{(1,1), (2,2), (3,7), (3,5)\}$ is a set of ordered pairs but it is not a function because (3,7) and (3,5) have the same first members. By contrast, $\{(1,2), (3,4), (4,4)\}$ is a set of ordered pairs which is a function with domain $\{1,3,4\}$ and range $\{2,4\}$. Note that in the latter example, there are two pairs which have the same second member; this does not violate the definition of a function.

Hence, if at least two different ordered pairs of a set have the same first member, the set is not a function. In other words, if a domain member appears with more than one range member, the set is not a function. Geometrically, this means that if the graph of a set of ordered pairs has more than one point with the same abscissa, the set is not a function.

EXAMPLE

Which of the following sets are functions? What is the domain and range of each? Graph each set.

a) $\{(1,2), (2,3), (3,4), (4,4), (5,6)\}$
b) $\{(x,y) | y = 3\}$
c) $\{(x,y) | x = 1\}$
d) $\{(x,y) | y = 3, x \in \{-1,0,2,3\}\}$
e) $\{(x,y) | y = 3x\}$

SOLUTION

a) $\{(1,2), (2,3), (3,4), (4,4), (5,6)\}$ is a function with domain $\{1,2,3,4,5\}$ and range $\{2,3,4,6\}$ (Figure 1).

Figure 1

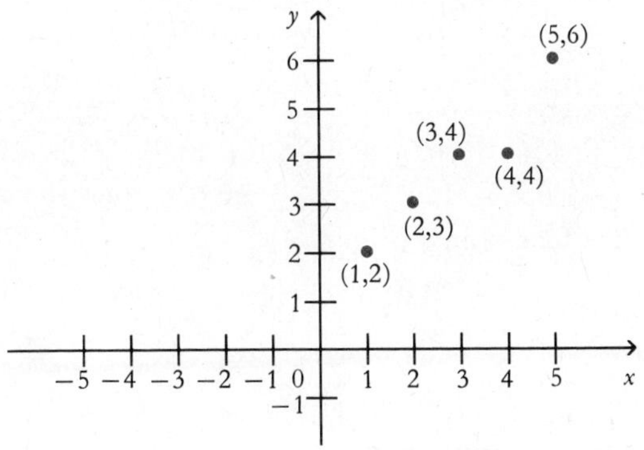

b) $\{(x,y) | y = 3\}$ is a function because no two different ordered pairs have the same first members. The domain is the set of all real numbers because there is no restriction on x, and the range is $\{3\}$. The graph is the line parallel to the x axis (Figure 2).

Figure 2

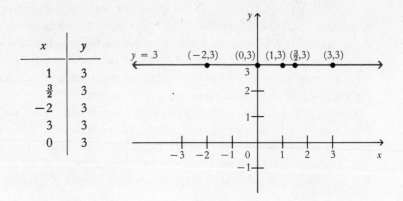

x	y
1	3
$\frac{3}{2}$	3
-2	3
3	3
0	3

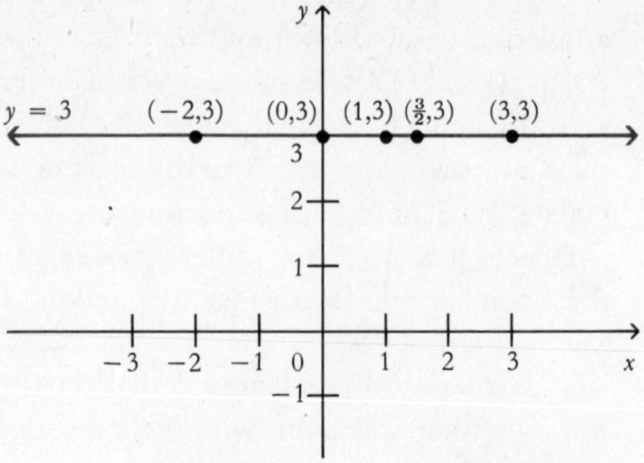

c) $\{(x,y)|x = 1\}$ is not a function because (1,0) and (1,2) are members of the set and they both have the same first element. The graph is the line parallel to the y axis (Figure 3).

Figure 3

x	y
1	0
1	2
1	-1
1	-2

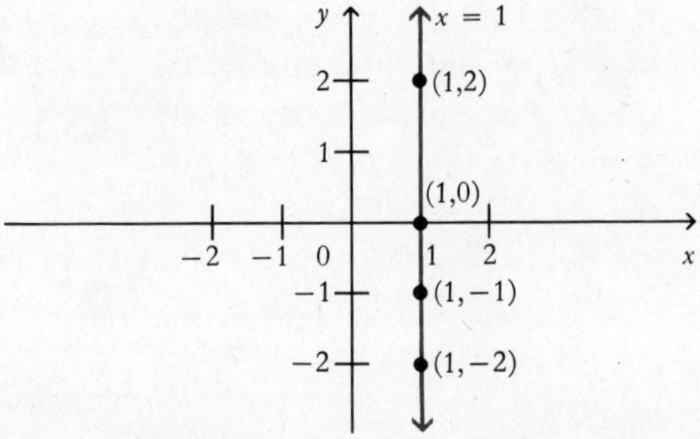

d) $\{(x,y)|y = 3x \text{ with } x \in \{-1,0,2,3\}\}$ represents the ordered pairs $\{(-1,-3), (0,0), (2,6), (3,9)\}$. It is a function with domain $\{-1,0,2,3\}$ and range $\{-3,0,6,9\}$, and the graph is composed of the four points (Figure 4).

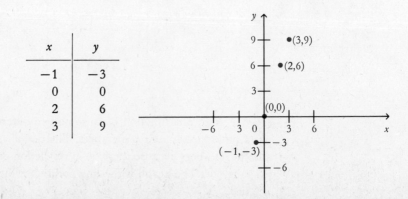

Figure 4

x	y
-1	-3
0	0
2	6
3	9

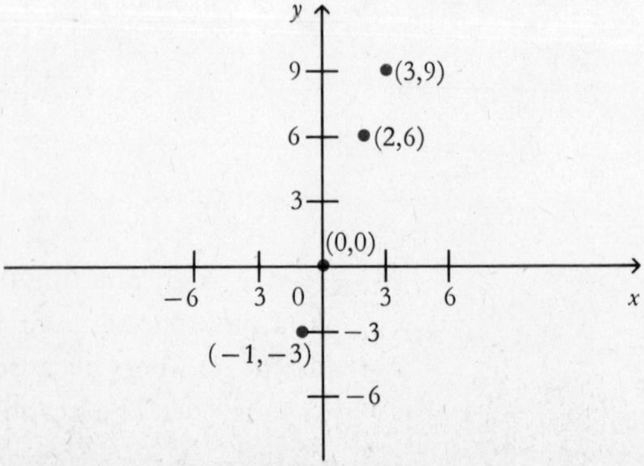

e) $\{(x,y) \mid y = 3x\}$ is a function and has as both its domain and range the set of all real numbers (Figure 5). (It can be seen geometrically that this set of ordered pairs is a function since no two points have the same abscissa.)

Figure 5

x	y
0	0
−1	−3
2	6
3	9

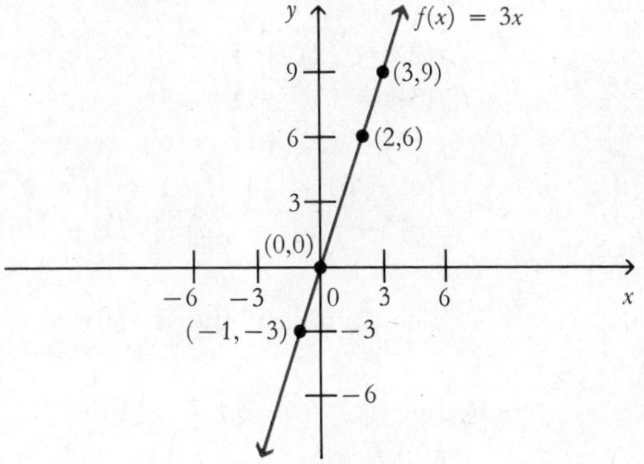

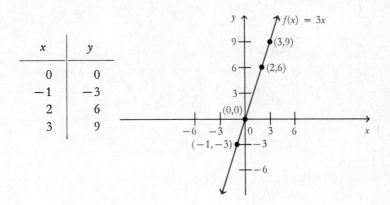

If x represents members of the domain of $y = x^2$, then y is considered to be a function of x and we indicate this by writing $f(x) = x^2$ (which reads "f of x equals x squared"). Hence, $f(x) = x^2$ is another way of writing the function $\{(x,y) \mid y = x^2\}$. In other words, $y = f(x)$ means that (x,y) is a member of the function f. For example, if $4x - 2y = 1$ and $y = f(x)$, then after solving for y in terms of x, we could write the function either as $\{(x,y) \mid y = 2x - \frac{1}{2}\}$ or as $f(x) = 2x - \frac{1}{2}$.

Now, it is important to realize that letters other than x, y, or f can be used to denote functions. For example, $h(r) = r^2 - 1$ is the function given by $\{(r,h(r)) \mid h(r) = r^2 - 1\}$; $3r + 5t = 3$ with $t = g(r)$ is the function given by $\left\{(r,t) \mid t = \dfrac{3 - 3r}{5}\right\}$; and $c(d) = \pi d$ is the function given by $\{(d,c(d)) \mid c(d) = \pi d\}$.

When functional notation is used, sometimes it is helpful to think of the variable which represents the members of the domain as a "missing blank." For example, $g(t) = t^2$ can be thought of as $g(\ ) = (\ )^2$; hence, if any expression (representing a real number) is to be used to represent a member of the domain, the corresponding member of the range can be found by substitution in the blank. For example, $g(3)$ can be determined by first writing the function as

$$g(\ ) = (\ )^2$$

so that by substitution

$$g(3) = (3)^2 = 9$$

Similarly,

$$g(3 - 5x) = (3 - 5x)^2 = 9 - 30x + 25x^2$$

EXAMPLES

1 Let $f(x) = x + 1$. Find $f(2), f(3), f(5), f(a-6)$, and $f(a) - f(6)$. What is the domain of f?

SOLUTION

$$f(2) = (2) + 1 = 3$$
$$f(3) = (3) + 1 = 4$$
$$f(5) = (5) + 1 = 6$$
$$f(a-6) = (a-6) + 1 = a - 5$$
$$f(a) - f(6) = [(a) + 1] - [(6) + 1] = a - 6$$

The domain is the set of real numbers R.

2 Let $f(x) = -3x + 1$. Find $f(0), f(1), f(-1), f(2), f(-2), f(3)$, and $f(-3)$.

SOLUTION

$$f(0) = -3(0) + 1 = 1$$
$$f(1) = -3(1) + 1 = -2$$
$$f(-1) = -3(-1) + 1 = 4$$
$$f(2) = -3(2) + 1 = -5$$
$$f(-2) = -3(-2) + 1 = 7$$
$$f(3) = -3(3) + 1 = -8$$
$$f(-3) = -3(-3) + 1 = 10$$

3 Let $f(x) = x^2 - 1$. What is the domain and the range of f? Find each of the following expressions.

a) $f(3) + f(5)$
b) $f(x+1)$
c) $f(2a+4)$
d) $2f(a-2)$

SOLUTION. The domain of the function f is the set of real numbers R. Since $x^2 \geq 0$, $x^2 - 1 \geq -1$; hence the range of f is the set $\{y | y \geq -1\}$.

a) $f(3) = (3)^2 - 1 = 8$

and

$$f(5) = (5)^2 - 1 = 24$$

so that

$$f(3) + f(5) = 8 + 24 = 32$$

b) $f(x+1) = (x+1)^2 - 1 = x^2 + 2x$
c) $f(2a+4) = (2a+4)^2 - 1 = 4a^2 + 16a + 15$
d) $2f(a-2) = 2[(a-2)^2 - 1] = 2[a^2 - 4a + 3] = 2a^2 - 8a + 6$

4 Given the *identity function* $f(x) = x$, find $f(-2)$, $f(-1)$, $f(0)$, $f(1)$, and $f(2)$. Sketch the graph of f.

SOLUTION. $f(-2) = -2$, $f(-1) = -1$, $f(0) = 0$, $f(1) = 1$, and $f(2) = 2$. The domain is the set of real numbers R, and the range is also the set of real numbers R (Figure 6).

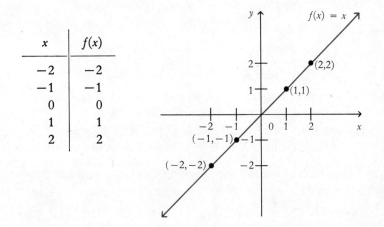

Figure 6

x	$f(x)$
−2	−2
−1	−1
0	0
1	1
2	2

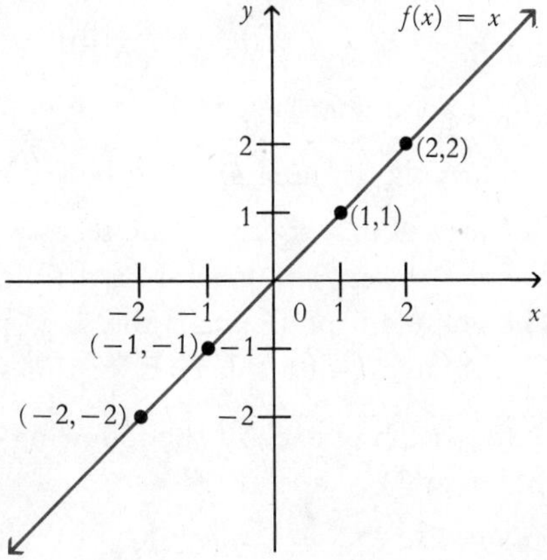

5 Given the *absolute value function* $f(x) = |x|$. Find $f(-2)$, $f(-1)$, $f(0)$, $f(1)$, and $f(2)$. Describe the domain and range of f. Sketch the graph of f.

SOLUTION. $f(-2) = |-2| = 2$, $f(-1) = |-1| = 1$, $f(0) = |0| = 0$, $f(1) = |1| = 1$, and $f(2) = |2| = 2$. The domain is the set of real numbers R and the range is the set of all nonnegative real numbers (Figure 7).

Figure 7

x	$f(x)$
−2	2
−1	1
0	0
1	1
2	2

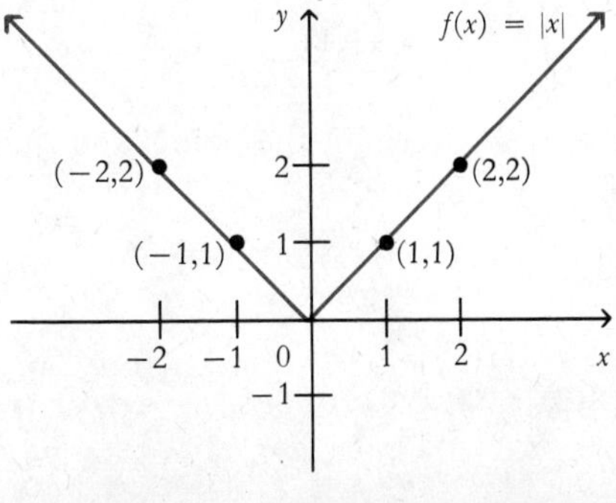

PROBLEM SET 7.5

1 Which of the following sets are functions?

a) $\{(2,4), (3,6), (7,2), (9,-3)\}$
b) $\{(1,1), (21,1), (3,1)\}$
c) $\{(-8,0), (-7,0), (-6,2), (-7,3), (5,1)\}$
d) $\{(x,y) \mid y = 2x - 1\}$

2 Find $f(1)$, $f(-1)$, $f(2)$, $f(-2)$, $f(5)$, and $f(-5)$.

a) $f(x) = 3x + 1$
b) $f = \{(1,2), (3,4), (5,7)\}$
c) $f(x) = 5x + 1$
d) $f(x) = \dfrac{|x|}{x}$
e) $f(x) = |x + 1|$
f) $f(x) = -7x + 2$
g) $f(x) = |x| + x$
h) $f(x) = -5x + 11$

3 Find the domain and range of each function in Problem 2.

4 Write a formula defining each of the following functions.

a) The area A of a square is the second power of its side x.
b) The perimeter P of a square is 4 times the side x.
c) The volume V of a cube is the third power of its edge x.
d) The volume V of a sphere is $\frac{4}{3}\pi$ times the cube of the radius r.

5 Sketch the graph of each of the following functions; find the domain and the range.

a) $f(x) = 2x$
b) $f(x) = |-2x|$
c) $f(x) = -2x + 1$
d) $f(x) = -3x + 7$
e) $f(x) = 3x + 5$
f) $f(x) = -2|x|$
g) $f(x) = \frac{3}{4}x - 5$
h) $f(x) = -\frac{1}{2}x + 3$

REVIEW PROBLEM SET

In problems 1–4, find the value of x and y so that the ordered pairs are equal.

1 $(2,y) = (2,3)$

2 $(x + 1, y - 1) = (-3,4)$

3 $(2x + 1, 3) = (5, y - 1)$

4 $(2 - x, 3 - y) = (1,1)$

In problems 5–12, plot the points and indicate which quadrant, if any, contains the points.

5 $(-1,3)$

6 $(3.1,-1.7)$

7 $(-3,-2)$

8 $(-\sqrt{3},-2)$

9 $(3,-9)$

10 $(-2,-4)$

11 $(3,0)$

12 $(-3,7)$

In problems 13–16, graph each set.

13 $\{(x,y)|x > 1\}$

14 $\{(x,y)|y < 2\}$

15 $\{(x,y)|y < -x + 2\}$

16 $\{(x,y)|y \geq 2x - 3\}$

In problems 17–20, determine which equations are linear.

17 $3x - 7y = 8$

18 $2xy + 7 = 3$

19 $\frac{4}{x} + \frac{5}{y} = 1$

20 $\frac{x}{7} + \frac{y}{3} = 8$

In problems 21–26, determine if any of the given ordered pairs are solutions of $2x - 3y = 6$.

21 $(0,2)$

22 $(-3,0)$

23 $(0,-2)$

24 $(-3,-4)$

25 $(3,0)$

26 $(5,4)$

In problems 27–30, list the elements of the sets.

27 $\{(x,y)|y = 2x + 7, x \in \{-2,-1,0,1,2\}\}$

28 $\{(x,y)|y = 4 - 3x, x \in \{0,1,2,3\}\}$

29 $\{(x,y)|2x + 3y = 9, x \in \{-1,0,1\}\}$

30 $\{(x,y)|-3x + 5y = 10, x \in \{-3,-2,-1,0\}\}$

In problems 31–36, graph each linear equation. Determine the x and y intercepts.

31 $2x + 3y = 6$

32 $5x + y = 7$

33 $-x + 4y = 8$

34 $3x - 4y = 12$

35 $3x - 2y = 5$

36 $5x + 7y = 8$

In problems 37–44, find the x intercept, the y intercept, and sketch the graph of each line.

37 $4x + 3y = 12$

38 $3x - 2y = 10$

39 $4x - 5y = 7$

40 $3x - 5y = 10$

41 $x - 2y = 5$

42 $\frac{x}{2} + \frac{y}{-1} = 1$

43 $3x + 2y + 5 = 0$

44 $y = -\frac{3}{2}x + 1$

45 Given the equation of the line $3x - 2y + 6 = 0$, find its y intercept and the ordinate of the point on the line whose abscissa is 4.

In problems 46–49, indicate which is a function. Indicate the domain and the range of each function.

46 $f(x) = 7x - 2$

47 $f(x) = 25 - x$

48 $\{(1,2), (2,3), (1,5), (6,7)\}$

49 $\{(x,y) \mid |y| = 2x - 1\}$

50 Which of the two graphs in Figure 1 is the graph of a function?

In problems 51–58, let $f(x) = 3x^2 + 2$. Find each of the following.

51 $f(-1)$

52 $f(0)$

Figure 1

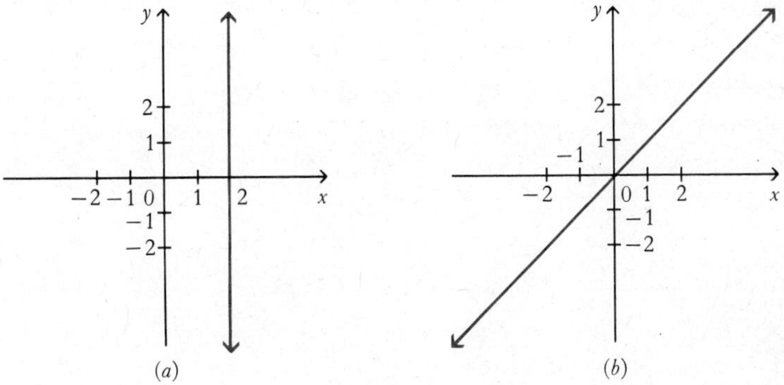

53 $f(1)$

54 $f(\frac{2}{3})$

55 $f(a)$

56 $f(b)$

57 $f(a + b)$

58 $f(a + b) - f(a)$

In problems 59–61, let $f(x) = |x|$ and $g(x) = 2x - 3$. Find each of the following.

59 x if $f(x + 3) = 7$

60 x if $f(x - 2) < 1$

61 x if $g(x - 5) > 2$

In problems 62–68, indicate which sets are functions. Indicate the domain and range, and graph each set.

62 $\{(x,y) | y = |2x - 1|\}$

63 $\{(x,y) | y = 2\}$

64 $\{(x,y) | x = -3\}$

65 $\{(x,y) | y \geq x - 1\}$

66 $\{(x,y) | y < -\frac{1}{3}x + 2\}$

67 $\{(x,y) | y \leq 3x + 1\}$

68 $\{(x,y) | y \leq 2x - 3\}$

CHAPTER 8

Systems of Equations and Inequalities

8 SYSTEMS OF EQUATIONS AND INEQUALITIES

Introduction

Chapter 7 presented graphical methods for showing the relationship between two variables expressed by equations of first-degree. It is our objective in this chapter to discuss the geometrical situation that arises when systems of two equations of first-degree are drawn on the same set of axes. In addition, we shall explore a variety of ways of solving such systems.

8.1 Solution Sets

A set of equations such as

$$\begin{aligned} 6x - 5y &= 8 \\ 3x + 10y &= -6 \end{aligned}$$

is called a system of equations. We know from Chapter 7 that the solution set of a first-degree equation of two variables consists of a set of ordered pairs of numbers. The *solution set* for the system consists of the ordered pairs of numbers that satisfy both equations simultaneously. In order to find the solution set of the above system, we let $A = \{(x,y)|$

$6x - 5y = 8\}$ and $B = \{(x,y)|x + 10y = -6\}$ so that the solution set of the system of two equations is their intersection $A \cap B$. Since A and B are infinite sets, we shall have to rely on methods to be developed later in the chapter to show that $A \cap B$ does not contain any other points.

EXAMPLES

Determine which of the given ordered pairs belongs to the solution set of the given system of equations.

1 (2,5), (3,−1), (4,0)

$$\begin{aligned} 2x - y &= 7 \\ 3x + 2y &= 7 \end{aligned}$$

SOLUTION. For (2,5) we have

$$2(2) - (5) \stackrel{?}{=} 7 \quad \text{and} \quad 3(2) + 2(5) \stackrel{?}{=} 7$$
$$-1 \neq 7 \quad \text{and} \quad 16 \neq 7$$

Hence, (2,5) is not a solution.

For (3,−1), we have

$$2(3) - (-1) \stackrel{?}{=} 7 \quad \text{and} \quad 3(3) + 2(-1) \stackrel{?}{=} 7$$
$$7 = 7 \quad \text{and} \quad 7 = 7$$

Hence, (3,−1) is a solution.

For (4,0) we have

$$2(4) - (0) \stackrel{?}{=} 7 \quad \text{and} \quad 3(4) + 2(0) \stackrel{?}{=} 7$$
$$8 \neq 7 \quad \text{and} \quad 12 \neq 7$$

Hence, (4,0) is not a solution.

2 (0,4), ($\frac{9}{2}$,1), (3,0)

$$\begin{aligned} 4x + 3y &= 12 \\ 2x - 3y &= 6 \end{aligned}$$

SOLUTION. For (0,4), we have

$$4(0) + 3(4) \stackrel{?}{=} 12 \quad \text{and} \quad 2(4) - 3(0) \stackrel{?}{=} 6$$
$$12 = 12 \quad \text{and} \quad 8 \neq 6$$

Hence, (0,4) is not a solution.

For $(\frac{9}{2},1)$ we have

$$4(\tfrac{9}{2}) + 3(1) \stackrel{?}{=} 12 \quad \text{and} \quad 2(\tfrac{9}{2}) - 3(1) \stackrel{?}{=} 6$$
$$21 \neq 12 \quad \text{and} \quad 6 = 6$$

Hence, $(\frac{9}{2},1)$ is not a solution.

For (3,0) we have

$$4(3) + 3(0) \stackrel{?}{=} 12 \quad \text{and} \quad 2(3) - 3(0) \stackrel{?}{=} 6$$
$$12 = 12 \quad \text{and} \quad 6 = 6$$

Hence, (3,0) is a solution.

3 (5,1), (4,3), (0,4)

$$2x + y = 11$$
$$x + 4y = 16$$

For (5,1), we have

$$2(5) + 1 \stackrel{?}{=} 11 \quad \text{and} \quad 5 + 4(1) \stackrel{?}{=} 16$$
$$11 = 11 \quad \text{and} \quad 9 \neq 16$$

Hence, (5,1) is not a solution.

For (4,3) we have

$$2(4) + 3 \stackrel{?}{=} 11 \quad \text{and} \quad 4 + 4(3) \stackrel{?}{=} 16$$
$$11 = 11 \quad \text{and} \quad 16 = 16$$

Hence, (4,3) is a solution.

For (0,4) we have

$$2(0) + 4 \stackrel{?}{=} 11 \quad \text{and} \quad 0 + 4(4) \stackrel{?}{=} 16$$
$$4 \neq 11 \quad \text{and} \quad 16 = 16$$

Hence, (0,4) is not a solution.

PROBLEM SET 8.1

In problems 1–10, determine which of the given ordered pairs belongs to the solution set of each of the given systems of equations.

1 $x + y = 7$, $x - y = 1$ (5,2), (6,5), (4,3)

2 $2x - y = -5$, $-x + y = 3$ (2,4), (−2,1), (0,3)

3 $3x - 2y = 4$, $x - y = 0$ $\quad (2,2), (5,2), (4,4)$

4 $5x + y = 5$, $3x + 2y = 10$ $\quad (0,5), (1,0), (1,3)$

5 $9x - 8y = -1$, $x + y = -2$ $\quad (0,\frac{1}{8}), (-3,2), (-1,-1)$

6 $\frac{x}{2} + \frac{y}{3} = 3$, $-\frac{x}{2} + \frac{v}{2} = 2$ $\quad (2,6), (-4,4), (-4,0)$

7 $3y - 2 = x$, $5 - x = -2$ $\quad (1,1), (7,3), (10,4)$

8 $-x + 2y = 7$, $2x - y = -8$ $\quad (-3,2), (1,10), (9,8)$

9 $y = 2x + 1$, $y = -\frac{1}{2}x - 3$ $\quad (-\frac{8}{5}, -\frac{11}{5}), (0,1), (2,-4)$

10 $y = 3x - 3$, $y = 7x - 7$ $\quad (1,0), (0,1), (2,3)$

8.2 Graphical Solution of Systems of Equations

We saw in Chapter 7 that the graph of a linear equation in two variables is a straight line. Therefore, if we graph the equations of a system of two linear equations, using the same coordinate axes for both, we get two straight lines in the plane. The solution of a system of linear equations is represented by the point of intersection of the two lines representing each equation. For example, to solve the system of linear equations

$$3x - 4y = 5$$
$$2x + 5y = 11$$

by the graphical method, construct tables for both equations:

$3x - 4y = 5$

x	y
0	$-\frac{5}{4}$
$\frac{8}{3}$	$\frac{3}{4}$
4	$\frac{7}{4}$

$2x + 5y = 11$

x	y
0	$\frac{11}{5}$
5	$\frac{1}{5}$
4	$\frac{3}{5}$

Next plot the points determined by the corresponding values in each table and draw a straight line for each of these tables (Figure 1) to get the point of intersection (3,1). Hence the solution set is $\{(3,1)\}$.

Figure 1

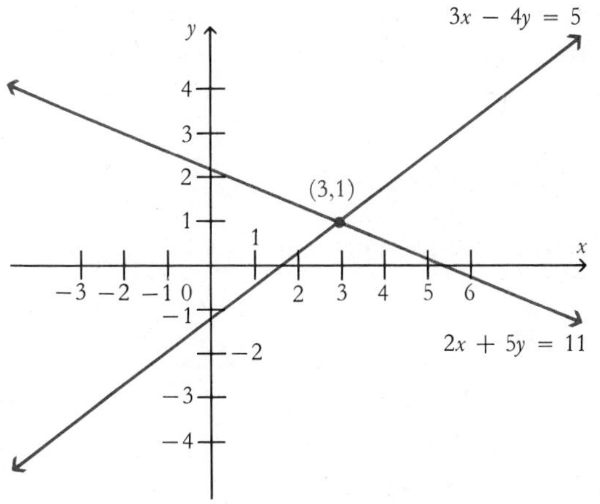

Now consider the system of equations

$$y = 3x + 2$$
$$y = 3x - 1$$

Constructing tables for both equations, we have

$y = 3x + 2$

x	y
−1	−1
0	2
1	5

$y = 3x - 1$

x	y
−1	−4
0	−1
1	2

Plotting the points determined by the corresponding values in each table and drawing two straight lines, we notice the lines are parallel (Figure 2).

Figure 2

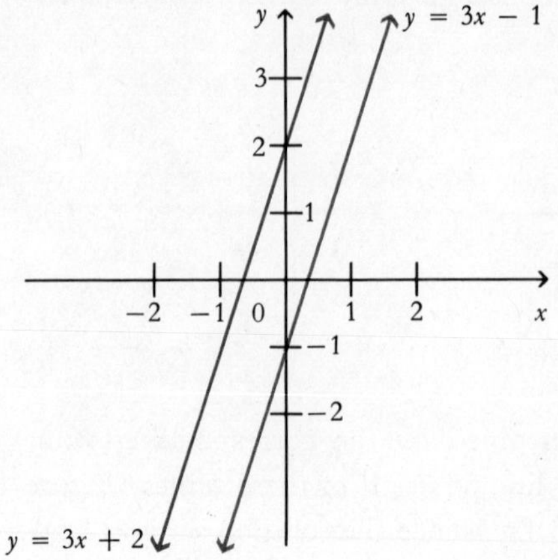

Hence, they have no points in common, and the solution set is ∅.

Using the same approach for the system of equations

$$y = 2x + 1$$
$$2y - 4x = 2$$

we see that the graphs of the two lines coincide (Figure 3):

Figure 3

$y = 2x + 1$

x	y
−2	−3
0	1
2	5

$2y - 4x = 2$

x	y
−2	−3
0	1
2	5

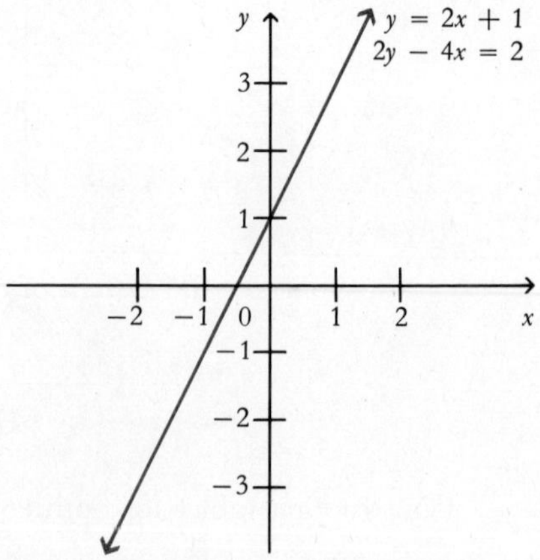

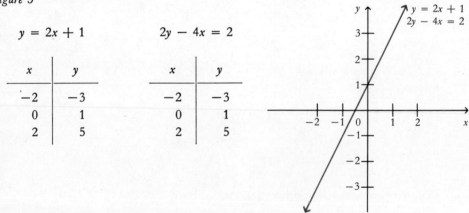

Hence, this system does not have a unique solution. That is, every ordered pair that satisfies one equation also satisfies the other equation.

When systems of two linear equations in two variables are graphed on the same coordinate plane, three general relationships between the two lines are evident: Either they are a pair of intersecting lines (Figure 1); or they are parallel lines (Figure 2); or they coincide (Figure 3). Hence, we have the following possibilities:

1 If the two lines intersect at one point (Figure 1), then we call the system *independent*.

2 If the two lines are parallel (Figure 2), they have no points in common. Hence, the solution set is the empty set, and the system is called *inconsistent.*

3 If the two lines coincide (Figure 3), then the coordinates of every point on the first line are solutions of the given system, and all points on one line are points on the second line. In this case, we say that the system is *dependent.*

EXAMPLES

In each of the following examples, graph the equations of each system on the same coordinate system, decide whether or not the linear system has a solution, and then find the solution.

1 $x + y = 8$
$x - y = 2$

SOLUTION. We construct the following tables and graph both equations on the same coordinate system:

Figure 4

$x + y = 8$

x	y
0	8
4	4
8	0

$x - y = 2$

x	y
2	0
4	2
6	4

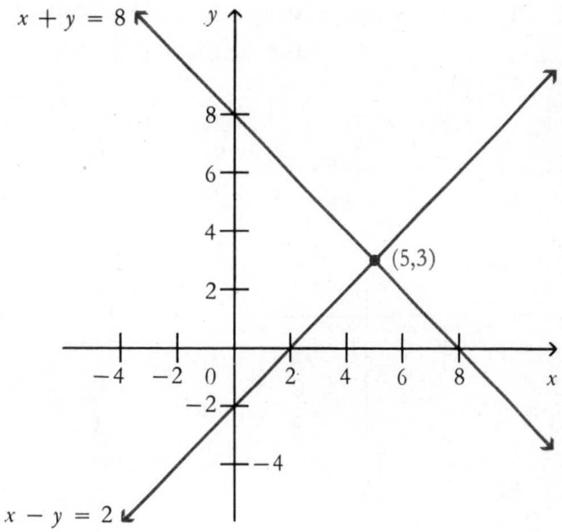

From the graph (Figure 4), we notice that the point of intersection is (5,3), and so the solution set is $\{(5,3)\}$.

2 $3x + 2y = 11$
$-2x + y = 2$

SOLUTION. Graphing the two equations on the same coordinate axes (Figure 5), we find the solution set from their point of intersection to be $\{(1,4)\}$.

Figure 5

$3x + 2y = 11$

x	y
0	$\frac{11}{2}$
1	4
$\frac{11}{3}$	0

$-2x + y = 2$

x	y
−1	0
0	2
1	4

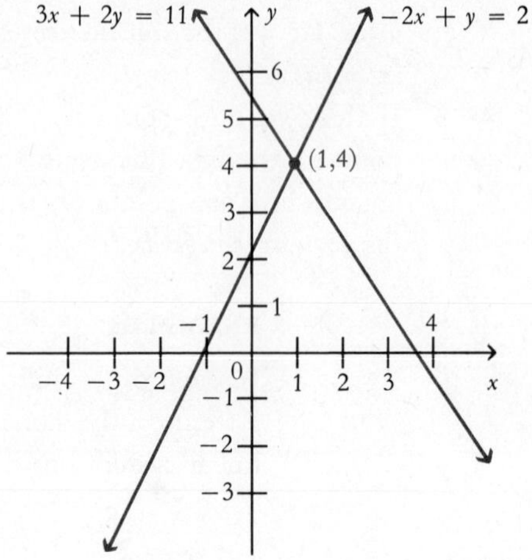

3 $2x - 3y = 1$
$5x + 2y = 12$

SOLUTION. Graphing these two equations (Figure 6), we find the solution set to be $\{(2,1)\}$.

Figure 6

$2x - 3y = 1$

x	y
0	$-\frac{1}{3}$
$\frac{1}{2}$	0
1	$\frac{1}{3}$

$5x + 2y = 12$

x	y
0	6
$\frac{12}{5}$	0
1	$\frac{7}{2}$

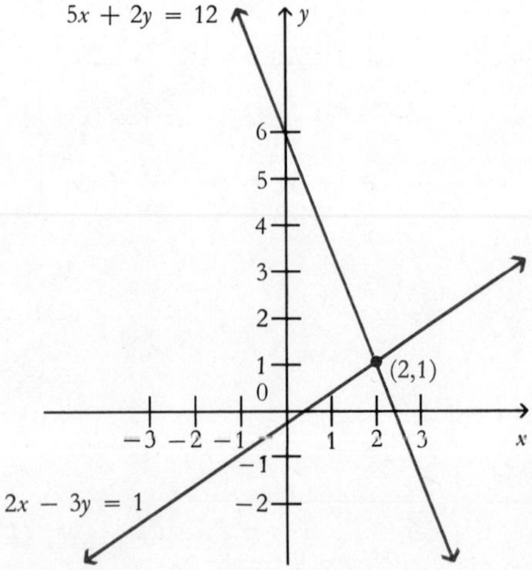

4 $2x + 3y = 12$
$x - 2y = -1$

SOLUTION. Graphing these two equations (Figure 7), we find the solution set to be $\{(3,2)\}$.

Figure 7

$2x + 3y = 12$

x	y
0	4
3	2
6	0

$x - 2y = -1$

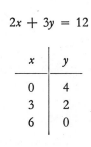

x	y
-1	0
0	$\frac{1}{2}$
1	1

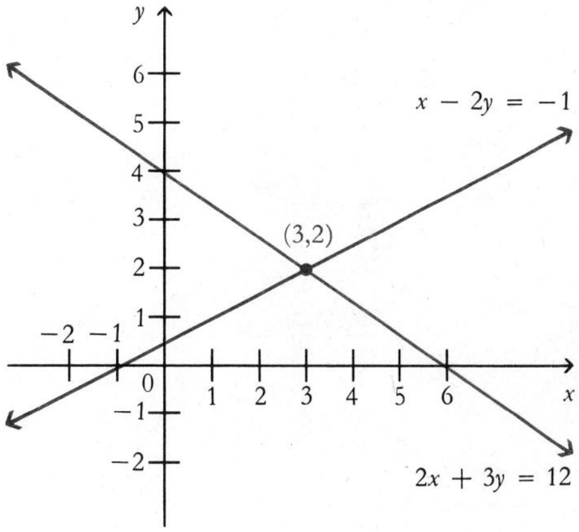

5 $y = -x + 3$
$y = -x + \frac{1}{5}$

SOLUTION. Graphing these two equations (Figure 8), we find that they do not intersect. Hence there is no solution. That is, the system is inconsistent.

Figure 8

$y = -x + 3$

x	y
0	3
1	2
2	1

$y = -x + \frac{1}{5}$

x	y
0	$\frac{1}{5}$
1	$-\frac{4}{5}$
2	$-\frac{9}{5}$

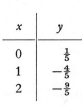

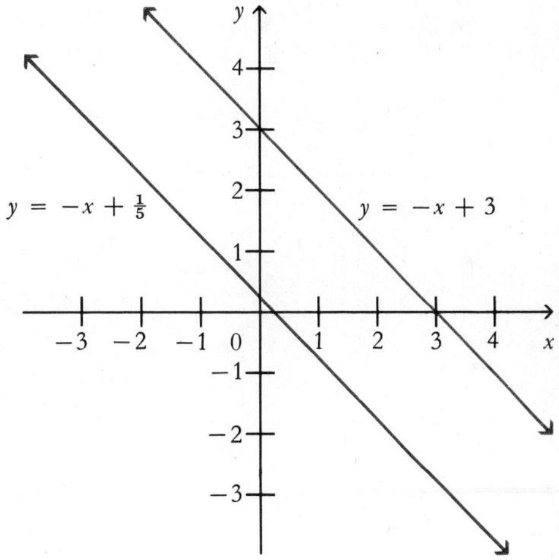

6 $2x + 3y = 1$
$2x + 3y = 5$

SOLUTION. Graphing these two equations (Figure 9), we find that they do not intersect. Hence, there is no solution. That is, the system is inconsistent.

Figure 9

$2x + 3y = 1$

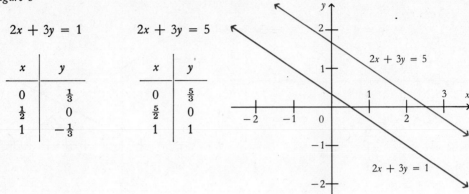

x	y
0	$\frac{1}{3}$
$\frac{1}{2}$	0
1	$-\frac{1}{3}$

$2x + 3y = 5$

x	y
0	$\frac{5}{3}$
$\frac{5}{2}$	0
1	1

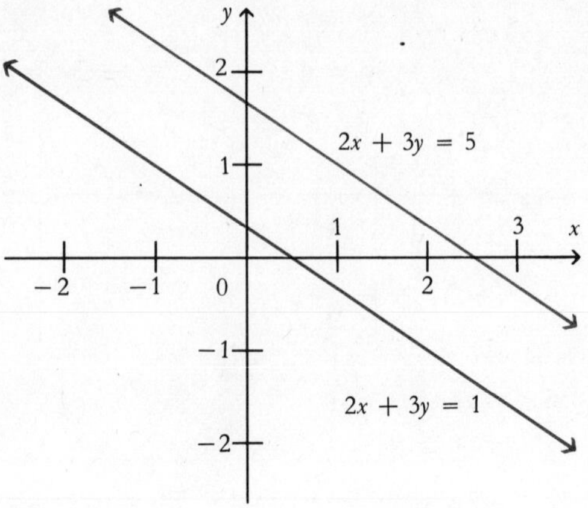

7 $\begin{aligned} 4x - 7y &= 2 \\ -8x + 14y &= -4 \end{aligned}$

SOLUTION. Graphing these two equations (Figure 10), we find that they coincide. That is, the system is dependent and there are an infinite number of solutions.

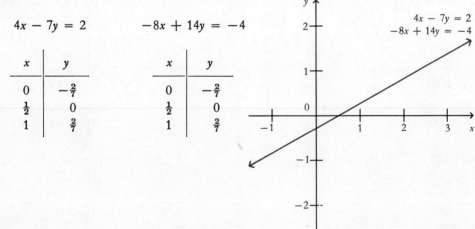

Figure 10

$4x - 7y = 2$

x	y
0	$-\frac{2}{7}$
$\frac{1}{2}$	0
1	$\frac{2}{7}$

$-8x + 14y = -4$

x	y
0	$-\frac{2}{7}$
$\frac{1}{2}$	0
1	$\frac{2}{7}$

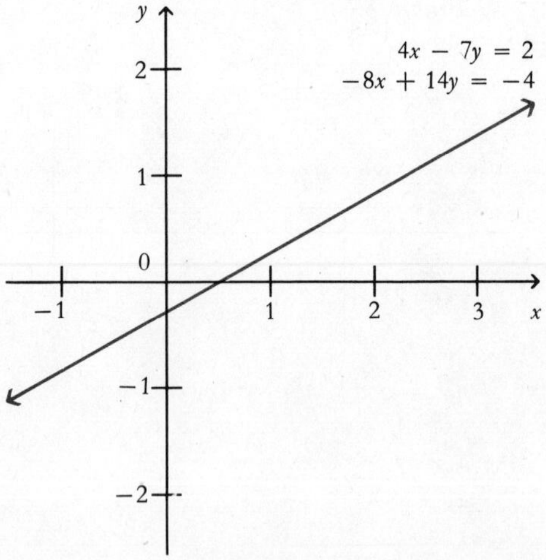

PROBLEM SET 8.2

In problems 1–22, graph the equations of each system on the same coordinate system, decide whether or not the linear system has a solution, and then find the solution set.

1 $\begin{aligned} x - 3y &= 3 \\ x + y &= 7 \end{aligned}$

2 $x - y = -2$
$x + y = 10$

3 $x + 2y = 12$
$x + y = 21$

4 $x - 2y = 2$
$2x + 3y = 11$

5 $x - 2y = -1$
$-x + 2y = 3$

6 $3x - y = 4$
$-3x + y = 1$

7 $-2x + y = 1$
$4x - 2y = -2$

8 $\frac{x}{2} + \frac{y}{3} = 1$
$3x + 2y = 6$

9 $4x - y = 5$
$x + 2y = 8$

10 $3x + 2y = 13$
$2x - 3y = 0$

11 $5x + 6y = 1$
$-3x - 2y = 1$

12 $\frac{x}{5} + \frac{y}{6} = 2$
$2x + y = 16$

13 $y = 2x - 1$
$y = 2x + 1$

14 $2y - x = 3$
$4y - 2x = 5$

15 $x + y = 5$
$x - y = 1$

16 $4x - y = -5$
$-3x + y = 3$

17 $2x + y = 23$
$2x - y = 5$

18 $3x - 5y = 0$
$2x + 5y = 25$

19 $2x - 3y = 6$
$x + y = -7$

20 $x + y = -2$
$-x + y = 6$

21 $x + y = 10$
$x - y = 0$

22 $y = \frac{1}{2}x + 3$
$y = \frac{1}{2}x - 3$

8.3 Solving Systems of Equations by the Addition or Subtraction Method

The solution of systems of equations by graphing is not an accurate method. The results of this method depend on how accurately the graphs were constructed. In order to obtain exact solutions, we must use algebraic methods. One such method, which involves the addition property of the equality relation, is called the *addition or subtraction method.* For example, consider the following system:

$$x + y = 8$$
$$x - y = 2$$

Adding the corresponding members of the given equations, we have

$$x + y + x - y = 8 + 2$$
$$2x = 10$$
$$x = 5$$

Substituting 5 for x in the first of the given equations, we obtain $5 + y = 8$ or $y = 3$.

Check: Substituting 5 for x and 3 for y in both original equations, we obtain $5 + 3 = 8$ and $5 - 3 = 2$. Therefore, the solution set is $\{(5,3)\}$.

EXAMPLES

Solve each of the following systems by the addition or subtraction method.

1 $x + y = 8$
$x - y = 4$

SOLUTION. Adding the corresponding members of the equations, we get

$$x + y + x - y = 8 + 4,$$

so that

$$2x = 12$$
$$x = 6$$

Substituting 6 for x in the first equation, we obtain

$$6 + y = 8$$
$$y = 2$$

Hence, the solution set is $\{(6,2)\}$.

2 $2x + y = 7$
$2x - y = 13$

SOLUTION. Adding the corresponding members of the equations, we get

$$2x + y + 2x - y = 7 + 13,$$

so that

$$4x = 20$$
$$x = 5$$

Therefore,

$$2(5) + y = 7$$
$$10 + y = 7$$
$$y = -3$$

Hence, the solution set is $\{(5,-3)\}$

3 $6x - 5y = 8$
$3x + 10y = -6$

SOLUTION. Neither addition nor subtraction will eliminate one variable from the equations as they stand. However, if we multiply the first equation by 2, we obtain

$$12x - 10y = 16$$
$$3x + 10y = -6$$

Adding the corresponding members of the equations, we get

$$15x = 10$$
$$x = \frac{10}{15} = \frac{2}{3}$$

Therefore,

$$6(\tfrac{2}{3}) - 5y = 8$$
$$4 - 5y = 8$$
$$-5y = 4$$
$$y = -\tfrac{4}{5}$$

Hence, the solution set is $\{(\frac{2}{3}, -\frac{4}{5})\}$.

4 $5x - 3y = 32$
$5x + 7y = -8$

SOLUTION. Subtracting the second equation from the first, we get

$$(5x - 3y) - (5x + 7y) = 32 - (-8),$$

so that

$$-10y = 40$$
$$y = -4$$

Therefore,

$$5x - 3(-4) = 32$$
$$5x + 12 = 32$$
$$5x = 20$$
$$x = 4$$

Hence, the solution set is $\{(4, -4)\}$.

5 $3x + 2y = 8$
$9x - 4y = 9$

SOLUTION. Multiplying the first equation by 2, we get

$$6x + 4y = 16$$
$$9x - 4y = 9$$

Adding the corresponding members of the equations, we obtain

$$15x = 25$$
$$x = \frac{25}{15} = \frac{5}{3}$$

Therefore,

$$\begin{aligned} 3(\tfrac{5}{3}) + 2y &= 8 \\ 5 + 2y &= 8 \\ 2y &= 3 \\ y &= \tfrac{3}{2} \end{aligned}$$

Hence, the solution set is $\{(\frac{5}{3},\frac{3}{2})\}$.

6 $\begin{aligned} 4x + 3y &= 46 \\ 2x - 3y &= 14 \end{aligned}$

SOLUTION. Adding the corresponding members of the equations, we get

$$\begin{aligned} 6x &= 60 \\ x &= 10 \end{aligned}$$

Therefore,

$$\begin{aligned} 2(10) - 3y &= 14 \\ 20 - 3y &= 14 \\ -3y &= -6 \\ y &= 2 \end{aligned}$$

Hence, the solution set is $\{(10,2)\}$.

7 $\begin{aligned} 2x + 3y &= 4 \\ 3x + 5y &= 5 \end{aligned}$

SOLUTION. Multiplying the first equation by 3 and the second equation by 2, we obtain

$$\begin{aligned} 6x + 9y &= 12 \\ 6x + 10y &= 10 \end{aligned}$$

Subtracting the corresponding members of the second equation from those of the first equation, $-y = 2$ or $y = -2$; then $3x + 5(-2) = 5$ or $3x - 10 = 5$, so that $3x = 15$ or $x = 5$. Hence, the solution set is $\{(5,-2)\}$.

PROBLEM SET 8.3

In problems 1–22, solve each system of equations by using the addition or subtraction method.

1 $\begin{aligned} 3x - y &= 6 \\ 2x + y &= 4 \end{aligned}$

2 $x + y = 10$
$x - y = 12$

3 $3x - 2y = 17$
$x + y = 9$

4 $-x + 3y = 9$
$x - y = 1$

5 $x - 2y = 2$
$2x + 3y = 11$

6 $3x + 4y = 29$
$2x + 3y = 21$

7 $3x + 2y = 12$
$x + y = 5$

8 $11x + 3y = 12$
$4x + y = 5$

9 $2x + y = 7$
$x + 2y = 5$

10 $7x - 5y = 8$
$-x + y = -2$

11 $4x - 5y = 10$
$-2x + 3y = 11$

12 $4x + 9y = 10$
$-7x + 6y = 12$

13 $7x + 3y = 4$
$-2x + 5y = 6$

14 $13x + 12y = 50$
$11x - 9y = 4$

15 $x - 5y = -8$
$5x + y = 12$

16 $8x + 10y = 3$
$3x - 5y = 2$

17 $11x + 3y = 14$
$22x - 5y = 17$

18 $\frac{2x}{3} - \frac{3y}{4} = \frac{7}{12}$
$8x + 5y = 9$

19 $-14x + 3y = -5$
$7x - 2y = 1$

20 $$\begin{aligned} -15x + 12y &= 9 \\ 3x - 2y &= 1 \end{aligned}$$

21 $$\begin{aligned} \frac{x}{2} + \frac{y}{5} &= 4 \\ \frac{x}{3} + \frac{y}{5} &= \frac{10}{3} \end{aligned}$$

22 $$\begin{aligned} 2x - 3y &= 4 \\ x + 6y &= 2 \end{aligned}$$

8.4 Solving Systems of Equations by Substitution

Another algebraic method for solving systems of linear equations is called the *method of substitution.* Let us see how to apply this method by considering the following example:

$$\begin{aligned} y &= 2x - 1 \\ y &= -3x + 1 \end{aligned}$$

Geometrically, solving this linear system means finding the coordinates of the point of intersection of the two lines (Figure 1). Algebraically, this means that we are to use the equations to find a pair of numbers (x,y) which satisfy both equations simultaneously.

Figure 1

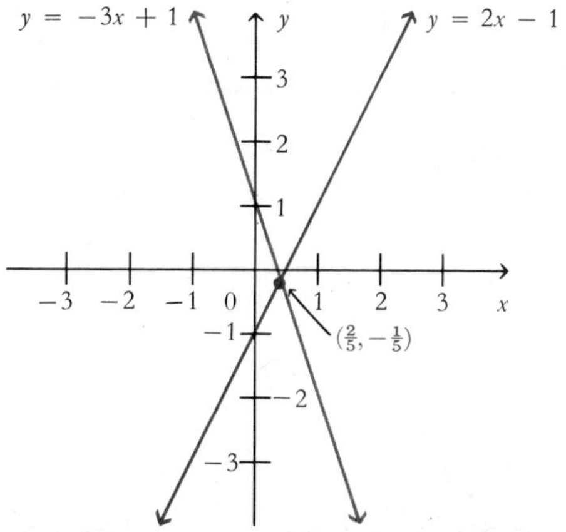

Since we are trying to find one value for x that yields the same y value in both equations, we assume that the variable y represents the same number in each equation, or

$$y = y$$

Now

$$2x - 1 = -3x + 1$$

from which we get

$$5x = 2 \quad \text{or} \quad x = \tfrac{2}{5}$$

Hence,

$$y = 2x - 1 = 2(\tfrac{2}{5}) - 1 = \frac{-1}{5}$$

so that the point of intersection is $\left(\frac{2}{5}, \frac{-1}{5}\right)$, and the solution set is $\{(\frac{2}{5}, -\frac{1}{5})\}$.

Notice that once the value of x was determined, the value of y was computed by using $y = 2x - 1$. If $y = -3x + 1$ had been used, that is, $y = -3(\frac{2}{5}) + 1$, we would have also obtained $y = \frac{-1}{5}\cdot$ This is not surprising because of the initial assumption that the y values were the same in order to get $x = \frac{2}{5}$. Thus, once the common value of x is determined, either equation can be used to compute the corresponding value of y.

EXAMPLES

In examples 1–4, graph the equations of each system on the same coordinate system, decide whether or not the linear system has a solution, and solve the system by substitution. Check the solution.

1 $y = 3x$
$y = 5x + 1$

SOLUTION. From the graph of these equations (Figure 2), we see that the lines intersect. Assuming the y values to be the same, we get

$$\begin{aligned} y &= y \\ 5x + 1 &= 3x \\ 2x &= -1 \\ x &= -\tfrac{1}{2} \end{aligned}$$

and

$$y = 3(-\tfrac{1}{2}) = -\tfrac{3}{2}$$

Figure 2

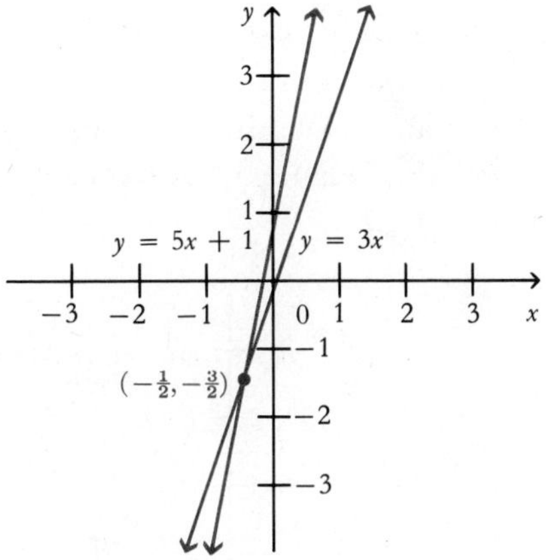

Check:

$y = 3x$ and $y = 5x + 1$

$-\frac{3}{2} \stackrel{?}{=} 3(-\frac{1}{2})$ $\quad -\frac{3}{2} = 5(-\frac{1}{2}) + 1$

$-\frac{3}{2} = -\frac{3}{2}$ $\quad -\frac{3}{2} = -\frac{3}{2}$

Hence, $\{(-\frac{1}{2}, -\frac{3}{2})\}$ is the solution set.

2 $x + y = 3$
$5x + 5y = 1$

SOLUTION. First we graph both equations of the linear system on the same coordinate axes. Note that the lines are parallel and therefore do not intersect, so that the given system does not have a solution (Figure 3). The solution set is the empty set $\emptyset$.

Figure 3

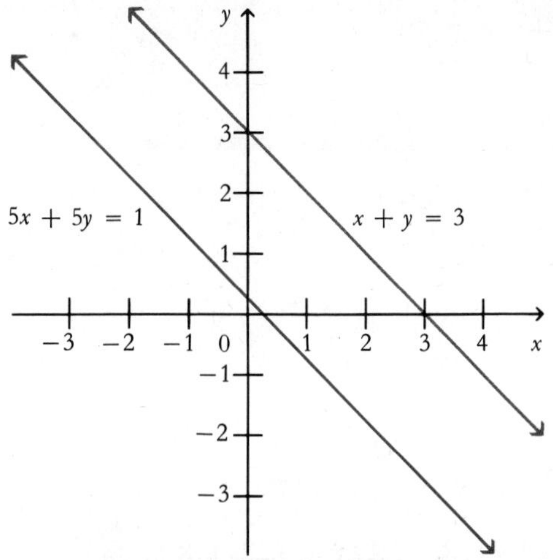

3 $x - 6y = 1$
$3x - 3 = 18y$

SOLUTION. We graph both equations of the linear system on the same coordinate axes. Note that the graphs of the two equations are on the same line (Figure 4). The solution of the system, the

Figure 4

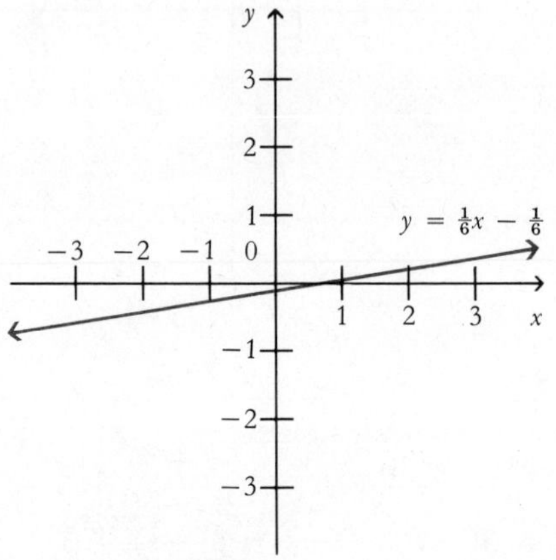

intersection of the lines, is the infinite set $\{(x,y) | y = \frac{1}{6}x - \frac{1}{6}\}$ since all points on this line satisfy the equations of the system.

4 $5x + 2y = 1$
$3x - y = 2$

SOLUTION. The graph of these equations show that a solution exists (Figure 5). The second equation can be rewritten as $y = 3x - 2$, and by substituting this latter equation into the first equation, $5x + 2y = 1$, we get

$$5x + 2(3x - 2) = 1$$
$$5x + 6x - 4 = 1$$
$$11x = 5$$

so that

$$x = \frac{5}{11}$$

Hence, $y = 3(\frac{5}{11}) - 2 = \frac{15}{11} - \frac{22}{11} = -\frac{7}{11}$.

Check:

$$5(\tfrac{5}{11}) + 2(-\tfrac{7}{11}) = \frac{25 - 14}{11} = 1$$

and

$$3(\tfrac{5}{11}) - (-\tfrac{7}{11}) = \frac{15 + 7}{11} = 2$$

Therefore, $\{(\frac{5}{11}, -\frac{7}{11})\}$ is the solution set.

Figure 5

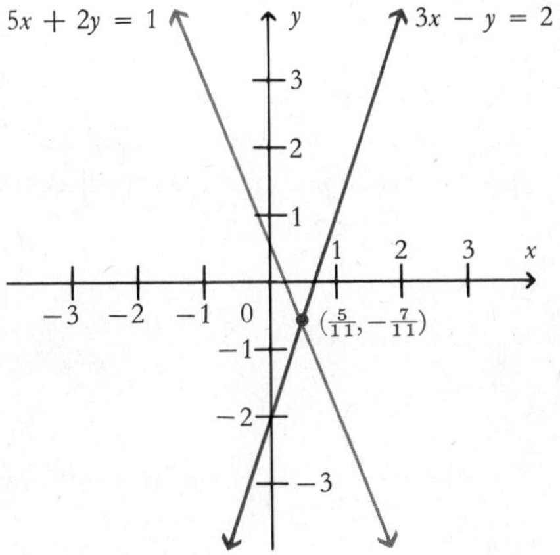

EXAMPLES

In examples 5 and 6, solve the systems by the substitution method and check the results.

5 $3x + 4y = 7$
$x - 4y = 5$

SOLUTION. Solving the equation $x - 4y = 5$ for x, we get $x = 4y + 5$. Substituting this expression for x in the first equation and solving for y, we obtain

$$\begin{aligned} 3(4y + 5) + 4y &= 7 \\ 12y + 15 + 4y &= 7 \\ 16y &= -8 \\ y &= -\tfrac{1}{2} \end{aligned}$$

and

$$\begin{aligned} x &= 4y + 5 \\ x &= 4(-\tfrac{1}{2}) + 5 \\ x &= 3 \end{aligned}$$

Check:

$$3(3) + 4(-\tfrac{1}{2}) \stackrel{?}{=} 7 \qquad 3 - 4(-\tfrac{1}{2}) \stackrel{?}{=} 5$$
$$9 - 2 = 7 \qquad 3 + 2 = 5$$

Hence, the solution set is $\{(3, -\tfrac{1}{2})\}$.

6 $x - y = 3a + b$
$2x + 3y = 2a$

where a and b are constant real numbers.

SOLUTION. Solving for x in the first equation, we get

$$x = y + 3a + b$$

Substituting for x in the second equation, we obtain

$$2(y + 3a + b) + 3y = 2a$$

so that

$$\begin{aligned} 2y + 6a + 2b + 3y &= 2a \\ 5y &= -4a - 2b \\ y &= -\frac{4a + 2b}{5} \end{aligned}$$

so that

$$x = y + 3a + b = -\frac{4a + 2b}{5} + 3a + b = \frac{11a + 3b}{5}$$

Check:

$$\left(\frac{11a + 3b}{5}\right) - \left(-\frac{4a + 2b}{5}\right) = \frac{15a + 5b}{5} = 3a + b$$

$$2\left(\frac{11a + 3b}{5}\right) + 3\left(-\frac{4a + 2b}{5}\right) = \frac{10a + 0b}{5} = 2a$$

Hence, the solution set is $\left\{\left(\frac{11a + 3b}{5}, -\frac{4a + 2b}{5}\right)\right\}$.

PROBLEM SET 8.4

In problems 1–22, graph the equations of each system on the same coordinate system; decide whether or not each linear system has a solution; then solve the system by substitution. Finally, check the solution.

1 $2x - y = 3$
$x + y = 3$

2 $7x + y = 5$
$y - 2x = 3$

3 $-3x + y = -1$
$-2x - y = -4$

4 $x + y + 8 = 0$
$2x - 7y = 5$

5 $x + y = 0$
$x - y = 0$

6 $3y + 2z = 5$
$2y = -3z + 1$

7 $3x + y = 1$
$9x + 3y = -4$

8 $13r + 5s = 2$
$2 - 2s = 6r$

9 $x - y = -7$
$-5x + 5y = 35$

10 $x = 5 - y$
$y = 5 - x$

11 $2x - y = 9$
$5x - 3y = 14$

12 $5x - 2y = 35$
$x + 4y = 25$

13 $4x - 3y = 1$
$3x - 4y = 6$

14 $8 = x + 3y$
$x + 8y = 53$

15 $2x - y = 5$
$x + 2y = 25$

16 $\frac{1}{3}x - \frac{1}{4}y = 2$
$\frac{1}{4}x - \frac{1}{2}y = 7$

17 $\frac{1}{2}x + \frac{1}{3}y = 13$
$-x - y = 5$

18 $4x + 3y = 5a$
$x + 4y = 6b$

19 $3x - 2y = 6a$
$6x - 3y = b$

20 $3x + y = a$
$x - 3y = b$

21 $3ax + 2by = 6$
$2ax - 5by = 7$

22 $4ax + 3by = 1$
$5ax - 2by = 2$

8.5 Word Problems

We have seen how to work problems containing one variable. In this section we will illustrate how word problems that contain more than one unknown quantity can be more easily solved if more than one variable is introduced. However, before the problem can be solved, the number of equations formed must be equal to the number of variables used. Since most story problems are not solved by trial and error methods, the following steps are suggested as a guide:

1 Read the problem carefully, determining what facts are given and precisely what is to be found.
2 Draw diagrams if possible to help interpret the given information and the nature of the solution.
3 Express the given information and the solution in mathematical symbols and/or numbers.
4 Express the problem in the form of equations.
5 Solve the system of equations.
6 Check the solution to see if it satisfies the equations and to see if it makes sense with regard to the story problem.

EXAMPLES

1 The sum of two numbers is 16 and their difference is 4. Find the numbers.

SOLUTION. Let x and y represent the two numbers. Thus, the two equations which express the given conditions are:

$$x + y = 16$$
$$x - y = 4$$

Solving this system by the addition method, we have

$$\begin{aligned} 2x &= 20 \\ x &= 10 \\ 10 + y &= 16 \\ y &= 6 \end{aligned}$$

Hence, the two numbers are 10 and 6.

Check:

$$10 + 6 = 16$$
$$10 - 6 = 4$$

2 Find two numbers such that 4 times the first plus the second equals 16, while 12 times the first minus twice the second equals 2.

SOLUTION. Let x = the first number and y = the second number. The equations are

$$\begin{aligned} 4x + y &= 16 \\ 12x - 2y &= 2 \end{aligned}$$

Solving this system, we have

$$\begin{aligned} 8x + 2y &= 32 \\ 12x - 2y &= 2 \\ 20x &= 34 \\ x &= \tfrac{34}{20} = \tfrac{17}{10} \\ 4(\tfrac{17}{10}) + y &= 16 \\ \tfrac{34}{5} + y &= 16 \\ y &= \tfrac{46}{5} \end{aligned}$$

Check:

$$\begin{aligned} 4(\tfrac{17}{10}) + \tfrac{46}{5} &= \tfrac{34}{5} + \tfrac{46}{5} = \tfrac{80}{5} = 16 \\ 12(\tfrac{17}{10}) - 2(\tfrac{46}{5}) &= \tfrac{102}{5} - \tfrac{92}{5} = \tfrac{10}{5} = 2 \end{aligned}$$

3 Jammie is 1 year more than twice Gus's age. The sum of their ages is 7. How old is each?

SOLUTION. Let x = Jammie's age and y = Gus's age. The equations are

$$\begin{aligned} x &= 2y + 1 \\ x + y &= 7 \end{aligned}$$

Solving this system by the substitution method, we have

$$\begin{aligned} 2y + 1 + y &= 7 \\ 3y &= 6 \\ y &= 2 \\ x &= 2(2) + 1 \\ x &= 5 \end{aligned}$$

Hence, Jammie is 5 years old and Gus is 2 years old.

4 Charlotte has \$1.95 in dimes and quarters. She has 2 more dimes than quarters. How many of each type of coin does she have?

SOLUTION. Let x = number of dimes and y = number of quarters. The equations are

$$0.10x + 0.25y = 1.95$$

so that

$$10x + 25y = 195$$

and

$$x - y = 2$$

Solving this system, we have

$$\begin{aligned} x &= y + 2 \\ 10(y + 2) + 25y &= 195 \\ 10y + 20 + 25y &= 195 \\ 35y &= 175 \\ y &= 5 \end{aligned}$$

Therefore,

$$x = 5 + 2 = 7$$

That is, there are 7 dimes and 5 quarters.

5 A man has 136 bills in singles and fives. Upon counting his money, he finds he has a total of $200. How many bills of each type does he have?

SOLUTION. Let x = number of singles and y = number of fives. The equations are

$$\begin{aligned} x + y &= 136 \\ x + 5y &= 200 \end{aligned}$$

Solving by subtracting the first equation from the second, we get

$$\begin{aligned} 4y &= 64 \\ y &= 16 \\ x + 16 &= 136 \\ x &= 120 \end{aligned}$$

Hence, there are 120 singles and 16 fives.

6 If the sum of the digits of a two-digit number is added to the number, the sum is 49. The units' digit is 5 greater than the tens' digit. Find the number.

SOLUTION. Let x = the tens' digit and y = the units' digit. Thus, the number is $10x + y$. The equations are

$$x + y + 10x + y = 49$$
$$y = x + 5$$
$$11x + 2y = 49$$
$$y = x + 5$$

Solving, we have

$$11x + 2(x + 5) = 49$$
$$11x + 2x + 10 = 49$$
$$13x = 39$$
$$x = 3$$
$$y = 3 + 5 = 8$$

Hence, the number $10x + y$ is $10(3) + 8$, or 38.

7 A farmer sold two grades of wheat, one grade at 57 cents a bushel and another at 68 cents a bushel. How many bushels of each kind did he sell if he received \$1,095.20 for 1,850 bushels?

SOLUTION. Let x = number of bushels of 57-cent wheat and y = number of bushels of 68-cent wheat. The equations are

$$0.57x + 0.68y = 1{,}095.20$$
$$57x + 68y = 109{,}520$$
$$x + y = 1{,}850$$

Solving by the subtraction method, we have

$$57x + 68y = 109{,}520$$
$$57x + 57y = 105{,}450$$
$$11y = 4{,}070$$
$$y = 370$$
$$x + 370 = 1{,}850$$
$$x = 1{,}480$$

Hence, there were 1,480 bushels sold at 57 cents per bushel, and 370 bushels sold at 68 cents per bushel.

8 Roy invested \$40,000, part of it at $4\frac{1}{2}$ percent interest and the rest at 6 percent. The interest from the whole investment is \$2,130. How much did he invest at each rate?

SOLUTION. Let x = amount invested at $4\frac{1}{2}$ percent and y = amount invested at 6 percent. The equations are

$$0.045x + 0.06y = 2{,}130$$

so that

$$45x + 60y = 2{,}130{,}000$$

and

$$x + y = 40{,}000$$

Solving this system, we have

$$\begin{aligned} 45x + 60y &= 2{,}130{,}000 \\ 45x + 45y &= 1{,}800{,}000 \\ 15y &= 330{,}000 \\ y &= 22{,}000 \\ x + 22{,}000 &= 40{,}000 \\ x &= 18{,}000 \end{aligned}$$

Hence, he invested \$18,000 at $4\frac{1}{2}$ percent and \$22,000 at 6 percent.

PROBLEM SET 8.5

In problems 1–40, use a system of equations to solve the given word problem.

1 The sum of two numbers is 12. If one of the numbers is multiplied by 5 and the other is multiplied by 8, the sum of the products is 75. Find the numbers.

2 Find two numbers such that twice the first plus 5 times the second is 20 and 4 times the first less 3 times the second is 14.

3 In a number the tens' digit is 5 greater than the units' digit and the number is 63 greater than the sum of its digits. Find the number.

4 The sum of the digits of a two-place number is 13 and the number is 1 less than 10 times the difference of the digits, the units' digit being the greater. Find the number.

5 If the digits of a two-place number are interchanged, the number is increased by 27. The sum of the digits is 11. What is the number?

6 To a two-place number is added 3 times the tens' digit and twice the units' digit, giving 51 as a sum. The sum of the digits is 7. Find the number.

7 From a two-place number 5 times the tens' digit and 3 times the units' digit are subtracted, leaving a remainder of 32. The tens' digit less the units' digit is 4. Find the number.

8 If the sum of the digits of a number is subtracted from the number, the remainder is 54. Find all that you can about the number.

9 In a two-place number the sum of the digits is 12. If 6 is subtracted from the tens' digit and 5 added to the units' digit, the result is 38. Find the number.

10 The sum of the digits of a certain two-digit number is equal to 15. The number with the digits reversed is 27 less than the original number. Find the number.

11 The tens' digit of a certain two-digit number exceeds the units' digit by 4. The number itself is 7 times the sum of the digits. Find the number.

12 Of two partners, A and B, A has $\frac{2}{3}$ as much invested in a business as B. If the profits from the business are \$3,600, how much should each of the partners receive?

13 A man takes 5 hours and 20 minutes to row 6 miles up a river and back. If he can row 3 miles downstream in the same time that he can row 1 mile upstream, find the rate at which he rows and the rate of the current.

14 Separate 72 into two numbers so that the larger number will be 8 more than the smaller.

15 The difference of two numbers is 20 and the larger of the two is six times the smaller. Find the numbers.

16 At the first home basketball game, a total of 470 student tickets and 280 adult tickets were sold. At the next game, 320 adult tickets and 500 student tickets were sold. If the receipts were \$585 and \$650 respectively, find the cost of each type of ticket.

17 In a certain rectangle the perimeter is 84 feet. If the length is increased by 9 feet and the width is halved, the perimeter becomes 12 feet larger. Find the original dimensions of the figure.

18 The sum of the digits of a two-digit number is 14. By interchanging the tens digit and the units digit, the number is increased by 18. Find the original number.

19 A man bought 13 stamps at one price and 32 stamps at a higher price, and paid \$2.63. If he had bought twice as many of the lower priced stamps and 6 fewer of the more expensive stamps, he would have spent 3 cents less. What was the price of each type of stamp?

20 The rowing team of a certain school can row downstream at the rate of 7 miles an hour and upstream at the rate of 3 miles an hour. Find the team's rate in still water and the rate of the current.

21 A man invested a total of \$4,000 in securities. Part was invested at 5 percent, and the rest was invested at 4 percent. His annual income from both investments was \$180. What amount was invested at each rate?

22 Wilma has 11 coins consisting of nickels and dimes. She has 3 more dimes than nickels. How many of each does she have if the total amount is 90¢?

23 Joan is 3 years older than Doreen. Four years ago she was twice as old as Doreen. How old is each now?

24 Two cars start from the same point, at the same time, and travel in opposite directions. One car travels at a rate that is 10 miles per hour faster than the other. After 3 hours, they are 330 miles apart. How fast is each traveling?

25 A car that traveled at an average rate of speed of 40 miles per hour went 35 miles farther than one that went 55 miles per hour. The slower car traveled 2 hours longer than the faster one. How long did each car travel?

26 The sum of the digits of a two-digit number is 11. The ones' digit is 1 less than twice the tens' digit. Find the number.

27 The sum of the digits of a two-digit number is 7. If the digits are reversed, the new number thus formed is 9 more than the original number. Find the number.

28 The perimeter of an isosceles triangle is 26 inches. The base is 2 inches longer than one of the equal sides of the triangle. Find the length of each of the sides of the triangle.

29 Two cars start from points 400 miles apart and travel toward each other. They meet after 5 hours. Find the average rate of speed of each car if one travels 10 miles per hour faster than the other.

30 A committee is to be formed consisting of 21 students with the stipulation that the number of boys on the committee must be more than twice as many as the number of girls. How many of each may be on the committee?

31 Find two numbers whose sum is 2 such that the larger number is 2 more than the smaller number.

32 A storekeeper made up a 5-pound box of candy to sell at 80 cents a pound. He did this by mixing two varieties that sell at 60 cents and 90 cents a pound respectively. How much of each of these kinds did he use to make his mixture?

33 The sum of the length and the width of a rectangle is 22 feet. If the length is increased by 4 feet and the width decreased by 2 feet, its area is increased by 8 square feet. What are the original dimensions?

34 A rectangular field is 40 rods longer than it is wide and the length of the fence around it is 560 rods. What are the dimensions of the field?

35 Find two numbers whose sum is 64, such that the smaller divided by their difference gives $\frac{1}{2}$ as a quotient.

36 In a collection of dimes and quarters there are 51 more dimes than quarters, and the value of the collection is \$10.70. What is the number of each kind of coin? Solve by using one unknown and also by using two unknowns.

37 Of an investment of \$125,000, part yields 5 percent interest and part yields 7 percent. How much is invested at each rate if the total income is \$7,050?

38 A merchant bought 800 pounds of tea for \$390. For part of it he paid 45 cents a pound and for the rest 55 cents a pound. How much did he buy at each price?

39 How much cream containing 22 percent butterfat and milk containing 4 percent butterfat must be mixed to make 40 gallons of cream containing 20 percent butterfat?

40 A chemist has in his laboratory the same acid in two strengths. Six parts of the first mixed with four parts of the second gives a mixture 86 percent pure, and four parts of the first mixed with six parts of the second gives a mixture 84 percent pure. What is the percent of purity of each acid?

8.6 Systems of Linear Inequalities

We discussed the graphical solutions of systems of linear equations in Section 8.2, and we discussed the graphing of linear inequalities in two variables in Chapter 7. In this section we will discuss the graphical method for obtaining the solution sets of systems of linear inequalities such as:

$$\begin{aligned} x + y &> 3 \\ 3x - y &> 6 \end{aligned}$$

First, draw the graphs of the equation $x + y = 3$ and $3x - y = 6$ on the same axes (Figure 1). The graph of the solution set of $x + y > 3$ consists of all points in the plane above the graph of $x + y = 3$. Similarly the graph of the solution set $3x - y > 6$ or $y < 3x - 6$ consists

Figure 1

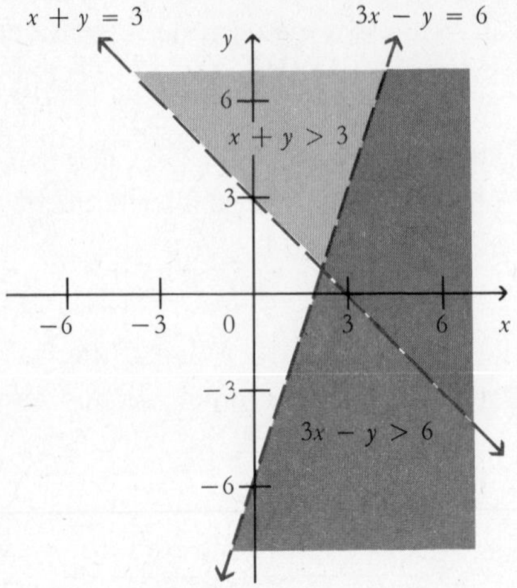

of all points in the plane below the graph of $3x - y = 6$. Therefore, the graph of the solution set of the system of inequalities is the middle colored region which contains all points with coordinates satisfying both inequalities. The boundary lines are broken to indicate that they are not part of the graph.

EXAMPLES

Describe the solution sets of the given systems of linear inequalities.

1 $x + y \geq 1$
$x - y \geq 1$

SOLUTION. First, draw the graphs of the equations $x + y = 1$ and $x - y = 1$ on the same coordinate axes (Figure 2). The graph of

Figure 2

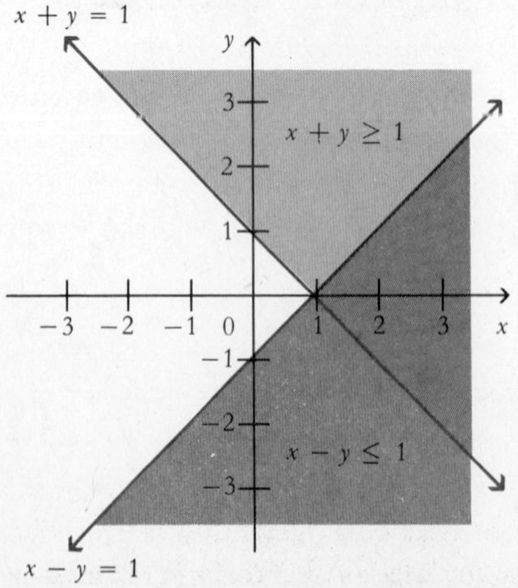

the solution set of $x + y \geq 1$ consists of all points in the plane above the graph of $x + y = 1$, including the line $x + y = 1$. Similarly, the graph of the solution set of $x - y \geq 1$ consists of all points in the plane below the graph of $x - y = 1$, including the line $x - y = 1$. Therefore, the graph of the solution sets of the system of inequalities is the middle colored region which contains all points with coordinates satisfying both inequalities.

2 $y - 2x \geq 0$
$y - 2 < 0$

SOLUTION. First, draw the graphs of $y = 2x$ and $y = 2$. The graph of $y - 2x \geq 0$ consists of all points on or above the line $y = 2x$. The graph of $y - 2 < 0$ consists of all points that lie below the line $y = 2$. The solution set of the system consists of all points that satisfy both conditions, as illustrated by the colored region (Figure 3).

Figure 3

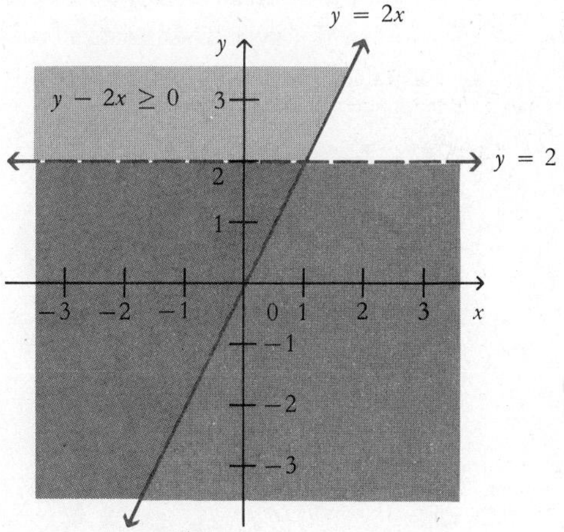

3 $y \leq 2 - x$
$y \geq 3 + x$

SOLUTION. The solution of $y \leq 2 - x$ is the set of points that lie on or below the line $y = 2 - x$, and the solution of $y \geq 3 + x$ is the set of points that lie on or above the line $y = 3 + x$. The solution set of the system consists of all points that satisfy both conditions, as indicated by the middle colored region (Figure 4).

Figure 4

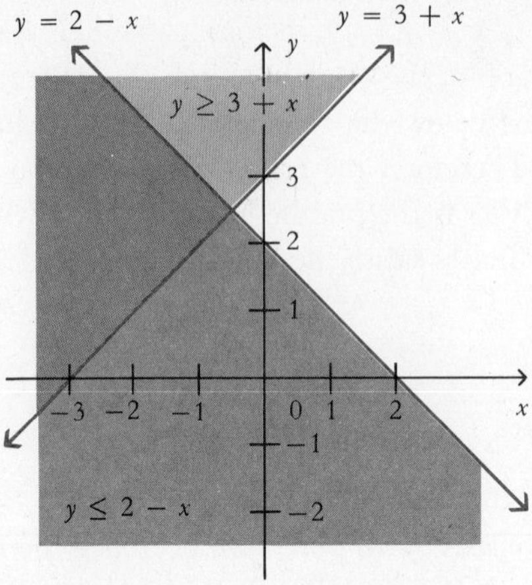

4 $x + 2y \geq 3$
$2y \leq 5 - x$

SOLUTION. The solution of $x + 2y \geq 3$ consists of all points that lie on or above the line $x + 2y = 3$, and the solution of $2y \leq 5 - x$ consists of all points that lie on or below the line $2y = 5 - x$. The solution of the system consists of all points that satisfy both conditions, as indicated by the middle colored region (Figure 5).

Figure 5

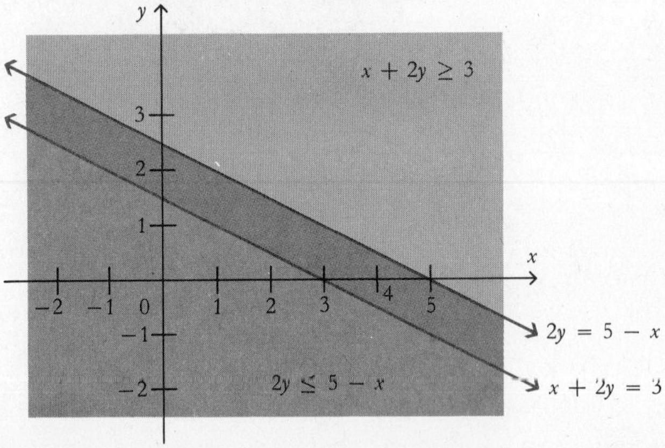

PROBLEM SET 8.6

In problems 1–16, describe the simultaneous solution sets of the given pairs of linear inequalities. Shade the region on a graph.

1 $x < 2$
$y \geq 1$

2 $x < 2$
$y \geq -3$

3 $x \geq -1$
$y \leq 3$

4 $3x \leq 6$
$-2y \leq 4$

5 $x + y \leq 2$
$y - 1 > 2x$

6 $y - x \leq -1$
$3y - x > x$

7 $y \geq x$
$y \leq 2$

8 $y \leq x$
$y \geq -1$

9 $x + y \leq 3$
$x - y \geq 3$

10 $x + y > 5$
$x - y < 9$

11 $x + 2y < 12$
$x - y > 6$

12 $x + 2y > 12$
$3x - y < 1$

13 $3x + y \geq 6$
$x - y \geq 1$

14 $2x - y \leq -2$
$x - 2y \geq -2$

15 $2x - 3y \leq -3$
$5x - 2y \geq 9$

16 $y \leq 2x + 4$
$y \geq 3 - 2x$

REVIEW PROBLEM SET

In problems 1–6, show that the given ordered pair is the solution of the given systems.

1 (2,3)
$$4x - y = 5$$
$$-3x + 2y = 0$$

2 (−4,7)
$$5x + 2y = -6$$
$$-8x + 4y = -4$$

3 $(\frac{1}{2}, -\frac{3}{2})$ $\quad 4x + 8y = -10$
$5x - 3y = 7$

4 $(5,9)$ $\quad x + y = 14$
$3x - y = 6$

5 $(14,3)$ $\quad x + 2y = 20$
$x + y = 17$

6 $(30,40)$ $\quad 4x + 3y = 240$
$-x + y = 10$

In problems 7–24, solve the systems of linear equations by the graphical method.

7 $2x + 3y = 8$
$3x - y = 1$

8 $2x + 3y = 8$
$2x + y = 6$

9 $-x + 2y = 3$
$3x - y = -4$

10 $3x + 2y = 6$
$7x + 2y = 10$

11 $x + y = 5$
$x - y = 31$

12 $3x + y = 4$
$3x - y = -10$

13 $2x + y = 13$
$3x + y = 17$

14 $6x - y = 22$
$2x + y = 2$

15 $x + 4y = 10$
$x + 2y = 4$

16 $7x + 3y = 40$
$x + 3y = 4$

17 $2x + 3y = -3$
$3x - 4y = 38$

18 $4x - 7y = 7$
$6x - 5y = 27$

19 $15x + 17y = 32$
$8x - 19y = -11$

20 $6x - 9y = 11$
$8x - 11y = 3$

21 $13x + 20y = 79$
$-2x + 10y = 14$

22 $2x - 8y = 7$
$3x + 2y = 9$

23 $3x - 4y = -19$
$2x + y = 2$

24 $4x + 3y = 12$
$3x + y = 7$

In problems 25–32, solve the systems of linear equations by the substitution method.

25 $4x + 3y = 10$
$2x + y = 4$

26 $x + 3y = 5$
$3x - y = 5$

27 $x - 5 = 2y$
$3 - y = x$

28 $x + y = 8$
$x - y = 4$

29 $y = 3x + 7$
$y = -2x + 5$

30 $6x - 5y = 8$
$3x + 10y = -5$

31 $2x + 3y = 4$
$5x - 2y = 8$

32 $2x + y = 4$
$4x + 5y = 2$

In problems 33–44, solve each system of linear equations by the addition or subtraction method.

33 $5x - y = 7$
$-x + y = 5$

34 $3x - 2y = 9$
$5x + 2y = 7$

35 $11x - 7y = 5$
$14x - 7y = 11$

36 $x - 3y = 7$
$x + y = -1$

37 $2x - 5y = -7$
$3x - 5y = 2$

38 $x + 2y = 3$
$2x + y = 5$

39 $3x + 5y = 9$
$-2x + 3y = 13$

40 $4x + y = 17$
$3x - 5y = -10$

41 $4x + 3y = 12$
$3x - 2y = 8$

42 $7x - 2y = 9$
$2x + 5y = 7$

43 $x - 2y = 4$
$2x + y = 2$

44 $x - y = -3$
$x + y = 2$

In problems 45–50, find the solution of the systems of linear inequalities by graphing.

45 $x + y \leq 3$
$-x + y \geq 2$

46 $x - 2y + 4 > 0$
$2x + y - 2 > 0$

47 $y > x - 2$
$y < x - 3$

48 $x - 2y + 4 > 0$
$2x + y < 2$

49 $2x + 3y \leq 5$
$3x - 2y \geq 4$

50 $x - y + 3 \geq 0$
$x + y - 2 \leq 0$

In problems 51–56, solve by using systems of linear equations.

51 The sum of two numbers is 33 and their difference is 7. Find the numbers.

52 The sum of Tom's and John's ages is 27. Twice Tom's age is 15 greater than John's age. Find their ages.

53 Joe invests $10,000 in two different banks. If one pays an annual interest rate of 4 percent and the other pays 4.5 percent, how much was invested in each bank if the interest earned the first year was $430.

54 A two-digit number is 4 times the sum of its digits. If the number is multiplied by 2 and decreased by 12, the result will be the number with its digits interchanged. Find the number.

55 A college mailed 120 letters, some requiring 8 cents postage and the rest requiring 10 cents. If the total bill is $11.20, find the number sent at each rate.

56 In a special showing at a certain movie theater, admission tickets are 65 cents for adults and 25 cents for children. If the receipts from 740 tickets are $385, how many tickets of each kind were sold?

CHAPTER 9

Roots and Radical Expressions

9 ROOTS AND RADICAL EXPRESSIONS

Introduction

The laws of exponents, where the powers are positive integers, were studied in Chapter 2. In this chapter, we will extend the idea of a power and give a meaning to such expressions as $\sqrt[n]{a}$, called the "nth root of a." One consequence of this extension is that we obtain properties of roots similar to those of exponents.

9.1 Roots

Consider the two numbers 3 and 9. We know from Chapter 2 that these two numbers are related by the equation $3^2 = 9$. We say that 9 is the second power of 3 or 3 is the second root (*square root*) of 9. Similarly, in the equation $3^3 = 27$, we say that 3 is the third root (*cube root*) of 27. In general, if a is a real number and n is a positive integer greater than 1, then the number x, such that $x^n = a$, is called the nth *root of* a. To indicate the nth root of a, we use the expression $\sqrt[n]{a}$. In the expression $\sqrt[n]{a}$ the symbol $\sqrt[n]{}$ is called a *radical*, n is called the *index*, and a is called the *radicand*. Since n is a positive integer greater than or equal to 2, we only show the index for those values of n greater than 2. Thus, $\sqrt{4} = 2$ since $4 = 2^2$, $\sqrt[3]{8} = 2$ since $8 = 2^3$, and $\sqrt[5]{-32} = -2$ since $-32 = (-2)^5$.

When we square a positive or negative number, we get a positive number. For example, $(3)^2 = 9$ and $(-3)^2 = 9$. Thus, every positive number has two square roots, one positive and the other negative.

We use the expression $\sqrt{9}$ to indicate the positive root 3 (the principal square root), $-\sqrt{9}$ to indicate the negative root -3, and $\pm\sqrt{9}$ to represent both roots. (Zero, however, has only one square root, zero.) For example, $\sqrt{16} = 4$, while $-\sqrt{16} = -4$ and $\pm\sqrt{16} = \pm 4$. In general, $\sqrt[n]{a}$ represents the principal nth root of a, and is written as $\sqrt[n]{a}$.

If $a > 0$ and n is any positive integer, then $\sqrt[n]{a}$ is positive; for example, $\sqrt[3]{64} = 4$. Here $a = 64$ and $n = 3$.

If $a < 0$ and n is any positive odd integer, then $\sqrt[n]{a}$ is negative; for example, $\sqrt[3]{-125} = -5$. Here $a = -125$ and $n = 3$.

If $a < 0$ and n is any positive even integer, then $\sqrt[n]{a}$ is not defined in the real numbers. For example, $\sqrt{-4}$ is not defined. Here $a = -4$ and $n = 2$.

EXAMPLES

In examples 1–4, change each exponential equation to radical form.

1 $5^2 = 25$

SOLUTION

$\sqrt{25} = 5$

2 $12^2 = 144$

SOLUTION

$\sqrt{144} = 12$

3 $5^3 = 125$

SOLUTION

$\sqrt[3]{125} - 5$

4 $3^4 = 81$

SOLUTION

$\sqrt[4]{81} = 3$

In examples 5–13, compute the value of each expression. Assume all variables to be positive.

5 $\sqrt{121}$

SOLUTION

$\sqrt{121} = 11 \qquad \text{since } 11^2 = 121$

6 $\sqrt{0.01}$

SOLUTION

$\sqrt{0.01} = 0.1 \qquad \text{since } (0.1)^2 = 0.01$

7 $\sqrt{0.49}$

SOLUTION

$\sqrt{0.49} = 0.7 \qquad \text{since } (0.7)^2 = 0.49$

8 $\sqrt[4]{-16}$

SOLUTION. Since the index is even and the radicand is negative, $\sqrt[4]{-16}$ is not defined as a real number.

9 $\sqrt[3]{-64}$

SOLUTION

$\sqrt[3]{-64} = -4 \qquad \text{since } (-4)^3 = -64$

10 $\sqrt[3]{125}$

SOLUTION

$\sqrt[3]{125} = 5 \qquad \text{since } 5^3 = 125$

11 $\sqrt[3]{216}$

SOLUTION

$\sqrt[3]{216} = 6 \qquad \text{since } 6^3 = 216$

12 $\sqrt[4]{x^8}$

SOLUTION

$\sqrt[4]{x^8} = \sqrt[4]{(x^2)^4} = x^2 \qquad \text{since } (x^2)^4 = x^8$

13 $\sqrt[3]{8x^6}$

SOLUTION

$\sqrt[3]{8x^6} = \sqrt[3]{(2x^2)^3} = 2x^2 \qquad \text{since } (2x^2)^3 = 2^3(x^2)^3 = 8x^6$

PROBLEM SET 9.1

In problems 1–10, change each equation from exponential form to radical form.

1 $2^4 = 16$

2 $(-3)^3 = -27$

3 $3^4 = 81$

4 $5^3 = 125$

5 $x^4 = 7$

6 $13^5 = x$

7 $x^3 = 64$

8 $(2x)^6 = 3y$

9 $(-3)^5 = -243$

10 $(5x)^4 = 2y$

In problems 11–37, compute the value of each expression. Assume all variables to be positive.

11 $\sqrt{9}$

12 $\sqrt{25}$

13 $-\sqrt{49}$

14 $\sqrt{121}$

15 $-\sqrt{144}$

16 $-\sqrt{400}$

17 $-\sqrt{81}$

18 $\sqrt{169}$

19 $-\sqrt{441}$

20 $\sqrt{x^6}$

21 $\sqrt{x^{10}}$

22 $\sqrt{16x^2}$

23 $\sqrt{121x^4}$

24 $\sqrt{2.56}$

25 $\sqrt{0.0361}$

26 $\sqrt{0.81}$

27 $\sqrt{0.0289}$

28 $-\sqrt{0.0001}$

29 $\sqrt[3]{-8}$

30 $\sqrt[3]{512}$

31 $\sqrt[3]{-27}$

32 $\sqrt[3]{-0.125}$

33 $\sqrt[3]{-1331}$

34 $-\sqrt[3]{-x^3}$

35 $\sqrt[4]{16}$

36 $\sqrt[4]{81x^4}$

37 $\sqrt[4]{625x^8}$

9.2 Roots of Products

We know that $\sqrt{625} = 25$, since the number 625 can be written as a product of 25 and 25. Also $\sqrt{25} \cdot \sqrt{25} = 5 \cdot 5 = 25$ so that $\sqrt{625} = \sqrt{25}\sqrt{25}$. But $\sqrt{625}$ can also be written as $\sqrt{25 \cdot 25}$, and so we can conclude that $\sqrt{25 \cdot 25} = \sqrt{25}\sqrt{25}$. Similarly, $\sqrt[3]{64} = 4$, since the number 64 can be written as the product of 8 and 8. Also, $\sqrt[3]{8} \cdot \sqrt[3]{8} = 2 \cdot 2 = 4$ so that $\sqrt[3]{64} = \sqrt[3]{8}\sqrt[3]{8}$. But $\sqrt[3]{64}$ can also be written as $\sqrt[3]{8 \cdot 8}$; hence we can conclude that $\sqrt[3]{8 \cdot 8} = \sqrt[3]{8}\sqrt[3]{8}$. This property can be generalized as follows:

Property 1

If a and b are real numbers and n is a positive integer, then $\sqrt[n]{ab} = \sqrt[n]{a}\sqrt[n]{b}$. For example, $\sqrt{500} = \sqrt{5}\sqrt{100} = 10\sqrt{5}$ and $\sqrt[3]{81} = \sqrt[3]{27 \cdot 3} = \sqrt[3]{27}\sqrt[3]{3} = 3\sqrt[3]{3}$.

EXAMPLES

Use Property 1 to find the simplest form of each of the following expressions. Assume all variables to be positive.

1 $\sqrt{320}$

SOLUTION

$$\sqrt{320} = \sqrt{64 \cdot 5} = \sqrt{64}\sqrt{5} = 8\sqrt{5}$$

2 $\sqrt{396}$

SOLUTION

$$\sqrt{396} = \sqrt{36 \cdot 11} = \sqrt{36}\sqrt{11} = 6\sqrt{11}$$

3 $\sqrt{75}$

SOLUTION

$$\sqrt{75} = \sqrt{25 \cdot 3} = \sqrt{25}\sqrt{3} = 5\sqrt{3}$$

4 $\sqrt{72}$

SOLUTION

$$\sqrt{72} = \sqrt{36 \cdot 2} = \sqrt{36}\sqrt{2} = 6\sqrt{2}$$

5 $\sqrt[3]{-250}$

SOLUTION

$$\sqrt[3]{-250} = \sqrt[3]{(-125)2} = \sqrt[3]{-125}\sqrt[3]{2} = -5\sqrt[3]{2}$$

6 $\sqrt[3]{200}$

SOLUTION

$$\sqrt[3]{200} = \sqrt[3]{8 \cdot 25} = \sqrt[3]{8}\sqrt[3]{25} = 2\sqrt[3]{25}$$

7 $\sqrt[3]{64}$

SOLUTION

$$\sqrt[3]{64} = 4, \text{ since } 4^3 = 64$$

8 $\sqrt[3]{-9}$

SOLUTION

$$\sqrt[3]{-9} = \sqrt[3]{(-1)9} = \sqrt[3]{-1}\sqrt[3]{9} = -\sqrt[3]{9}$$

9 $\sqrt[3]{48}$

SOLUTION

$$\sqrt[3]{48} = \sqrt[3]{8 \cdot 6} = \sqrt[3]{8}\sqrt[3]{6} = 2\sqrt[3]{6}$$

10 $\sqrt[3]{8x^4}$

SOLUTION

$$\sqrt[3]{8x^4} = \sqrt[3]{8x^3 \cdot x} = \sqrt[3]{8x^3}\sqrt[3]{x} = 2x\sqrt[3]{x}$$

11 $\sqrt{9x^3}$

SOLUTION

$$\sqrt{9x^3} = \sqrt{9x^2 \cdot x} = \sqrt{9x^2}\sqrt{x} = 3x\sqrt{x}$$

12 $\sqrt[5]{128x^8}$

SOLUTION

$$\sqrt[5]{128x^8} = \sqrt[5]{32x^5 \cdot 4x^3} = \sqrt[5]{32x^5}\,\sqrt[5]{4x^3} = 2x\sqrt[5]{4x^3}$$

13 $\sqrt[4]{x^3y^5}$

SOLUTION

$$\sqrt[4]{x^3y^5} = \sqrt[4]{y^4 \cdot x^3y} = \sqrt[4]{y^4}\,\sqrt[4]{x^3y} = y\sqrt[4]{x^3y}$$

PROBLEM SET 9.2

In problems 1–34, use Property 1 to find the simplest form of each expression. Assume all variables to be positive.

1 $\sqrt{32}$

2 $\sqrt{18}$

3 $\sqrt{45}$

4 $\sqrt{27}$

5 $\sqrt{160}$

6 $\sqrt{325}$

7 $\sqrt{150}$

8 $\sqrt{125}$

9 $\sqrt{925}$

10 $\sqrt{1,452}$

11 $\sqrt[3]{72}$

12 $\sqrt[3]{108}$

13 $\sqrt[3]{320}$

14 $\sqrt[3]{-56}$

15 $\sqrt[3]{27x^6}$

16 $\sqrt[3]{-875x^3}$

17 $\sqrt[4]{32}$

18 $\sqrt[4]{162}$

19 $\sqrt[5]{224}$

20 $\sqrt[6]{256}$

21 $\sqrt[5]{224x^{10}}$

22 $\sqrt[6]{x^{12}y^{18}}$

23 $\sqrt[6]{x^9y^{12}}$

24 $\sqrt{80x^6}$

25 $\sqrt[4]{32y^8}$

26 $\sqrt{8x^2y^8}$

27 $\sqrt[3]{125x^5}$

28 $\sqrt[7]{-128x^{21}}$

29 $\sqrt[5]{-32x^6y^{12}}$

30 $\sqrt[3]{-8x^3}$

31 $\sqrt[4]{48x^7y^{11}}$

32 $\sqrt[4]{256x^{12}}$

33 $\sqrt[3]{-243z^{10}}$

34 $\sqrt{625x^{15}y^{21}}$

9.3 Roots of Quotients

In finding the square root of $\frac{121}{400}$, we write $\sqrt{\frac{121}{400}} = \frac{11}{20}$ or $\dfrac{\sqrt{121}}{\sqrt{400}} = \dfrac{11}{20}$; hence we have $\sqrt{\dfrac{121}{400}} = \dfrac{\sqrt{121}}{\sqrt{400}}$. Also, since $\sqrt[3]{\frac{8}{27}} = \frac{2}{3}$ and $\dfrac{\sqrt[3]{8}}{\sqrt[3]{27}} = \dfrac{2}{3}$, we have $\sqrt[3]{\dfrac{8}{27}} = \dfrac{\sqrt[3]{8}}{\sqrt[3]{27}}$. In general, we have the following property:

Property 1

If a and b are real numbers and n is a positive integer, then

$$\sqrt[n]{\frac{a}{b}} = \frac{\sqrt[n]{a}}{\sqrt[n]{b}}$$

For example,

$$\sqrt{\frac{15}{25}} = \frac{\sqrt{15}}{\sqrt{25}} = \frac{\sqrt{15}}{5}$$

EXAMPLES

Use Property 1 to simplify each of the following roots. Assume all variables to be positive.

1 $\sqrt{\frac{36}{16}}$

SOLUTION

$$\sqrt{\frac{36}{16}} = \frac{\sqrt{36}}{\sqrt{16}} = \frac{6}{4} = \frac{3}{2}$$

2 $\sqrt{\frac{49}{9}}$

SOLUTION

$$\sqrt{\frac{49}{9}} = \frac{\sqrt{49}}{\sqrt{9}} = \frac{7}{3}$$

3 $\sqrt{\frac{144}{64}}$

SOLUTION

$$\sqrt{\frac{144}{64}} = \frac{\sqrt{144}}{\sqrt{64}} = \frac{12}{8} = \frac{3}{2}$$

4 $\sqrt[3]{\frac{27}{64}}$

SOLUTION

$$\sqrt[3]{\frac{27}{64}} = \frac{\sqrt[3]{27}}{\sqrt[3]{64}} = \frac{3}{4}$$

5 $\sqrt[3]{\frac{125}{27}}$

SOLUTION

$$\sqrt[3]{\frac{125}{27}} = \frac{\sqrt[3]{125}}{\sqrt[3]{27}} = \frac{5}{3}$$

6 $\dfrac{\sqrt{1{,}000}}{\sqrt{40}}$

SOLUTION

$$\frac{\sqrt{1{,}000}}{\sqrt{40}} = \sqrt{\frac{1{,}000}{40}} = \sqrt{25} = 5$$

7 $\dfrac{\sqrt[3]{64}}{\sqrt[3]{8}}$

SOLUTION

$$\frac{\sqrt[3]{64}}{\sqrt[3]{8}} = \sqrt[3]{\frac{64}{8}} = \sqrt[3]{8} = 2$$

8 $\dfrac{\sqrt{3x}}{\sqrt{12}}$

SOLUTION

$$\frac{\sqrt{3x}}{\sqrt{12}} = \sqrt{\frac{3x}{12}} = \sqrt{\frac{x}{4}} = \frac{\sqrt{x}}{2}$$

9 $\sqrt[3]{\dfrac{27x^6}{8x^9}}$

SOLUTION

$$\sqrt[3]{\frac{27x^6}{8x^9}} = \sqrt[3]{\frac{27}{8x^3}} = \frac{\sqrt[3]{27}}{\sqrt[3]{8x^3}} = \frac{3}{2x}$$

10 $\sqrt[5]{\dfrac{32x^{15}}{243x^{10}}}$

SOLUTION

$$\sqrt[5]{\frac{32x^{15}}{243x^{10}}} = \sqrt[5]{\frac{32x^5}{243}} = \frac{\sqrt[5]{32x^5}}{\sqrt[5]{243}} = \frac{2x}{3}$$

PROBLEM SET 9.3

In problems 1–24, use Property 1 to find the simplest form of each expression. Assume all variables to be positive.

1 $\sqrt{\frac{3}{25}}$

2 $\sqrt{\frac{7}{16}}$

3 $\sqrt{\frac{5}{9}}$

4 $\sqrt{\frac{7}{81}}$

5 $\sqrt{\frac{8}{49}}$

6 $\sqrt{\frac{7}{121}}$

7 $\sqrt[3]{-\frac{1}{8}}$

8 $\sqrt[3]{\frac{5}{27}}$

9 $\sqrt[3]{-\frac{27}{64}}$

10 $\sqrt[3]{\frac{8}{125}}$

11 $\sqrt{\frac{x}{y^2}}$

12 $\sqrt[3]{\frac{x}{y^3}}$

13 $\sqrt[6]{\frac{3}{64}}$

14 $\sqrt[4]{\frac{100}{10{,}000}}$

15 $\sqrt[3]{\frac{3}{1{,}000}}$

16 $\sqrt[3]{\frac{-27}{64x^3}}$

17 $\sqrt[3]{\frac{125}{8x^6}}$

18 $\frac{\sqrt[3]{54}}{\sqrt[3]{18}}$

19 $\frac{\sqrt[3]{32}}{\sqrt[3]{4}}$

20 $\sqrt[4]{\frac{64}{81x^8}}$

21 $\frac{\sqrt{324}}{\sqrt{81x^2}}$

22 $\frac{\sqrt{324x^4}}{\sqrt{25x^2}}$

23 $\dfrac{\sqrt[3]{27x^3}}{\sqrt[3]{8x^6}}$

24 $\dfrac{\sqrt{1{,}000x^2}}{\sqrt{40x^2}}$

9.4 Addition and Subtraction of Radicals

Expressions containing radicals can be added and subtracted by using the distributive property. For example, $3\sqrt{2} + 4\sqrt{2} - \sqrt{2} = (3 + 4 - 1)\sqrt{2} = 6\sqrt{2}$, and $3\sqrt{5} + 7\sqrt{5} - 3\sqrt{5} = (3 + 7 - 3)\sqrt{5} = 7\sqrt{5}$. *Similar radicals* are those radicals which can be expressed with the same index and radicand. Thus $\sqrt{50}$ and $\sqrt{72}$ are similar radicals, since $\sqrt{50} = 5\sqrt{2}$ and $\sqrt{72} = 6\sqrt{2}$. On the other hand, $\sqrt{50}$ and $\sqrt{75}$ are not similar radicals, for $\sqrt{50} = 5\sqrt{2}$ whereas $\sqrt{75} = 5\sqrt{3}$. Hence, expressions such as $7\sqrt{3} + 3\sqrt{3}$ and $8\sqrt[4]{x} - 3\sqrt[4]{x}$ can be combined by applying the distributive property of real numbers. Expressions such as $\sqrt{72} + \sqrt{50}$ must be simplified before they can be combined. For example, $\sqrt{72} + \sqrt{50} = 6\sqrt{2} + 5\sqrt{2} = (6 + 5)\sqrt{2} = 11\sqrt{2}$.

EXAMPLES

Simplify each of the following expressions by combining similar terms.

1 $\sqrt{2} + 3\sqrt{2}$

SOLUTION

$$\begin{aligned} \sqrt{2} + 3\sqrt{2} &= 1\sqrt{2} + 3\sqrt{2} \\ &= (1 + 3)\sqrt{2} \\ &= 4\sqrt{2} \end{aligned}$$

2 $\sqrt{8} + \sqrt{32}$

SOLUTION

$$\begin{aligned} \sqrt{8} + \sqrt{32} &= \sqrt{4 \cdot 2} + \sqrt{16 \cdot 2} \\ &= \sqrt{4}\sqrt{2} + \sqrt{16}\sqrt{2} \end{aligned}$$

$$\begin{aligned} &= 2\sqrt{2} + 4\sqrt{2} \\ &= (2 + 4)\sqrt{2} \\ &= 6\sqrt{2} \end{aligned}$$

3 $\sqrt{3} + \sqrt{27} + \sqrt{243}$

SOLUTION

$$\begin{aligned} \sqrt{3} + \sqrt{27} + \sqrt{243} &= \sqrt{3} + \sqrt{9 \cdot 3} + \sqrt{81 \cdot 3} \\ &= \sqrt{3} + \sqrt{9}\sqrt{3} + \sqrt{81}\sqrt{3} \\ &= \sqrt{3} + 3\sqrt{3} + 9\sqrt{3} \\ &= (1 + 3 + 9)\sqrt{3} \\ &= 13\sqrt{3} \end{aligned}$$

4 $4\sqrt{12} + 5\sqrt{8} - \sqrt{50}$

SOLUTION

$$\begin{aligned} 4\sqrt{12} + 5\sqrt{8} - \sqrt{50} &= 4\sqrt{4 \cdot 3} + 5\sqrt{4 \cdot 2} - \sqrt{25 \cdot 2} \\ &= 4\sqrt{4}\sqrt{3} + 5\sqrt{4}\sqrt{2} - \sqrt{25}\sqrt{2} \\ &= 4 \cdot 2\sqrt{3} + 5 \cdot 2\sqrt{2} - 5\sqrt{2} \\ &= 8\sqrt{3} + 10\sqrt{2} - 5\sqrt{2} \\ &= 8\sqrt{3} + (10 - 5)\sqrt{2} \\ &= 8\sqrt{3} + 5\sqrt{2} \end{aligned}$$

5 $\sqrt{50} + \sqrt{8}$

SOLUTION

$$\begin{aligned} \sqrt{50} + \sqrt{8} &= \sqrt{25 \cdot 2} + \sqrt{4 \cdot 2} \\ &= 5\sqrt{2} + 2\sqrt{2} \\ &= 7\sqrt{2} \end{aligned}$$

PROBLEM SET 9.4

In problems 1–24, simplify by combining similar terms. Assume that all variables are positive.

1 $\sqrt{8} + \sqrt{18}$

2 $\sqrt{18} + \sqrt{2}$

3 $\sqrt{72} - \sqrt{2}$

4 $\sqrt{50} + \sqrt{128}$

5 $5\sqrt{125} + 6\sqrt{45}$

6 $\sqrt{48} - \sqrt{12}$

7 $\sqrt{98} - \sqrt{50}$

8 $4\sqrt{18} - 2\sqrt{72}$

9 $2\sqrt{108} - 3\sqrt{27}$

10 $\sqrt{108} - 2\sqrt{48}$

11 $\sqrt{18} - \sqrt{8} + \sqrt{32}$

12 $\sqrt{20} + \sqrt{45} - \sqrt{80}$

13 $\sqrt{294} + \sqrt{486} - \sqrt{24}$

14 $\sqrt{75} - \sqrt{3} - \sqrt{12}$

15 $2\sqrt[3]{3} + \sqrt[3]{24} + 5\sqrt[3]{192}$

16 $\sqrt[3]{16x^4} - x\sqrt[3]{54x}$

17 $\sqrt[3]{16} - \sqrt[3]{54} + 2\sqrt[3]{2}$

18 $\sqrt[3]{8} - \sqrt[3]{125}\sqrt[3]{64}$

19 $4\sqrt{3} - 5\sqrt{12} + 2\sqrt{75}$

20 $\sqrt[3]{2} + \sqrt[3]{16} - \sqrt[3]{54}$

21 $\sqrt{48x} - \sqrt{12x} + \sqrt{300x}$

22 $2\sqrt{108x} - \sqrt{27x} + \sqrt{363x}$

23 $\sqrt{x^3} + \sqrt{25x^3} + \sqrt{9x}$

24 $\sqrt{x^3} - 2x\sqrt{x^5} + 3x\sqrt{x^7}$

9.5 Products of Roots

In this section, we will use Property 1 of Section 9.2, represented by the formula $\sqrt[n]{ab} = \sqrt[n]{a}\sqrt[n]{b}$, as a basis for the multiplication of radicals. This property can be extended to handle cases where the multiplicand or multiplier or both are indicated sums of radicals. For example, the products $\sqrt{3}(\sqrt{3} + \sqrt{2})$ and $(\sqrt{5} + \sqrt{2})(\sqrt{5} + 3\sqrt{2})$ can be found by using the distributive property, as is the case when multiplying polynomials. Thus, $\sqrt{3}(\sqrt{3} + \sqrt{2}) = \sqrt{3}\sqrt{3} + \sqrt{3}\sqrt{2} = 3 + \sqrt{6}$ and

$$\begin{aligned}
&(\sqrt{5} + \sqrt{2})(\sqrt{5} + 3\sqrt{2}) \\
&\quad = \sqrt{5}(\sqrt{5} + 3\sqrt{2}) + \sqrt{2}(\sqrt{5} + 3\sqrt{2}) \\
&\quad = \sqrt{5}\sqrt{5} + 3\sqrt{5}\sqrt{2} + \sqrt{2}\sqrt{5} + 3\sqrt{2}\sqrt{2} \\
&\quad = 5 + 3\sqrt{10} + \sqrt{10} + 6 \\
&\quad = 11 + 4\sqrt{10}
\end{aligned}$$

EXAMPLES

Find the products, and express each in simplest form. Assume all variables to be positive.

1 $\sqrt{5}(\sqrt{15} + \sqrt{25})$

SOLUTION

$$\begin{aligned}\sqrt{5}(\sqrt{15} + \sqrt{25}) &= \sqrt{5}\sqrt{15} + \sqrt{5}\sqrt{25} \\ &= \sqrt{75} + 5\sqrt{5} \\ &= \sqrt{25 \cdot 3} + 5\sqrt{5} \\ &= \sqrt{25}\sqrt{3} + 5\sqrt{5} \\ &= 5\sqrt{3} + 5\sqrt{5}\end{aligned}$$

2 $\sqrt{6}(\sqrt{8} + \sqrt{18})$

SOLUTION

$$\begin{aligned}\sqrt{6}(\sqrt{8} + \sqrt{18}) &= \sqrt{6}\sqrt{8} + \sqrt{6}\sqrt{18} \\ &= \sqrt{48} + \sqrt{108} \\ &= \sqrt{16 \cdot 3} + \sqrt{36 \cdot 3} \\ &= \sqrt{16}\sqrt{3} + \sqrt{36}\sqrt{3} \\ &= 4\sqrt{3} + 6\sqrt{3} \\ &= 10\sqrt{3}\end{aligned}$$

3 $\sqrt{x}(\sqrt{x} + \sqrt{xy})$

SOLUTION

$$\begin{aligned}\sqrt{x}(\sqrt{x} + \sqrt{xy}) &= \sqrt{x}\sqrt{x} + \sqrt{x}\sqrt{xy} \\ &= x + \sqrt{x}\sqrt{x}\sqrt{y} \\ &= x + x\sqrt{y}\end{aligned}$$

4 $(\sqrt{10} + \sqrt{2})(\sqrt{10} - \sqrt{2})$

SOLUTION

$$\begin{aligned}&(\sqrt{10} + \sqrt{2})(\sqrt{10} - \sqrt{2}) \\ &\quad = \sqrt{10}\sqrt{10} - \sqrt{10}\sqrt{2} + \sqrt{2}\sqrt{10} - \sqrt{2}\sqrt{2} \\ &\quad = 10 - 2 \\ &\quad = 8\end{aligned}$$

5 $(\sqrt{3} - \sqrt{2})(2\sqrt{3} + \sqrt{2})$

SOLUTION

$$\begin{aligned}(\sqrt{3}-\sqrt{2})(2\sqrt{3}+\sqrt{2}) &= 2\sqrt{3}\sqrt{3}+\sqrt{3}\sqrt{2}-2\sqrt{2}\sqrt{3}-\sqrt{2}\sqrt{2}\\ &= 2\cdot 3+\sqrt{6}-2\sqrt{6}-2\\ &= 4-\sqrt{6}\end{aligned}$$

6 $(\sqrt{x}+y)(\sqrt{x}+2y)$

SOLUTION

$$\begin{aligned}(\sqrt{x}+y)(\sqrt{x}+2y) &= \sqrt{x}\sqrt{x}+2y\sqrt{x}+y\sqrt{x}+2y^2\\ &= x+3y\sqrt{x}+2y^2\end{aligned}$$

7 $(\sqrt{x}+2\sqrt{y})^2$

SOLUTION

$$\begin{aligned}(\sqrt{x}+2\sqrt{y})^2 &= (\sqrt{x}+2\sqrt{y})(\sqrt{x}+2\sqrt{y})\\ &= \sqrt{x}\sqrt{x}+2\sqrt{x}\sqrt{y}+2\sqrt{x}\sqrt{y}+4\sqrt{y}\sqrt{y}\\ &= x+4\sqrt{x}\sqrt{y}+4y\\ &= x+4\sqrt{xy}+4y\end{aligned}$$

PROBLEM SET 9.5

In problems 1–32, express each product in simplest form. Assume all variables to be positive.

1 $\sqrt{6}\sqrt{3}$

2 $\sqrt{18}\sqrt{2}$

3 $\sqrt{22}\sqrt{11}$

4 $\sqrt{2}\sqrt{3}\sqrt{6}$

5 $4\sqrt{5}\,2\sqrt{10}$

6 $\sqrt{6}\sqrt{2}\sqrt{8}$

7 $(3\sqrt{2})^2$

8 $(\sqrt{5}x)^2$

9 $(\sqrt{2}-1)^2$

10 $(\sqrt{3}+\sqrt{2})^2$

11 $(4\sqrt{2}-6\sqrt{3})^2$

12 $(3\sqrt{5}-5\sqrt{3})^2$

13 $\sqrt{3}(\sqrt{18}-\sqrt{2})$

14 $\sqrt{5}(4\sqrt{5}-3\sqrt{2})$

15 $\sqrt[3]{9}\sqrt[3]{6}$

16 $\sqrt[3]{21}\sqrt[3]{49}$

17 $\sqrt[4]{2}\sqrt[4]{8}\sqrt[4]{5}$

18 $\sqrt[5]{9}\sqrt[5]{40}$

19 $(1 + \sqrt{2})(2 - \sqrt{2})$

20 $(\sqrt{3} + 1)(2\sqrt{3} - 1)$

21 $(3\sqrt{2} + 1)(\sqrt{2} - 3)$

22 $(3\sqrt{5} + 2)(\sqrt{5} - 2)$

23 $(2\sqrt{2} - \sqrt{3})(3\sqrt{2} + 2\sqrt{3})$

24 $(3\sqrt{3} + \sqrt{5})(2\sqrt{3} - 2\sqrt{5})$

25 $\sqrt{x}(\sqrt{x} - \sqrt{y})$

26 $(\sqrt{x} + \sqrt{y})(\sqrt{x} - \sqrt{y})$

27 $(\sqrt{x} - 1)(2\sqrt{x} + 3)$

28 $(2\sqrt{y} + 1)(3\sqrt{y} + 2)$

29 $(\sqrt{x} + 2\sqrt{y})(3\sqrt{x} - \sqrt{y})$

30 $(2\sqrt{x} - 3\sqrt{y})(4\sqrt{x} + 2\sqrt{y})$

31 $(2\sqrt{x} - y)^2$

32 $(\sqrt{x} + 3\sqrt{y})^2$

9.6 Rationalizing Denominators

As we discussed in Section 1.3, $\sqrt{2}$ is an irrational number. In many problems in mathematics, numbers, for example, $\frac{1}{\sqrt{2}}$, have irrational denominators. This form is acceptable unless it is desirable to approximate the result by a rational number. Since $\sqrt{2}$ is approximately 1.414, then $\frac{1}{\sqrt{2}}$ is approximately $\frac{1}{1.414}$. However, $\frac{1}{\sqrt{2}}$ can be written as $\frac{\sqrt{2}}{2}$ simply by multiplying the numerator and the denominator of the fraction by $\sqrt{2}$. The form $\frac{\sqrt{2}}{2}$ is easier to approximate than the form $\frac{1}{\sqrt{2}}$, since $\frac{1.414}{2} = 0.707$, while $\frac{1}{1.414}$ requires a more involved division.

Thus,

$$\frac{5}{\sqrt{3}} = \frac{5\sqrt{3}}{\sqrt{3}\sqrt{3}} = \frac{5\sqrt{3}}{3}$$

and

$$\frac{4}{\sqrt{5}} = \frac{4\sqrt{5}}{\sqrt{5}\sqrt{5}} = \frac{4\sqrt{5}}{5}$$

The process of eliminating radicals from the denominator of a fraction is called *rationalizing the denominator.* To rationalize a denominator that is an indicated sum of terms involving square roots, we use the identity

$$(a + b)(a - b) = a^2 - b^2$$

For example, let us rationalize the denominator of $\dfrac{2 + \sqrt{3}}{\sqrt{5} - \sqrt{3}}$. Now, we see from above that $(\sqrt{5} - \sqrt{3})(\sqrt{5} + \sqrt{3}) = (\sqrt{5})^2 - (\sqrt{3})^2 = 5 - 3 = 2$. Thus,

$$\begin{aligned}\frac{2 + \sqrt{3}}{\sqrt{5} - \sqrt{3}} &= \frac{(2 + \sqrt{3})(\sqrt{5} + \sqrt{3})}{(\sqrt{5} - \sqrt{3})(\sqrt{5} + \sqrt{3})} \\ &= \frac{2\sqrt{5} + 2\sqrt{3} + \sqrt{3}\sqrt{5} + \sqrt{3}\sqrt{3}}{2} \\ &= \frac{2\sqrt{5} + 2\sqrt{3} + \sqrt{15} + 3}{2}\end{aligned}$$

EXAMPLES

Rationalize the denominator of each fraction. Assume all variables to be positive.

1 $\dfrac{2}{\sqrt{3}}$

SOLUTION

$$\frac{2}{\sqrt{3}} = \frac{2\sqrt{3}}{\sqrt{3}\sqrt{3}} = \frac{2\sqrt{3}}{3}$$

2 $\sqrt{\frac{2}{5}}$

SOLUTION

$$\sqrt{\frac{2}{5}} = \frac{\sqrt{2}}{\sqrt{5}} = \frac{\sqrt{2}\sqrt{5}}{\sqrt{5}\sqrt{5}} = \frac{\sqrt{10}}{5}$$

3 $\dfrac{\sqrt{8}}{\sqrt{10}}$

SOLUTION

$$\frac{\sqrt{8}}{\sqrt{10}} = \frac{\sqrt{8}\sqrt{10}}{\sqrt{10}\sqrt{10}} = \frac{\sqrt{80}}{10} = \frac{\sqrt{16 \cdot 5}}{10} = \frac{\sqrt{16}\sqrt{5}}{10} = \frac{4\sqrt{5}}{10} = \frac{2\sqrt{5}}{5}$$

4 $\dfrac{\sqrt{x}}{\sqrt{y}}$

SOLUTION

$$\frac{\sqrt{x}}{\sqrt{y}} = \frac{\sqrt{x}\sqrt{y}}{\sqrt{y}\sqrt{y}} = \frac{\sqrt{xy}}{y}$$

5 $\dfrac{\sqrt{6} - 2}{\sqrt{3}}$

SOLUTION

$$\begin{aligned}\frac{\sqrt{6} - 2}{\sqrt{3}} &= \frac{(\sqrt{6} - 2)\sqrt{3}}{\sqrt{3}\sqrt{3}} \\ &= \frac{\sqrt{6}\sqrt{3} - 2\sqrt{3}}{3} \\ &= \frac{\sqrt{18} - 2\sqrt{3}}{3} \\ &= \frac{\sqrt{9 \cdot 2} - 2\sqrt{3}}{3} \\ &= \frac{\sqrt{9}\sqrt{2} - 2\sqrt{3}}{3} \\ &= \frac{3\sqrt{2} - 2\sqrt{3}}{3}\end{aligned}$$

6 $\dfrac{5}{\sqrt{3} - 1}$

SOLUTION

$$\begin{aligned}\frac{5}{\sqrt{3} - 1} &= \frac{5(\sqrt{3} + 1)}{(\sqrt{3} - 1)(\sqrt{3} + 1)} \\ &= \frac{5\sqrt{3} + 5}{(\sqrt{3})^2 - 1^2} \\ &= \frac{5\sqrt{3} + 5}{3 - 1} \\ &= \frac{5\sqrt{3} + 5}{2}\end{aligned}$$

7 $\dfrac{5}{\sqrt[3]{16}}$

SOLUTION

$$\begin{aligned}\frac{5}{\sqrt[3]{16}} &= \frac{5}{\sqrt[3]{4^2}} \\ &= \frac{5\sqrt[3]{4}}{\sqrt[3]{4^2}\sqrt[3]{4}} \\ &= \frac{5\sqrt[3]{4}}{\sqrt[3]{4^3}} \\ &= \frac{5\sqrt[3]{4}}{4}\end{aligned}$$

8 $\dfrac{\sqrt{5}+\sqrt{2}}{\sqrt{5}-\sqrt{2}}$

SOLUTION

$$\begin{aligned}\frac{\sqrt{5}+\sqrt{2}}{\sqrt{5}-\sqrt{2}} &= \frac{(\sqrt{5}+\sqrt{2})(\sqrt{5}+\sqrt{2})}{(\sqrt{5}-\sqrt{2})(\sqrt{5}+\sqrt{2})} \\ &= \frac{(\sqrt{5})^2 + 2\sqrt{2}\sqrt{5} + (\sqrt{2})^2}{(\sqrt{5})^2 - (\sqrt{2})^2} \\ &= \frac{5 + 2\sqrt{10} + 2}{5 - 2} \\ &= \frac{7 + 2\sqrt{10}}{3}\end{aligned}$$

9 $\dfrac{3}{\sqrt{x}-1}$

SOLUTION

$$\begin{aligned}\frac{3}{\sqrt{x}-1} &= \frac{3(\sqrt{x}+1)}{(\sqrt{x}-1)(\sqrt{x}+1)} \\ &= \frac{3\sqrt{x}+3}{(\sqrt{x})^2 - (1)^2} \\ &= \frac{3\sqrt{x}+3}{x-1}\end{aligned}$$

10 $\dfrac{\sqrt{x}+\sqrt{y}}{\sqrt{x}-\sqrt{y}}$

SOLUTION

$$\frac{\sqrt{x}+\sqrt{y}}{\sqrt{x}-\sqrt{y}} = \frac{(\sqrt{x}+\sqrt{y})(\sqrt{x}+\sqrt{y})}{(\sqrt{x}-\sqrt{y})(\sqrt{x}+\sqrt{y})}$$
$$= \frac{(\sqrt{x})^2 + 2\sqrt{x}\sqrt{y} + (\sqrt{y})^2}{(\sqrt{x})^2 - (\sqrt{y})^2}$$
$$= \frac{x + 2\sqrt{xy} + y}{x - y}$$

PROBLEM SET 9.6

In problems 1–30, rationalize the denominator of each expression. Assume all variables to be positive.

1 $\frac{2}{\sqrt{3}}$

2 $\frac{\sqrt{2}}{\sqrt{7}}$

3 $\frac{9}{\sqrt{21}}$

4 $\frac{8}{\sqrt{11}}$

5 $\frac{\sqrt{6xy}}{\sqrt{2x}}$

6 $\frac{10x}{\sqrt{2x}}$

7 $\frac{5}{\sqrt{2x}}$

8 $\frac{5\sqrt{6}}{\sqrt{6xy}}$

9 $\frac{3\sqrt{x}}{\sqrt{4x}}$

10 $\frac{8\sqrt{2}}{\sqrt{8x}}$

11 $\frac{10}{\sqrt{5}-1}$

12 $\frac{36}{\sqrt{3}+1}$

13 $\dfrac{\sqrt{2}}{1 + \sqrt{2}}$

14 $\dfrac{1 + \sqrt{2}}{\sqrt{2}}$

15 $\dfrac{3}{\sqrt{x} + \sqrt{y}}$

16 $\dfrac{\sqrt{2} + 1}{\sqrt{2} - 1}$

17 $\dfrac{\sqrt{x} - \sqrt{y}}{\sqrt{x} + \sqrt{y}}$

18 $\dfrac{\sqrt{8} - \sqrt{x}}{2 + \sqrt{x}}$

19 $\dfrac{1 + \sqrt{y}}{1 - \sqrt{y}}$

20 $\dfrac{3\sqrt{2} - \sqrt{3}}{2\sqrt{3} - \sqrt{2}}$

21 $\dfrac{y}{\sqrt{y} + 1}$

22 $\dfrac{3}{2 + \sqrt{3}}$

23 $\dfrac{\sqrt{2} + \sqrt{3}}{\sqrt{2} - \sqrt{3}}$

24 $\dfrac{\sqrt{y}}{\sqrt{x} - \sqrt{y}}$

25 $\dfrac{x}{x + \sqrt{y}}$

26 $\dfrac{1}{\sqrt{2} + \sqrt{3}}$

27 $\dfrac{\sqrt{3} - 3\sqrt{5}}{4\sqrt{3} + 3\sqrt{5}}$

28 $\dfrac{\sqrt{3} - 2}{(1 + \sqrt{3})}$

29 $\dfrac{2\sqrt{3} + \sqrt{2}}{\sqrt{6} + 2\sqrt{2}}$

30 $\dfrac{5}{2 - \sqrt{5}}$

REVIEW PROBLEM SET

In problems 1–6, simplify each expression. Assume all variables to be positive.

1 $\sqrt{625}$

2 $\sqrt{225}$

3 $\sqrt[4]{1296}$

4 $\sqrt[5]{-96}$

5 $\sqrt[6]{729}$

6 $\sqrt[4]{243x^8}$

In problems 7–14, use Property 1 of Section 9.2 to find the simplest form of each expression. Assume all variables to be positive.

7 $\sqrt{150}$

8 $\sqrt{108}$

9 $\sqrt{63}$

10 $\sqrt{500}$

11 $\sqrt[3]{a^4b^5}$

12 $\sqrt[3]{8xy^4}$

13 $\sqrt[4]{y^5}$

14 $\sqrt{0.0081}$

In problems 15–22, use Property 1 of Section 9.3 to find the simplest form of each expression. Assume all variables to be positive.

15 $\sqrt{\frac{25}{64}}$

16 $\sqrt{\frac{11}{9}}$

17 $\sqrt[3]{\frac{125}{8}}$

18 $\sqrt[3]{\frac{5}{27}}$

19 $\sqrt[4]{\frac{625}{16}}$

20 $\sqrt{\frac{x}{y^2}}$

21 $\sqrt[5]{\frac{3x}{32}}$

22 $\sqrt[6]{\frac{5}{x^{12}}}$

In problems 23–32, simplify each expression by combining similar terms.

23 $3\sqrt{2} + \sqrt{2}$

24 $4\sqrt{7} - \sqrt{7} + 3\sqrt{7}$

25 $7\sqrt[3]{5} + \sqrt[3]{5} - 2\sqrt[3]{5}$

26 $5\sqrt{x} - 2\sqrt{x} \qquad x > 0$

27 $2\sqrt{50} + 50\sqrt{2}$

28 $\sqrt{8} + \sqrt{18} + \sqrt{98}$

29 $\sqrt{72} - \sqrt{2}$

30 $\sqrt[3]{2} + \sqrt[3]{16}$

31 $3\sqrt{48} - \sqrt{147} + \sqrt{243}$

32 $\sqrt{63} + 2\sqrt{112} - \sqrt{252}$

In problems 33–40, express each product in simplest form.

33 $(2\sqrt{3} - \sqrt{2})(2\sqrt{3} + \sqrt{2})$

34 $(\sqrt{21} - \sqrt{5})(\sqrt{21} + \sqrt{5})$

35 $(\sqrt{33} - \sqrt{6})(\sqrt{33} + \sqrt{6})$

36 $(6 - 2\sqrt{2})(6 + 2\sqrt{2})$

37 $(\sqrt{x} + \sqrt{y})(\sqrt{x} - \sqrt{y}) \qquad x > 0, y > 0$

38 $(2\sqrt{10} + 3\sqrt{5})(\sqrt{10} + \sqrt{5})$

39 $(2 + \sqrt{7})(3 - \sqrt{7})$

40 $(26\sqrt{2} - 2)(13\sqrt{2} + 1)$

In problems 41–48, rationalize the denominator of each expression. Assume all variables to be positive.

41 $\dfrac{4}{\sqrt{3}}$

42 $\dfrac{10}{\sqrt[3]{5x}}$

43 $\dfrac{6}{5\sqrt{3x}}$

44 $\dfrac{6x}{\sqrt{162xy}}$

45 $\dfrac{1}{\sqrt{2} + \sqrt{3}}$

46 $\dfrac{2\sqrt{x} - 1}{2\sqrt{x} + 1}$

47 $\dfrac{x}{\sqrt{x} - \sqrt{y}}$

48 $\dfrac{3 - \sqrt{3}}{\sqrt{2} + \sqrt{3}}$

CHAPTER 10

Quadratic Equations

10 QUADRATIC EQUATIONS

Introduction

First-degree equations were studied in Chapter 5. In this chapter we will use the concepts of roots and radicals (Chapter 9) to solve *second-degree equations*, usually referred to as *quadratic equations*. For example,

$$x^2 - 1 = 0 \quad 3x^2 - 12x + 4 = 0 \quad \text{and} \quad x^2 + 4x - 5 = 0$$

are quadratic equations. More generally, any equation that is equivalent to an equation of the form $ax^2 + bx + c = 0$, where a, b, and c are real numbers with $a \neq 0$, is called a second-degree equation or a quadratic equation.

EXAMPLES

Write each of the following equations in the form $ax^2 + bx + c = 0$. Identify a, b, and c.

1 $x(2x - 1) + 6 = 0$

SOLUTION

$$x(2x - 1) + 6 = 0$$
$$2x^2 - x + 6 = 0$$

so that

$a = 2 \quad b = -1 \quad c = 6$

2 $6x^2 - 3x + 2 = x^2 - 2$

SOLUTION

$$6x^2 - 3x + 2 = x^2 - 2$$

then

$$5x^2 - 3x + 4 = 0$$

so that

$$a = 5 \qquad b = -3 \qquad c = 4$$

10.1 Solving Quadratic Equations by Roots Extraction

Let us first consider a method for solving a quadratic equation of the form $ax^2 + c = 0$, $a \neq 0$. For example, consider the equation $4x^2 - 9 = 0$. We shall follow the method used in solving first-degree equations. First we solve for x^2:

$$4x^2 - 9 = 0$$

Adding 9 to both sides, we have

$$4x^2 = 9$$

Dividing both sides by 4, we have

$$x^2 = \tfrac{9}{4}$$

Now, since x is the number whose squares are $\frac{9}{4}$, we solve the equation for x by determining square roots of $\frac{9}{4}$, so that

$$x = \pm\tfrac{3}{2}$$

That is,

$$x = \tfrac{3}{2} \qquad \text{or} \qquad x = -\tfrac{3}{2}$$

Hence, the solution set is $\{-\frac{3}{2}, \frac{3}{2}\}$.

Notice that the difference in solving the quadratic equation $4x^2 - 9 = 0$ from solving a first-degree equation such as $4x - 9 = 0$ lies in the last step where we determined square roots. The similarities that exist between this method of solving a quadratic equation of the form $ax^2 + c = 0$ and the method of solving a first-degree equation of the form $ax + b = 0$ can be illustrated as follows:

Second degree (quadratic)	*First degree*
$4x^2 - 9 = 0$	$4x - 9 = 0$
$4x^2 = 9$	$4x = 9$
$x^2 = \frac{9}{4}$	$x = \frac{9}{4}$
$x = \pm\sqrt{\frac{9}{4}}$	
$x = -\frac{3}{2}$	
or	
$x = \frac{3}{2}$	

The solution set of the first-degree equation contains only the one element $\frac{9}{4}$, whereas the solution set of the quadratic equation contains the two elements $-\frac{3}{2}$ and $\frac{3}{2}$. Since there are always two square roots for every nonzero real number, the solution set of a quadratic equation of the form $ax^2 + c = 0$, where $a \neq 0$ and $c < 0$, will consist of exactly two different elements. Later in this chapter we shall notice that the solution sets of all quadratic equations always contain two elements, although not necessarily different.

EXAMPLES

Solve the following quadratic equations and check the results.

1 $x^2 - 4 = 0$

SOLUTION

$$x^2 - 4 = 0$$
$$x^2 = 4$$

so that

$$x = -\sqrt{4} \quad \text{or} \quad x = \sqrt{4}$$

That is,

$$x = -2 \quad \text{or} \quad x = 2$$

Check: For $x = -2$, we have

$$x^2 - 4 = 0$$
$$(-2)^2 - 4 \stackrel{?}{=} 0$$
$$4 - 4 = 0$$

For $x = 2$, we have

$$x^2 - 4 = 0$$
$$(2)^2 - 4 \stackrel{?}{=} 0$$
$$4 - 4 = 0$$

Therefore, the solution set is $\{-2,2\}$.

2 $3x^2 - 27 = 0$

SOLUTION

$$3x^2 - 27 = 0$$
$$3x^2 = 27$$

so that

$$x^2 = 9$$

Then

$$x = -\sqrt{9} \quad \text{or} \quad x = \sqrt{9}$$

That is,

$$x = -3 \quad \text{or} \quad x = 3$$

Check: For $x = -3$,

$$3(-3)^2 - 27 = 3(9) - 27 = 27 - 27 = 0$$

For $x = 3$,

$$3(3)^2 - 27 = 3(9) - 27 = 27 - 27 = 0$$

Hence, the solution set is $\{-3,3\}$.

3 $(x - 1)^2 - 9 = 0$

SOLUTION. First, we solve for $x - 1$:

$$(x - 1)^2 - 9 = 0$$

so that

$(x - 1)^2 = 9$

or

$x - 1 = -\sqrt{9}$ or $x - 1 = \sqrt{9}$

Therefore,

$x - 1 = -3$ or $x - 1 = 3$

so that

$x = -2$ or $x = 4$

Check: For $x = -2$,

$$(x - 1)^2 - 9 = 0$$
$$(-2 - 1)^2 - 9 = 0$$
$$(-3)^2 - 9 = 9 - 9 = 0$$

For $x = 4$,

$$(x - 1)^2 - 9 = 0$$
$$(4 - 1)^2 - 9 = 0$$
$$3^2 - 9 = 9 - 9 = 0$$

Hence, the solution set is $\{-2,4\}$.

4 $(3x - 2)^2 - 16 = 0$

SOLUTION

$(3x - 2)^2 - 16 = 0$

so that

$(3x - 2)^2 = 16$

or

$3x - 2 = -\sqrt{16}$ or $3x - 2 = \sqrt{16}$

Therefore,

$3x - 2 = -4$ or $3x - 2 = 4$

so that

$$3x = -2 \qquad \text{or} \qquad 3x = 6$$

That is,

$$x = \frac{-2}{3} \qquad \text{or} \qquad x = 2$$

Hence, the solution set is $\{-\frac{2}{3}, 2\}$.

PROBLEM SET 10.1

In problems 1–5, write each equation in the form $ax^2 + bx + c = 0$. Identify a, b, and c.

1 $6x^2 - x + 9 = 4 - x(x - 2)$

2 $x(x - 3) + 12 = (3x - 7)x$

3 $(x - 3)^2 + (3x - 2)^2 = 5x^2$

4 $x(3x + 1) + 10 = 2x(x - 5) - x^2 + 6$

5 $3x(x - 8) = 4(2x - 3) - 5$

6 How many solutions does the quadratic equation $4x^2 - 25 = 0$ have? How are they related to each other?

In problems 7–32, solve each quadratic equation by determining square roots, check the results, and write the solution set.

7 $x^2 = 9$

8 $3x^2 - 4 = 0$

9 $5x^2 - 3 = 3x^2 - 2$

10 $x^2 - c = 0$, where $c > 0$

11 $9x^2 - 4 = 0$

12 $3x^2 - 2 = 0$

13 $(x - 1)^2 = 4$

14 $(2x + 1)^2 - 25 = 0$

15 $(x - 6)^2 - 5 = 0$

16 $(2x + 5)^2 - 9 = 0$

17 $(x - 1)^2 - 36 = 0$

18 $(2x - 3)^2 - 5 = 0$

19 $(x + \frac{3}{2})^2 - 1 = 0$

20 $(x - \frac{3}{8})^2 - 1 = 0$

21 $(6x - 5)^2 - 4 = 0$

22 $(x - 2)^2 - 1 = 0$

23 $(x - \frac{8}{3})^2 - 9 = 0$

24 $(x - 5)^2 - 4 = 0$

25 $(4x - 3)^2 - 16 = 0$

26 $(3x - 2)^2 - 49 = 0$

27 $2x(x + 2) = 4x + 15$

28 $x^2 = 63 - 6x^2$

29 $12x^2 - 125 = 7x^2$

30 $(x + 5)^2 = 49$

31 $(4x - 3)^2 = 625$

32 $(2x + 5)^2 = 36$

10.2 Solving Quadratic Equations by Factoring

A method for solving quadratic equations of the form $ax^2 + c = 0$ was illustrated in Section 10.1. For example, the solution set $\{-2,2\}$ of the equation $x^2 - 4 = 0$ is obtained from solving the equation for x^2, that is $x^2 = 4$, and then determining the square roots of 4. However, this method is not practical for solving equations of the form $ax^2 + bx + c = 0$, $a, b \neq 0$. For example, we cannot apply this method to solve $x^2 - 6x + 5 = 0$.

We saw in Chapter 3 how to factor some second-degree polynomials, and we shall use factoring as a tool for solving factorable second-degree equations. Thus, the equation $x^2 - 6x + 5 = 0$ can also be written as $(x - 1)(x - 5) = 0$. Property 12 of Chapter 1 states that: *The product of two real numbers is zero whenever one or both of the two numbers is zero. That is, for a and b real numbers, $ab = 0$ if and only if $a = 0$, or $b = 0$, or both.*

This property of real numbers implies that the product $(x - 1)(x - 5) = 0$ whenever $x - 1 = 0$ or $x - 5 = 0$, or both. Solving these two first-degree equations, we have $x = 1$ or $x = 5$. Thus, both values of x satisfy the original equation, which can be verified by substitution. In summary, the factoring method consists of the following steps:

$$x^2 - 6x + 5 = 0$$

$$(x - 1)(x - 5) = 0 \qquad \text{(Factoring)}$$

$$\left.\begin{aligned} x - 1 &= 0 \\ x - 5 &= 0 \end{aligned}\right\} \qquad \text{(Setting each factor equal to 0)}$$

Hence, $x = 1$ or $x = 5$, and the solution set is $\{1,5\}$.

The factoring method can be applied to any quadratic equation of the form $ax^2 + bx + c = 0$ whenever $ax^2 + bx + c$ can be factored in the sense discussed in Sections 3.4 and 3.5.

EXAMPLES

In examples 1 and 2, find the solution set of each equation.

1 $x(x - 2) = 0$

SOLUTION. Setting each factor equal to zero, we have

$x = 0$ or $x - 2 = 0$

so that

$x = 0$ or $x = 2$

The solution set is $\{0,2\}$.

2 $(2x - 3)(x + 2) = 0$

SOLUTION. Setting each factor equal to zero, we have

$2x - 3 = 0$ or $x + 2 = 0$

Then

$x = \frac{3}{2}$ or $x = -2$

The solution set is $\{-2,\frac{3}{2}\}$.

In examples 3–6, solve each quadratic equation by factoring. Find the solution set.

3 $x^2 - 5x + 6 = 0$

SOLUTION. Factoring $x^2 - 5x + 6 = 0$, we have

$(x - 2)(x - 3) = 0$

Setting each factor equal to zero, we have

$x - 2 = 0 \quad \text{or} \quad x - 3 = 0$

Then

$x = 2 \quad \text{or} \quad x = 3$

The solution set is $\{2,3\}$.

4 $6x^2 + 5x - 4 = 0$

SOLUTION

$$6x^2 + 5x - 4 = 0$$
$$(2x - 1)(3x + 4) = 0$$
$$2x - 1 = 0 \quad \text{or} \quad 3x + 4 = 0$$
$$x = \tfrac{1}{2} \quad \text{or} \quad x = -\tfrac{4}{3}$$

The solution set is $\{\frac{1}{2}, -\frac{4}{3}\}$

5 $x^2 + 3x = 0$

SOLUTION

$$x^2 + 3x = 0$$
$$x(x + 3) = 0$$
$$x = 0 \quad \text{or} \quad x = -3$$

The solution set is $\{0, -3\}$. [Notice that both 0 and -3 satisfy the equation $x^2 + 3x = 0$, but if both sides of the equation had been divided by x, the solution $x = 0$ would have been lost. (Why?)]

6 $x^2 - 4x + 4 = 0$

SOLUTION

$$x^2 - 4x + 4 = 0$$
$$(x - 2)(x - 2) = 0$$
$$x - 2 = 0 \quad \text{or} \quad x - 2 = 0$$
$$x = 2 \quad \text{or} \quad x = 2$$

The solution set is $\{2\}$.

We obtained two elements in each solution set in the above examples; however, in Example 6, the solutions were identical. We shall still consider this case as having two solutions in order to distinguish between the solution sets of the quadratic equation $x^2 - 4x + 4 = 0$ and the first-degree equation $x - 2 = 0$.

If we are given the solutions of a quadratic equation, we can find an equation to which the solution set applies. This is illustrated in the following examples.

EXAMPLES

Find a quadratic equation for each of the following solution sets.

1 $\{1,3\}$

SOLUTION

$$x = 1 \qquad \text{or} \qquad x = 3$$

Then

$$x - 1 = 0 \qquad \text{or} \qquad x - 3 = 0$$

so that

$$(x - 1)(x - 3) = 0 \qquad \text{or} \qquad x^2 - 4x + 3 = 0$$

2 $\{-\sqrt{2},\sqrt{2}\}$

SOLUTION

$$x = -\sqrt{2} \qquad \text{or} \qquad x = \sqrt{2}$$

Then

$$x + \sqrt{2} = 0 \qquad \text{or} \qquad x - \sqrt{2} = 0$$

so that

$$(x + \sqrt{2})(x - \sqrt{2}) = 0 \qquad \text{or} \qquad x^2 - 2 = 0$$

PROBLEM SET 10.2

In problems 1–16, find the solution set of each equation.

1 $(x + 2)(x + 1) = 0$

2 $(x - 1)(2x - 1) = 0$

3 $(2x + 1)(2x - 1) = 0$

4 $(3x - 2)(3x + 2) = 0$

5 $(5x - 2)(5x + 2) = 0$

6 $(6x - 1)(6x + 1) = 0$

7 $(x + \sqrt{3})(x - \sqrt{3}) = 0$

8 $x(x + 2\sqrt{2}) = 0$

9 $(3x + 2)(3x + 7) = 0$

10 $(5x - 3)(-2x + 1) = 0$

11 $(3x - 7)(x + 2) = 0$

12 $(3x - 2)(-2x - 1) = 0$

13 $(7x - 1)(-3x - 2) = 0$

14 $(9x - 5)(-5x + 1) = 0$

15 $(8x - 5)(5x + 3) = 0$

16 $(3x - 1)(-x + \sqrt{2}) = 0$

In problems 17–36, solve each equation by factoring. Find the solution set.

17 $3x^2 - 7x = 0$

18 $5(x + 25) = 6x^2$

19 $x^2 - 6x = -8$

20 $4x^2 - 16 = 0$

21 $3x^2 - 2x = 5$

22 $x^2 + 6x - 16 = 0$

23 $2x^2 + x - 21 = 0$

24 $(x + 3)(2x + 5) = 11(x + 3)$

25 $15x^2 - 19x + 6 = 0$

26 $8x^2 - 16 = 0$

27 $x^2 + 2x - 3 = 0$

28 $10x^2 + x - 2 = 0$

29 $4x^2 - (x + 1)^2 = 0$

30 $2x^2 - 16x = 0$

31 $2x^2 - x - 3 = 0$

32 $12x^2 - 125 = 7x^2$

33 $x(x - 2) = 9 - 2x$

34 $4x^2 + 25x + 6 = 0$

35 $18x^2 + 61x - 7 = 0$

36 $10x^2 - 31x - 14 = 0$

In problems 37–42, find a quadratic equation for each solution set.

37 $\{5,6\}$

38 $\{-2,3\}$

39 $\{-1,-2\}$

40 $\{5,\frac{1}{3}\}$

41 $\{-\frac{2}{3},\frac{1}{3}\}$

42 $\{\sqrt{2},-\sqrt{3}\}$

10.3 Solving Quadratic Equations by Completing the Square

In Section 10.1, equations such as $(x - 1)^2 - 9 = 0$ were solved by the method of determining square roots. This equation can also be written as $x^2 - 2x - 8 = 0$. (Why?) Thus, another method for solving equations of the form $ax^2 + bx + c = 0$, $b \neq 0$, is suggested by the above equivalence. That is, to solve the equation $x^2 - 2x - 8 = 0$, first change it to the form $(x - 1)^2 - 9 = 0$, and then solve this equation by the method of determining square roots. Thus, the equation $x^2 - 2x - 8 = 0$ can be written as

$$x^2 - 2x + 1 - 9 = 0 \qquad \text{since} \qquad -8 = 1 - 9$$

or

$$(x - 1)^2 - 9 = 0$$

so that

$$(x - 1)^2 = 9$$

Therefore, $x - 1 = -3$ or $x - 1 = 3$. Hence the solution set is $\{4,-2\}$.

This technique is called *completing the square*. Its use involves finding the constant necessary to form a perfect quadratic square. From the identity $(x \pm d)^2 = x^2 \pm 2dx + d^2$, it appears that in order to complete the square when the two terms $x^2 \pm 2dx$ are given, we must add the number d^2. Since $d^2 = [\frac{1}{2}(2d)]^2$, the term to be added is the square of one-half the coefficient of x. For example, the quadratic expression $x^2 - 6x$ can be written as a complete quadratic square by adding $[\frac{1}{2}(6)]^2$ or $(3)^2 = 9$. Thus, $x^2 - 6x + 9 = (x - 3)^2$ is a perfect quadratic square.

EXAMPLES

Solve each quadratic equation by completing the square. Find the solution set.

1 $x^2 + 4x + 2 = 0$

SOLUTION

$$\begin{aligned} x^2 + 4x + 2 &= 0 \\ x^2 + 4x &= -2 \\ x^2 + 4x + [\tfrac{1}{2}(4)]^2 &= -2 + [\tfrac{1}{2}(4)]^2 \qquad \text{(Completing the square)} \\ x^2 + 4x + 4 &= -2 + 4 \end{aligned}$$

or

$$x^2 + 4x + 4 = 2$$

That is,

$$(x + 2)^2 = 2$$

Then

$$x + 2 = -\sqrt{2} \quad \text{or} \quad x + 2 = \sqrt{2}$$

so that

$$x = -2 + \sqrt{2} \quad \text{or} \quad x = -2 - \sqrt{2}$$

Hence the solution set is $\{-2 - \sqrt{2}, -2 + \sqrt{2}\}$.

2 $x^2 + 2x - 4 = 0$

SOLUTION

$$\begin{aligned} x^2 + 2x - 4 &= 0 \\ x^2 + 2x &= 4 \\ x^2 + 2x + [\tfrac{1}{2}(2)]^2 &= 4 + [\tfrac{1}{2}(2)]^2 \qquad \text{(Completing the square)} \\ x^2 + 2x + 1 &= 4 + 1 \\ (x + 1)^2 &= 5 \end{aligned}$$

so that

$$x + 1 = -\sqrt{5} \quad \text{or} \quad x + 1 = \sqrt{5}$$

Then

$$x = -1 - \sqrt{5} \quad \text{or} \quad x = -1 + \sqrt{5}$$

Hence, the solution set is $\{-1 - \sqrt{5}, -1 + \sqrt{5}\}$.

In cases in which the coefficient of the x^2 term is not 1, we first divide the equation by the coefficient of the x^2 term, and then proceed as before. For example, we solve the equation $3x^2 - 2x - 2 = 0$ as follows:

$$\begin{aligned}
3x^2 - 2x - 2 &= 0 \\
x^2 - \tfrac{2}{3}x - \tfrac{2}{3} &= 0 && \text{(Dividing by 3)} \\
x^2 - \tfrac{2}{3}x &= \tfrac{2}{3} \\
x^2 - \tfrac{2}{3}x + [\tfrac{1}{2}(\tfrac{2}{3})]^2 &= \tfrac{2}{3} + [\tfrac{1}{2}(\tfrac{2}{3})]^2 && \text{(Completing the square)} \\
x^2 - \tfrac{2}{3}x + \tfrac{1}{9} &= \tfrac{2}{3} + \tfrac{1}{9}
\end{aligned}$$

so that

$$(x - \tfrac{1}{3})^2 = \tfrac{7}{9}$$

Then

$$x - \tfrac{1}{3} = \sqrt{\tfrac{7}{9}} \quad \text{or} \quad x - \tfrac{1}{3} = -\sqrt{\tfrac{7}{9}}$$

so that

$$x = \frac{1}{3} + \frac{\sqrt{7}}{3} \quad \text{or} \quad x = \frac{1}{3} - \frac{\sqrt{7}}{3}$$

Hence, the solution set is $\left\{\dfrac{1 + \sqrt{7}}{3}, \dfrac{1 - \sqrt{7}}{3}\right\}$.

EXAMPLES

Solve the quadratic equations by the method of completing the square.

1 $2x^2 + 3x - 2 = 0$

SOLUTION

$$\begin{aligned}
2x^2 + 3x - 2 &= 0 \\
x^2 + \tfrac{3}{2}x - 1 &= 0 && \text{(Dividing both sides of the equation by 2)}
\end{aligned}$$

$$x^2 + \tfrac{3}{2}x = 1 \qquad \text{(Adding 1 to both sides of the equation)}$$
$$x^2 + \tfrac{3}{2}x + [\tfrac{1}{2}(\tfrac{3}{2})]^2 = 1 + [\tfrac{1}{2}(\tfrac{3}{2})]^2 \qquad \text{(Completing the square)}$$
$$x^2 + \tfrac{3}{2}x + \tfrac{9}{16} = 1 + \tfrac{9}{16}$$

That is,

$$x^2 + \tfrac{3}{2}x + \tfrac{9}{16} = \tfrac{25}{16}$$
$$(x + \tfrac{3}{4})^2 = \tfrac{25}{16} \qquad \text{(Factoring the left side of the equation)}$$

Hence,

$$x + \tfrac{3}{4} = -\tfrac{5}{4} \quad \text{or} \quad x + \tfrac{3}{4} = \tfrac{5}{4}$$

so that

$$x = -\tfrac{3}{4} - \tfrac{5}{4} = \frac{-8}{4} = -2 \quad \text{or} \quad x = -\tfrac{3}{4} + \tfrac{5}{4} = \tfrac{2}{4} = \tfrac{1}{2}$$

The solution set is $\{-2, \tfrac{1}{2}\}$

2 $3x^2 - 10x - 2 = 0$

SOLUTION. Divide both sides of the equation by 3, so that

$$x^2 - \tfrac{10}{3}x - \tfrac{2}{3} = 0$$

Adding $\tfrac{2}{3}$ to both sides of the equation, we have

$$x^2 - \tfrac{10}{3}x = \tfrac{2}{3}$$

Completing the square, we get

$$x^2 - \tfrac{10}{3}x + (\tfrac{5}{3})^2 = \tfrac{2}{3} + (\tfrac{5}{3})^2$$

so that

$$(x - \tfrac{5}{3})^2 = \tfrac{31}{9}$$

Therefore

$$x - \frac{5}{3} = -\frac{\sqrt{31}}{3} \quad \text{or} \quad x - \frac{5}{3} = \frac{\sqrt{31}}{3}$$

so that

$$x = \frac{5}{3} - \frac{\sqrt{31}}{3} \quad \text{or} \quad x = \frac{5}{3} + \frac{\sqrt{31}}{3}$$

That is,

$$x = \frac{5 - \sqrt{31}}{3} \qquad \text{or} \qquad x = \frac{5 + \sqrt{31}}{3}$$

The solution set is $\left\{\frac{5 - \sqrt{31}}{3}, \frac{5 + \sqrt{31}}{3}\right\}$.

PROBLEM SET 10.3

In problems 1–10, fill each blank so that the resulting polynomial expression is a perfect square. Then express each polynomial expression as a square of a binomial.

1 $x^2 + 4x +$ ______

2 $x^2 + 6x +$ ______

3 $x^2 - 4x +$ ______

4 $x^2 + 10x +$ ______

5 $x^2 +$ ______ $x + 9$

6 $x^2 +$ ______ $+ 25$

7 $x^2 - 15x +$ ______

8 $x^2 +$ ______ $+ 4$

9 $x^2 +$ ______ $+ 36$

10 $x^2 - 9x +$ ______

In problems 11–48, solve each equation by completing the square. Find the solution set.

11 $3x^2 - 8x + 2 = 0$

12 $x^2 + 3x - 10 = 0$

13 $3x^2 - 7x - 13 = 0$

14 $x^2 - 13x + 30 = 0$

15 $x^2 - 8x + 15 = 0$

16 $9x^2 - 27x + 14 = 0$

17 $16x^2 - 24x + 9 = 0$

18 $4x^2 + 16x + 15 = 0$

19 $x^2 - 8x = 5$

20 $2x^2 + 7x + 3 = 0$

21 $5x^2 + 2x - 7 = 0$

22 $y^2 - 37y + 322 = 0$

23 $5x^2 - 3x - 4 = 0$

24 $x^2 - 12x + 35 = 0$

25 $x^2 - 8x + 13 = 0$

26 $x^2 + 18x + 12 = 0$

27 $x^2 + 2x - 3 = 0$

28 $x^2 + 21x + 10 = 0$

29 $3x^2 + 8x - 4 = 0$

30 $x^2 - 12x = -25$

31 $25x^2 - 25x - 14 = 0$

32 $8x = 4 + 3x^2$

33 $7x^2 + 6x + 1 = 0$

34 $25x^2 - 25x = 14$

35 $12x^2 - 20x + 4 = 0$

36 $18x^2 + 11x - 8 = 0$

37 $5x^2 + 11x + 5 = 0$

38 $20x^2 - 9x - 14 = 0$

39 $12x^2 + 7x + 1 = 0$

40 $12x^2 + 7x - 1 = 0$

41 $13x^2 - 39x = 8$

42 $3y^2 + 12y - 2 = 0$

43 $9y^2 - 27y = 4$

44 $5x^2 - 3x - 1 = 0$

45 $2x^2 + 13x - 3 = 0$

46 $7x^2 + 8x + 2 = 0$

47 $4x^2 - 16x + 1 = 0$

48 $5x^2 + 20x + 1 = 0$

10.4 Solving Quadratic Equations by the Quadratic Formula

The method of completing the square can be generalized to develop a formula that enables us to solve any quadratic equation.

Consider the quadratic equation $ax^2 + bx + c = 0$, with a, b, and c real numbers and $a \neq 0$. By applying the method of completing the square, we can obtain its solution set in terms of a, b, and c as follows:

$$ax^2 + bx + c = 0$$

$$x^2 + \frac{b}{a}x + \frac{c}{a} = 0 \qquad \text{(Dividing by } a\text{)}$$

$$x^2 + \frac{b}{a}x = -\frac{c}{a}$$

Adding $\left[\frac{1}{2}\left(\frac{b}{a}\right)\right]^2 = \frac{b^2}{4a^2}$ to both sides of the equation, we have

$$x^2 + \frac{b}{a}x + \frac{b^2}{4a^2} = \frac{b^2}{4a^2} - \frac{c}{a}$$

or

$$x^2 + \frac{b}{a}x + \frac{b^2}{4a^2} = \frac{b^2 - 4ac}{4a^2}$$

The left-hand side of the equation is the square of $x + \frac{b}{2a}$; so that

$$\left(x + \frac{b}{2a}\right)^2 = \frac{b^2 - 4ac}{4a^2}$$

which implies that

$$x + \frac{b}{2a} = \pm\sqrt{\frac{b^2 - 4ac}{4a^2}} \qquad \text{or} \qquad x + \frac{b}{2a} = \frac{\pm\sqrt{b^2 - 4ac}}{2a}$$

so that

$$x = \frac{-b}{2a} \pm \frac{\sqrt{b^2 - 4ac}}{2a} = \frac{-b \pm \sqrt{b^2 - 4ac}}{2a}$$

Therefore

$$x = \frac{-b + \sqrt{b^2 - 4ac}}{2a} \qquad \text{or} \qquad x = \frac{-b - \sqrt{b^2 - 4ac}}{2a}$$

That is, the solution set is

$$\left\{\frac{-b + \sqrt{b^2 - 4ac}}{2a}, \frac{-b - \sqrt{b^2 - 4ac}}{2a}\right\}$$

The preceding result, which provides a formula for solving any quadratic equation, is called the *quadratic formula*. In summary, if $ax^2 + bx + c = 0$, $a \neq 0$, then

$$x = \frac{-b \pm \sqrt{b^2 - 4ac}}{2a}$$

which is an abbreviated way of writing the two possible solutions.

EXAMPLES

Solve each quadratic equation by using the quadratic formula. Find the solution set.

1 $3x^2 - 4x + 1 = 0$

SOLUTION

$3x^2 - 4x + 1 = 0$

We have $a = 3$, $b = -4$, and $c = 1$. Substituting into the quadratic formula, $x = \dfrac{-b \pm \sqrt{b^2 - 4ac}}{2a}$, we obtain

$$x = \frac{-(-4) \pm \sqrt{(-4)^2 - 4(3)(1)}}{2(3)} \quad \text{or} \quad x = \frac{4 \pm \sqrt{4}}{6}$$

That is,

$$x = \frac{4 \pm 2}{6}$$

Hence,

$$x = \frac{4 + 2}{6} = 1 \quad \text{or} \quad x = \frac{4 - 2}{6} = \frac{1}{3}$$

Hence, the solution set is $\{\frac{1}{3}, 1\}$.

2 $2x^2 + 3x - 1 = 0$

SOLUTION

$2x^2 + 3x - 1 = 0$

With $a = 2$, $b = 3$, and $c = -1$, we have

$$x = \frac{-(3) \pm \sqrt{(3)^2 - 4(2)(-1)}}{2(2)} \quad \text{or} \quad x = \frac{-3 \pm \sqrt{17}}{4}$$

The solution set is $\left\{\frac{-3-\sqrt{17}}{4}, \frac{-3+\sqrt{17}}{4}\right\}$.

3 $x^2 + 2x - 2 = 0$

SOLUTION

$x^2 + 2x - 2 = 0$

Since $a = 1$, $b = 2$, and $c = -2$, we have

$$x = \frac{-(2) \pm \sqrt{(2)^2 - 4(1)(-2)}}{2(1)} \quad \text{or} \quad x = \frac{-2 \pm \sqrt{12}}{2}$$

so that

$$x = \frac{-2 \pm 2\sqrt{3}}{2} \quad \text{or} \quad x = -1 \pm \sqrt{3}$$

The solution set is $\{-1 - \sqrt{3}, -1 + \sqrt{3}\}$.

PROBLEM SET 10.4

In problems 1–40, solve each quadratic equation for x by using the quadratic formula. Find the solution set.

1 $2x^2 - 5x + 1 = 0$

2 $20 = 12x - x^2$

3 $x^2 - 6x + 1 = 0$

4 $x^2 - 18x + 56 = 0$

5 $x^2 + 3x + 2 = 0$

6 $4x^2 + 16x + 15 = 0$

7 $6x^2 - 7x = 5$

8 $x^2 + 3x - 10 = 0$

9 $6x^2 + 37x + 6 = 0$

10 $37x = x^2 + 322$

11 $6x^2 + 17x - 14 = 0$

12 $3x^2 + 8x + 4 = 0$

13 $15x^2 + 2x - 8 = 0$

14 $6x^2 - 2 = x$

15 $21x^2 - 43x + 20 = 0$

16 $7x = 10x^2 - 12$

17 $2x^2 - 3x - 5 = 0$

18 $10x^2 - 29x + 10 = 0$

19 $x^2 + 16x + 48 = 0$

20 $6x^2 + 6 = 13x$

21 $x^2 - 2x - 63 = 0$

22 $2x^2 + 3x + 1 = 0$

23 $12x^2 + 29x - 11 = 0$

24 $5 + x = 6x^2$

25 $32x^2 - 4x - 21 = 0$

26 $4x^2 + 11x = 3$

27 $9x^2 - 16x + 5 = 0$

28 $7x^2 = 12x + 3$

29 $3x^2 - 2x - 8 = 0$

30 $12x^2 - 14x + 3 = 0$

31 $x(x + 3) + 8(x - 8) = 0$

32 $5x = 17 - 3x^2$

33 $(x - 4)^2 + (3 - x)(1 + 2x) = 0$

34 $6x^2 + 3x = 5$

35 $6x^2 + 5x = 2$

36 $2x^2 - x - 2 = 0$

37 $-2x^2 - x + 4 = 0$

38 $x^2 + 5x + 1 = 0$

39 $5x^2 - 12x + 2 = 0$

40 $x^2 + 8x + 4 = 0$

REVIEW PROBLEM SET

In problems 1–10, solve each quadratic equation by extracting roots.

1 $x^2 - 4 = 0$

2 $3x^2 = 27$

3 $(3x + 2)^2 - 5 = 0$

4 $(1 - 5x)^2 - 121 = 0$

5 $(-2x + 5)^2 - 9 = 0$

6 $(3x - 7)^2 - 16 = 0$

7 $(7x - 2)^2 - 25 = 0$

8 $9x^2 - 169 = 0$

9 $(3x - 1)^2 - 64 = 0$

10 $(-5x + 3)^2 - 144 = 0$

In problems 11–34, solve each quadratic equation by factoring. Find the solution set.

11 $6x^2 - x - 1 = 0$

12 $6x^2 - 11x - 2 = 0$

13 $x(3x + 11) = 20$

14 $7x^2 - 11x = 6$

15 $4x^2 - 8x + 3 = 0$

16 $10x^2 + 11x - 6 = 0$

17 $x(6x + 7) = 3$

18 $25x^2 - 10x + 1 = 0$

19 $x^2 - 5x - 84 = 0$

20 $x^2 - 14x - 15 = 0$

21 $4x^2 - 11x + 6 = 0$

22 $6x^2 - 21x - 45 = 0$

23 $(1 - 3x)^2 - 121 = 0$

24 $6x^2 - 13x + 6 = 0$

25 $x^2 + 17x - 38 = 0$

26 $3x^2 - 4x - 7 = 0$

27 $7 - 36x = -5x^2$

28 $x^2 = -20x - 99$

29 $25y^2 - 4 = 0$

30 $2x^2 + 11x + 15 = 0$

31 $3x^2 - 2x - 5 = 0$

32 $6x^2 + x - 2 = 0$

33 $6x^2 + 5x - 6 = 0$

34 $x - 6x^2 + 2 = 0$

In problems 35–60, solve each quadratic equation by completing the square.

35 $x^2 - 10x - 24 = 0$

36 $3x^2 + 4x - 1 = 0$

37 $x^2 + x - 2 = 0$

38 $6x^2 - 31x + 21 = 0$

39 $x^2 + 10x + 13 = 0$

40 $3x^2 - 8x + 1 = 0$

41 $4x^2 - x - 1 = 0$

42 $16 - 6x - 3x^2 = 0$

43 $2x^2 + 5x - 17 = 0$

44 $49x^2 - 98x + 3 = 0$

45 $8x^2 + 6x - 7 = 0$

46 $5 - 8x + x^2 = 0$

47 $3x^2 - 14x + 12 = 0$

48 $2x^2 - 17x + 4 = 0$

49 $4x^2 + 19x + 20 = 0$

50 $15x^2 - 8x + 1 = 0$

51 $3x^2 + 7x + 2 = 0$

52 $5x^2 - 8x + 1 = 0$

53 $x^2 + 3x - 10 = 0$

54 $3x^2 - 5x + 2 = 0$

55 $2x^2 - 5x + 1 = 0$

56 $5x^2 - x - 3 = 0$

57 $x^2 - 2x = 24$

58 $x^2 - 3x = 40$

59 $2x^2 + 3x = 7$

60 $x^2 - 4x = 21$

In problems 61–76, solve each quadratic equation by using the quadratic formula.

61 $x^2 + 12x - 13 = 0$

62 $15x^2 - 7x = 4$

63 $5x^2 - x - 2 = 0$

64 $x(x + 2) = x + 2$

65 $x^2 - x - 3 = 0$

66 $\frac{1}{2}x^2 - 3x = 1$

67 $6x^2 + x - 2 = 0$

68 $4x^2 + 11x - 3 = 0$

69 $x^2 + 3x + 2 = 0$

70 $x^2 - 3x = 1$

71 $x^2 - 8x + 16 = 0$

72 $3x^2 - 7x + 1 = 0$

73 $x^2 = 5x - 6$

74 $2x^2 + x = 3$

75 $x = x^2 - 5$

76 $3x^2 - 5x + 1 = 0$

Appendixes

APPENDIX A MEASUREMENT

Lengths in the Metric System

meter (m)	millimeter (mm)	centimeter (cm)	decimeter	decameter	hectometer	kilometer (km)
1 m =	1000	100	10	$0.1 = \frac{1}{10}$	$0.01 = \frac{1}{100}$	$0.001 = \frac{1}{1000}$

Weights in the Metric System

gram (g)	milligram (mg)	centigram (cg)	decigram	decagram	hectogram	kilogram (kg)
1 g =	1000	100	10	$0.1 = \frac{1}{10}$	$0.01 = \frac{1}{100}$	$0.001 = \frac{1}{1000}$

Lengths in the English System

foot (ft)	inch (in)	yard (yd)	mile
1 ft =	12	$\frac{1}{3}$	$\frac{1}{5{,}280}$

Weights in the English System

pound (lb)	ounce (oz)	ton
1 lb =	16	$\frac{1}{2{,}000}$

Conversion from the English System to the Metric System

1 inch = 2.54 centimeters; 1 meter = 39.37 inches; 1 pound = 0.4536 kilogram

APPENDIX B FORMULAS FROM GEOMETRY

I Areas of Plane Figures

1 Area of a Rectangle

The area of a rectangle is the number of basic units (squares) it contains. In Figure 1 we have shown the number of squares that rectangle

Figure 1

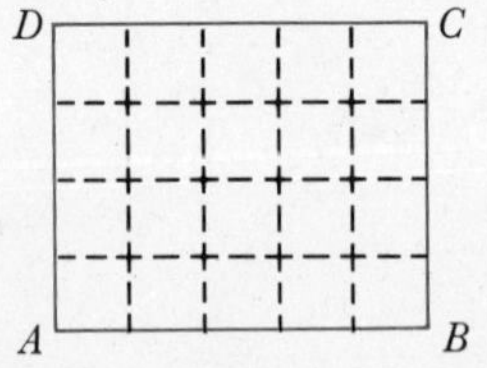

$ABCD$ contains. By actually counting, we find that there are 4 rows of squares, each containing 5 squares, for a total of 20 squares. Repeating this process with many other rectangles, we find that the number of squares contained in a rectangle is always *equal to the product of the base of the rectangle and the side of the rectangle, called the height* (in the example above, $20 = 5 \times 4$). In general, the area of a rectangle is the product of its length and width, or $A = lw$.

EXAMPLE

Find the area of a rectangle whose length is 5 centimeters and width is 7 centimeters.

SOLUTION

$$\begin{aligned} A &= lw \\ &= 5 \times 7 \\ &= 35 \end{aligned}$$

Therefore, the area is 35 square centimeters.

2 Area of a Square

The area of a square can be found by considering the fact that a square is also a rectangle. If we have a square of side s, then both its length and width are equal to s; hence, applying the formula for the area of a rectangle, we have

$$\begin{aligned} A &= lw \\ A &= ss \\ &= s^2 \end{aligned}$$

EXAMPLE

Find the area of a square whose side is 4.3 meters.

SOLUTION

$$\begin{aligned} A &= s^2 \\ &= (4.3)^2 \\ &= 18.49 \end{aligned}$$

Therefore, the area is 18.49 square meters.

3 Area of a Parallelogram

The area of a parallelogram (Figure 2) is equal to the product of the base and the height. In symbols, it is written $A = bh$, where b is the base and h is the height of the parallelogram. The *height* of a parallelogram, then, is the *perpendicular line segment drawn between two parallel sides, one of which is the base.*

Figure 2

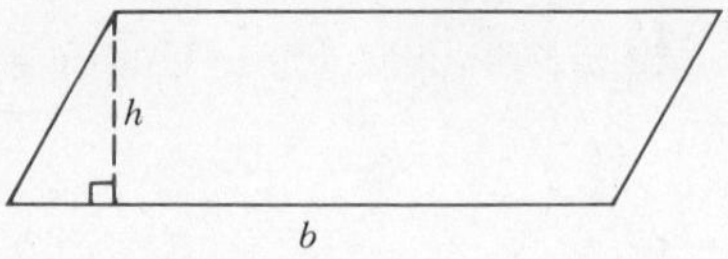

EXAMPLE

Find the area of a parallelogram whose base is 5 feet and whose height is 3 feet.

SOLUTION. The area of the parallelogram is expressed by $A = bh$ or $A = 5$ feet $\times$ 3 feet. Therefore, $A = 15$ square feet.

4 Area of a Triangle

The area of a triangle is equal to one-half the product of its base and height (Figure 3). Symbolically, it is written $A = \frac{1}{2}bh$.

Figure 3

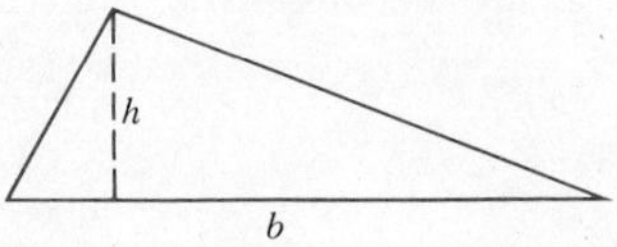

EXAMPLE

Find the area of a triangle whose base is 10 centimeters and whose height is 3.7 centimeters.

SOLUTION. The area of the triangle is $A = \frac{1}{2}bh$ or $A = (\frac{1}{2})(10)$ centimeters $\times$ 3.7 centimeters, or $A = 18.5$ square centimeters.

5 Area of a Trapezoid

The area of a trapezoid is equal to one-half the product of its height and the sum of its bases (Figure 4). Symbolically, $A = \frac{1}{2}h(b + b_1)$.

Figure 4

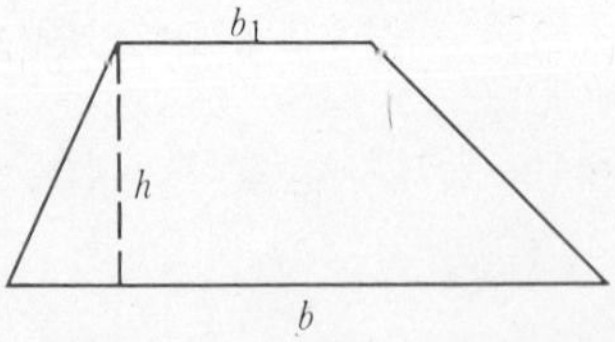

EXAMPLE

Find the area of a trapezoid whose bases are 12 inches and 8 inches and whose height is 5 inches.

SOLUTION. $A = \frac{1}{2}h(b + b_1)$ or $A = \frac{1}{2}$(5 inches) $\times$ (12 inches + 8 inches), or $A = 50$ square inches.

6 Area of a Circle

Given a circle of radius r (Figure 5), its area A is π times r^2, where $\pi =$ 3.1416 . . . . In symbols, $A = \pi r^2$.

Figure 5

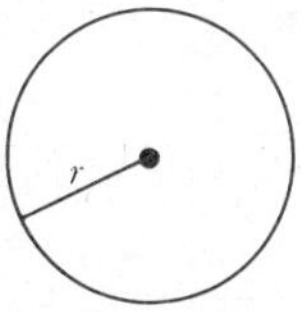

EXAMPLE

Find the area of a circle whose radius is 10 inches.

SOLUTION. $A = \pi r^2$ or $A = \pi(10 \text{ inches})^2$ or $A = 3.1416 \times 100$ or $A = 314.16$ square inches.

II Perimeters of Plane Figures

1 Perimeter of a Rectangle

The perimeter P of a rectangle is the sum of twice its length l and twice its width w (Figure 6). Symbolically, $P = 2l + 2w$.

Figure 6

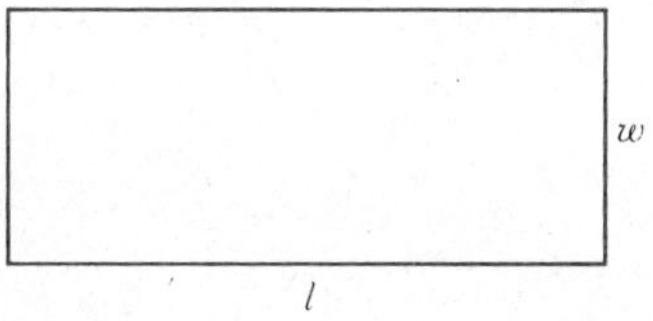

EXAMPLE

Find the perimeter of a rectangle whose length is 8 meters and whose width is 5 meters.

SOLUTION

$$\begin{aligned} P &= 2l + 2w \\ &= 2(8) + 2(5) \\ &= 16 + 10 \\ &= 26 \end{aligned}$$

Therefore, the perimeter is 26 meters.

2 Perimeter of a Square

The perimeter P of a square is four times its side (Figure 7). Symbolically, $P = 4s$.

Figure 7

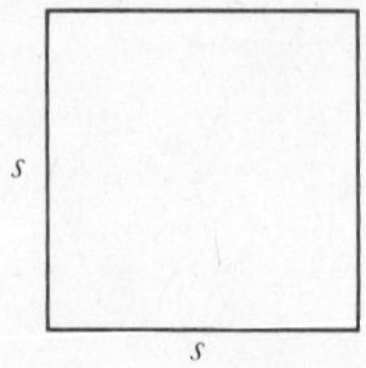

EXAMPLE

Find the perimeter of a square whose side is 8 centimeters.

SOLUTION

$$\begin{aligned} P &= 4s \\ &= 4(8) \\ &= 32 \end{aligned}$$

Therefore, the perimeter is 32 centimeters.

3 Perimeter of a Parallelogram

The perimeter P of a parallelogram is the sum of twice its length l and twice its width w (Figure 8). Symbolically, $P = 2l + 2w$.

Figure 8

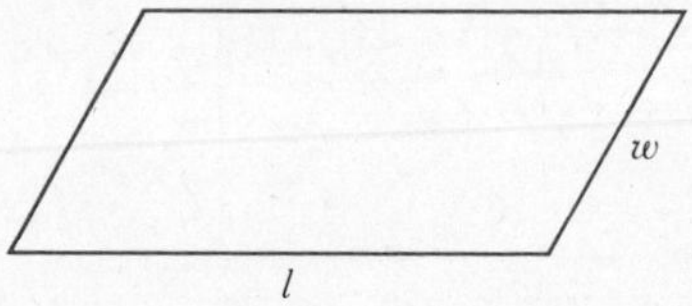

EXAMPLE

Find the perimeter of a parallelogram whose length is 10 inches and whose width is 6 inches.

SOLUTION

$$\begin{aligned} P &= 2l + 2w \\ &= 2(10) + 2(6) \\ &= 20 + 12 \\ &= 32 \end{aligned}$$

Therefore, the perimeter is 32 inches.

4 Perimeter of a Triangle

The perimeter P of a triangle is the sum of the length of its sides (Figure 9). Symbolically, $P = a + b + c$.

Figure 9

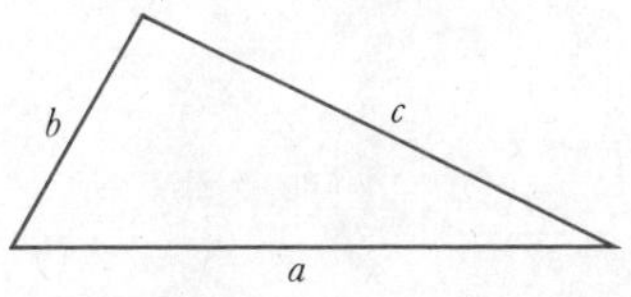

EXAMPLE

Find the perimeter of a triangle with sides 5 inches, 6 inches, and 8 inches.

SOLUTION

$$\begin{aligned} P &= a + b + c \\ &= 5 + 6 + 8 \\ &= 19 \end{aligned}$$

Therefore, the perimeter is 19 inches.

5 Perimeter of a Trapezoid

The perimeter P of a trapezoid is the sum of its sides (Figure 10). Symbolically, $P = a + b + c + d$.

Figure 10

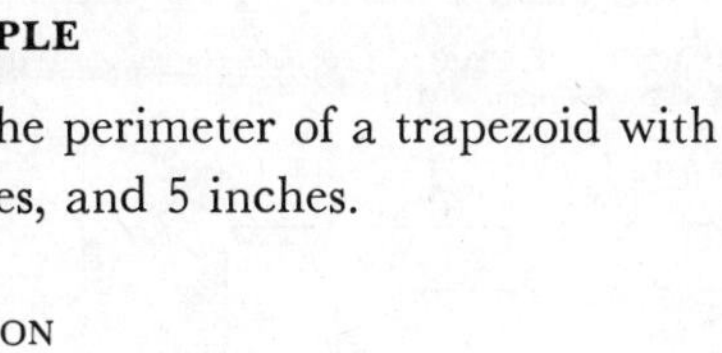

EXAMPLE

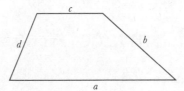

Find the perimeter of a trapezoid with sides 8 inches, 5 inches, 6 inches, and 5 inches.

SOLUTION

$$\begin{aligned} P &= a + b + c + d \\ &= 8 + 5 + 6 + 5 \\ &= 24 \end{aligned}$$

Therefore, the perimeter is 24 inches.

6 Circumference of a Circle

The circumference C of a circle of radius r is 2π times r (Figure 11). Symbolically, $C = 2\pi r$.

Figure 11

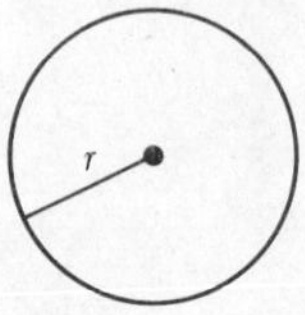

EXAMPLE

Find the circumference of a circle with a radius of 5 inches.

SOLUTION

$$\begin{aligned} C &= 2\pi r \\ &= 2\pi(5) \\ &= 10\pi \end{aligned}$$

Therefore, the circumference is 10π inches.

III Volumes and Surface Area

1 Volume and Surface Area of a Rectangular Parallelepiped

Denote the length of the rectangular solid by l, its width by w, and its height by h. Then its volume V is given by (Figure 12). The

Figure 12

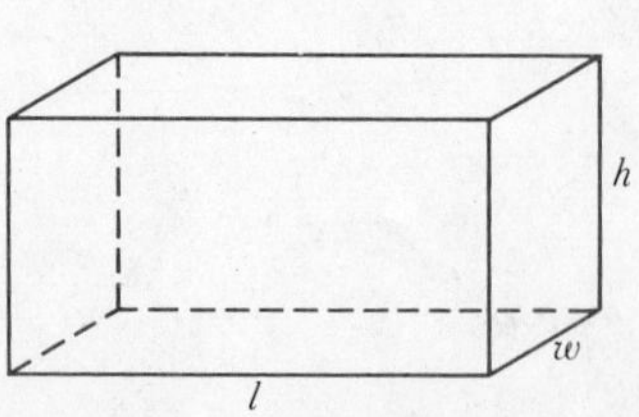

total surface area S of such a solid is the sum of the area of the faces and bases; that is, $S = 2(lw + lh + wh)$.

EXAMPLE

Find the volume and the total surface area of a rectangular box whose dimensions are 7, 5, and 3 centimeters.

SOLUTION. The volume is given by the formula $V = lwh = (7)(5)(3) =$ 105 cubic centimeters. The total surface area is given by the formula $S = 2(lw + lh + wh) = 2[(7)(5) + (7)(3) + (5)(3)] =$ $2(35 + 21 + 15) = 142$ square centimeters.

2 Volume and Surface Area of a Right Circular Cylinder

Denote the altitude of the *right circular cylinder* by h and the base by r. Its volume is given by $V = \pi r^2 h$, its lateral surface area. *L.S.* is given by $L.S. = 2\pi rh$, and its total surface area S is given by $S = 2\pi r^2 + 2\pi rh$ (Figure 13).

Figure 13

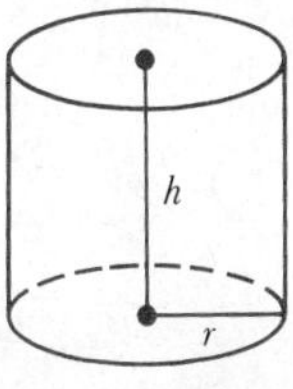

EXAMPLE

Find the lateral surface area, total surface area, and volume of a right circular cylinder of radius 3 centimeters and altitude 6 centimeters.

SOLUTION

$$\begin{aligned} L.S. &= 2\pi rh = 2\pi(3)(6) = 36\pi \text{ square centimeters} \\ S &= 2\pi r^2 + 2\pi rh \\ &= 2\pi(3)^2 + 36\pi \\ &= 18\pi + 36\pi \\ &= 54\pi \text{ square centimeters} \\ V &= \pi r^2 h = \pi(3)^2(6) = 54\pi \text{ cubic centimeters} \end{aligned}$$

3 Volume and Surface Area of the Pyramid

A *pyramid* is a three-dimensional figure whose base is a polygon and whose lateral faces are triangles (Figure 14). If the base of the pyramid

Figure 14

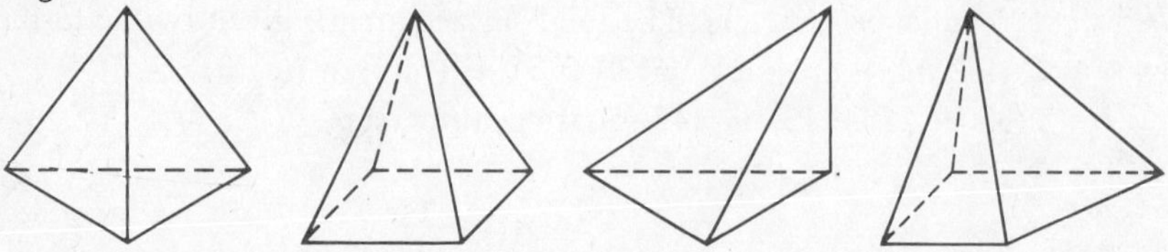

(*a*) (*b*) (*c*) (*d*)

is a regular polygon and the pyramid has equal lateral edges, it is called a *regular pyramid* (Figure 14*a* and *b*).

The volume V of a regular pyramid is expressed by the formula $V = \frac{1}{3}Ah$, where A is the area of the base and h is the altitude. Its lateral surface area L.S. is expressed by the formula L.S. $= \frac{1}{2}Pl$, where P is the perimeter of the base and l is the slant height (the height of a lateral face) (Figure 15).

Figure 15

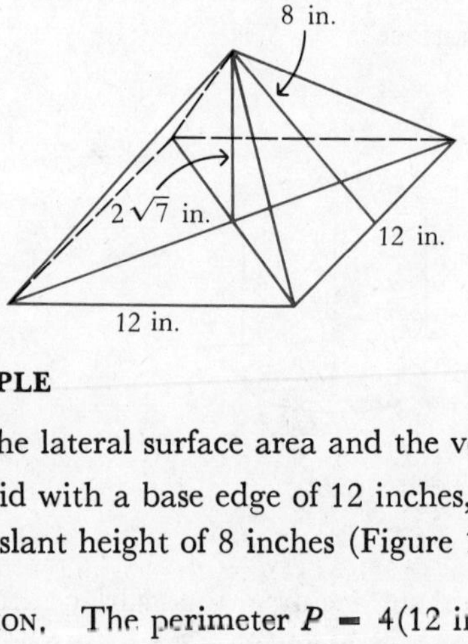

EXAMPLE

Find the lateral surface area and the volume of a regular square pyramid with a base edge of 12 inches, a height of $2\sqrt{7}$ inches, and a slant height of 8 inches (Figure 15).

SOLUTION. The perimeter $P = 4(12 \text{ inches}) = 48$ inches. Therefore, the lateral surface area is

$$\begin{aligned}\tfrac{1}{2}Pl &= \tfrac{1}{2}(48 \text{ inches})(8 \text{ inches})\\ &= 192 \text{ square inches}\end{aligned}$$

The volume of the pyramid is

$$\begin{aligned}V &= \tfrac{1}{3}Ah\\ &= \tfrac{1}{3}(144 \text{ square inches})(2\sqrt{7} \text{ inches})\\ &= 96\sqrt{7} \text{ cubic inches}\end{aligned}$$

4 Volume and Surface Area of a Right Circular Cone

The *right circular cone* resembles the pyramid except that the base of the cone is a circle (Figure 16). The volume of a right circular cone is given

Figure 16

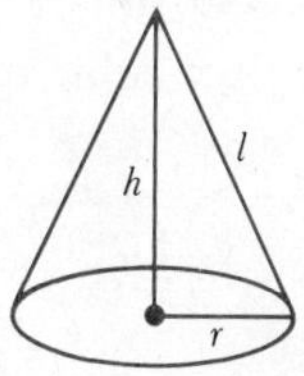

by the formula $V = \frac{1}{3}\pi r^2 h$, where r is the radius and h is the height. The lateral surface area is given by the formula L.S. $= \pi r l$, where l is the slant height.

EXAMPLE

Find the lateral surface area and the volume of a right circular cone of radius 6 inches, altitude 8 inches, and slant height 10 inches.

SOLUTION

$$\text{L.S.} = \pi r l = \pi(6 \text{ inches})(10 \text{ inches}) = 60\pi \text{ square inches}$$
$$V = \tfrac{1}{3}\pi r^2 h = \tfrac{1}{3}(36\pi \text{ square inches})(8 \text{ inches}) = 96\pi \text{ cubic inches}$$

5 Volume and Surface Area of a Sphere

A *sphere* is the set of all points in space at a given distance from a given point (Figure 17). Let r be the radius of the sphere; then the surface area

Figure 17

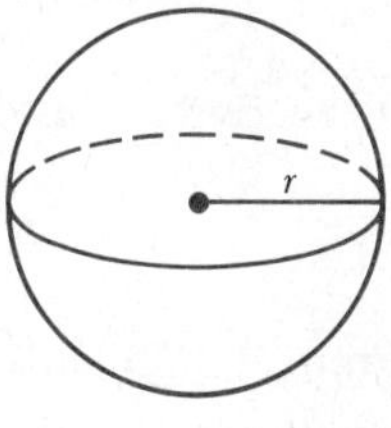

S is expressed by the formula $S = 4\pi r^2$ and the volume V is expressed by the formula $V = \frac{4}{3}\pi r^3$.

EXAMPLE

Find the surface area and the volume of a sphere of radius 3 meters.

SOLUTION

$$S = 4\pi r^2 = 4\pi(3)^2 = 36\pi \text{ square meters}$$
$$V = \tfrac{4}{3}\pi r^3 = \tfrac{4}{3}\pi(3)^3 = 36\pi \text{ cubic meters}$$

PROBLEM SET

1 Find the area of the following rectangles.

a) $h = 3$ inches, $b = 5$ inches
b) $h = 4.01$ centimeters, $b = 6.32$ centimeters

2 Find the area of the following parallelograms.

a) $h = 4.32$ feet, $b = 5.7$ feet
b) $h = 6\frac{3}{5}$ inches, $b = 4\frac{1}{5}$ inches

3 Find the area of the following triangles.

a) $h = 6$ inches, $b = 10$ inches
b) $h = 16$ centimeters, $b = 11$ centimeters

4 Find the area of the following squares.

a) $s = 3.5$ inches
b) $s = 16$ feet

5 Find the area of the following trapezoids.

a) $h = 6$ feet, $b = 11$ feet, $b_1 = 4$ feet
b) $h = 7.5$ inches, $b = 4.3$ inches, $b_1 = 5.2$ inches

6 Find the area of the following circles.

a) $r = 5$ inches
b) $r = 7$ feet

7 Find the perimeter of a rectangle with the indicated length and width.

a) $l = 8$ inches, $w = 2$ inches
b) $l = 7$ meters, $w = 4$ meters

8 Find the perimeter of a square with the given side.

a) $s = 10$ inches
b) $s = 5$ meters

9 Find the perimeter of the parallelogram $ABCD$ with the given dimensions.

a) $l = 10$ inches, $w = 6$ inches
b) $l = 8$ inches, $w = 4$ inches

10 Find the perimeter of a triangle with the given sides.

a) 6 inches, 4 inches, and 8 inches
b) 5 meters, 4 meters, and 3 meters

11 Find the perimeter of a trapezoid $ABCD$ (Figure 18) with the given dimensions.

a) $a = 10$ inches, $b = 4$ inches, $c = 6$ inches, and $d = 4$ inches
b) $a = \frac{1}{2}$ foot, $b = \frac{1}{4}$ foot, $c = \frac{1}{3}$ foot, and $d = \frac{1}{4}$ foot

Figure 18

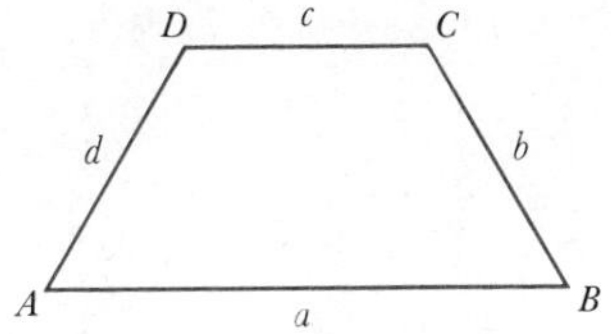

12 Find the circumference of a circle with the given radius.

a) $r = 4$ centimeters
b) $r = 5$ inches

13 Find the volume and the total surface area of the rectangular box with the following dimensions.

a) $l = 3$ meters, $w = 4$ meters, and $h = 5$ meters
b) $l = 2$ feet, $w = 1$ foot, and $h = 3$ feet

14 Find the lateral surface area, total surface area, and the volume of the following cylinders.

a) $r = 2$ inches, $h = 4$ inches
b) $r = 3$ feet, $h = 4$ feet

15 Find the total surface area and the volume of the following regular pyramids.

a) Area of the square base = 16 square inches, the slant height = $2\sqrt{10}$ inches, and the altitude = 6 inches.
b) The base is an equilateral triangle whose perimeter is 18 meters, the slant height is 10 meters, and the altitude = $\sqrt{97}$ meters.

16 Find the missing part for each of the following right circular cones.

a) $r = 3$ inches, $h = 4$ inches, $l = 5$ inches, L.S. = ?, S = ?, and V = ?
b) $r = 5$ centimeters, $h = 12$ centimeters, $l = 13$ centimeters, L.S. = ?, S = ?, and V = ?

17 Find the surface area and volume of a sphere with the given radius.

a) 5 centimeters
b) 2 inches

APPENDIX C

Powers and Roots

Number	Square	Square Root	Cube	Cube Root
1	1	1.000	1	1.000
2	4	1.414	8	1.260
3	9	1.732	27	1.442
4	16	2.000	64	1.587
5	25	2.236	125	1.710
6	36	2.449	216	1.817
7	49	2.646	343	1.913
8	64	2.828	512	2.000
9	81	3.000	729	2.080
10	100	3.162	1,000	2.154
11	121	3.317	1,331	2.224
12	144	3.464	1,728	2.289
13	169	3.606	2,197	2.351
14	196	3.742	2,744	2.410
15	225	3.873	3,375	2.466
16	256	4.000	4,096	2.520
17	289	4.123	4,913	2.571
18	324	4.243	5,832	2.621
19	361	4.359	6,859	2.668
20	400	4.472	8,000	2.714
21	441	4.583	9,261	2.759
22	484	4.690	10,648	2.802
23	529	4.796	12,167	2.844
24	576	4.899	13,824	2.884
25	625	5.000	15,625	2.924
26	676	5.099	17,576	2.962
27	729	5.196	19,683	3.000
28	784	5.292	21,952	3.037
29	841	5.385	24,389	3.072
30	900	5.477	27,000	3.107
31	961	5.568	29,791	3.141
32	1,024	5.657	32,768	3.175
33	1,089	5.745	35,937	3.208
34	1,156	5.831	39,304	3.240
35	1,225	5.916	42,875	3.271
36	1,296	6.000	46,656	3.302
37	1,369	6.083	50,653	3.332
38	1,444	6.164	54,872	3.362
39	1,521	6.245	59,319	3.391
40	1,600	6.325	64,000	3.420
41	1,681	6.403	68,921	3.448
42	1,764	6.481	74,088	3.476
43	1,849	6.557	79,507	3.503
44	1,936	6.633	85,184	3.530
45	2,025	6.708	91,125	3.557
46	2,116	6.782	97,336	3.583
47	2,209	6.856	103,823	3.609
48	2,304	6.928	110,592	3.634
49	2,401	7.000	117,649	3.659
50	2,500	7.071	125,000	3.684
51	2,601	7.141	132,651	3.708
52	2,704	7.211	140,608	3.733
53	2,809	7.280	148,877	3.756
54	2,916	7.348	157,464	3.780
55	3,025	7.416	166,375	3.803
56	3,136	7.483	175,616	3.826
57	3,249	7.550	185,193	3.849
58	3,364	7.616	195,112	3.871
59	3,481	7.681	205,379	3.893
60	3,600	7.746	216,000	3.915
61	3,721	7.810	226,981	3.936
62	3,844	7.874	238,328	3.958
63	3,969	7.937	250,047	3.979
64	4,096	8.000	262,144	4.000
65	4,225	8.062	274,625	4.021
66	4,356	8.124	287,496	4.041
67	4,489	8.185	300,763	4.062
68	4,624	8.246	314,432	4.082
69	4,761	8.307	328,509	4.102
70	4,900	8.367	343,000	4.121
71	5,041	8.426	357,911	4.141
72	5,184	8.485	373,248	4.160
73	5,329	8.544	389,017	4.179
74	5,476	8.602	405,224	4.198
75	5,625	8.660	421,875	4.217
76	5,776	8.718	438,976	4.236
77	5,929	8.775	456,533	4.254
78	6,084	8.832	474,552	4.273
79	6,241	8.888	493,039	4.291
80	6,400	8.944	512,000	4.309
81	6,561	9.000	531,441	4.327
82	6,724	9.055	551,368	4.344
83	6,889	9.110	571,787	4.362
84	7,056	9.165	592,704	4.380
85	7,225	9.220	614,125	4.397
86	7,396	9.274	636,056	4.414
87	7,569	9.327	658,503	4.431
88	7,744	9.381	681,472	4.448
89	7,921	9.434	704,969	4.465
90	8,100	9.487	729,000	4.481
91	8,281	9.539	753,571	4.498
92	8,464	9.592	778,688	4.514
93	8,649	9.644	804,357	4.531
94	8,836	9.695	830,584	4.547
95	9,025	9.747	857,375	4.563
96	9,216	9.798	884,736	4.579
97	9,409	9.849	912,673	4.595
98	9,604	9.899	941,192	4.610
99	9,801	9.950	970,299	4.626
100	10,000	10.000	1,000,000	4.642

ANSWERS TO SELECTED PROBLEMS

CHAPTER 1

PROBLEM SET 1.1, PAGE 8

1. $10 = 3 + 7$
3. $30 = 6 \times 5$
5. 36
7. 34
9. 34
11. 63
13. 19
15. 4
17. 12
19. 9
21. 3
23. $\frac{26}{25}$
25. 1
27. $\frac{1}{3}$
29. 180
31. $6(x + 5)$
33. $7(x + 5)$
35. $10x + 3 - 5x$
37. 7
39. 5
41. 6
43. 8
45. 11
47. 19

49. 13
51. 19
53. 2
55. 119

PROBLEM SET 1.2, PAGE 14

1. $4 \in A$ $6 \in B$ $9 \in C$ $1 \notin C$
$10 \notin A$ $7 \notin B$ $8 \in C$ $4 \notin C$
$6 \in A$ $5 \in B$ $10 \in C$ $3 \notin C$
3. {Tuesday, Thursday}
5. {2,3,4,5,6,7,8,9}
7. $\{x \mid x = 5n, n \in N\}$
9. $\{x \mid x$ is a counting number greater than 5 and less than 18$\}$
11. finite
13. infinite
15. infinite
17. true
19. true
21. −3, −1, 1, 3, 5, 7
23. 1, 3, 5, 7, 9, 11
25. Ø, {3}
27. Ø, {1}, {2}, {3}, {1,2}, {1,3}, {2,3}, {1,2,3}
29. {1,2,3,4,5,6,7}
31. {1,2,4,5,6,8}
33. {2,6}
35. {5,6}
37. {1,2,3,4,5,6,7}
39. {4}
41. {6}
43. {5,6,4,8}
45. Ø

PROBLEM SET 1.3, PAGE 21

1.
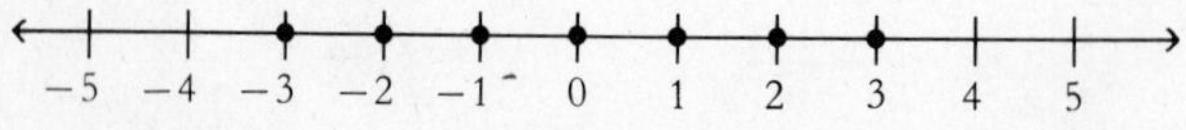

3.
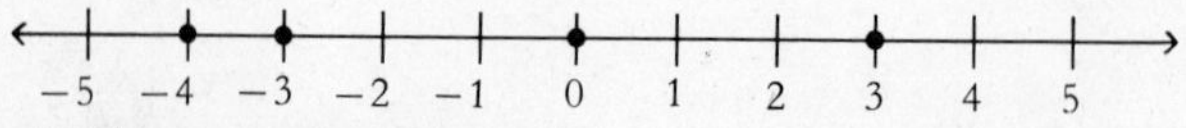

5.
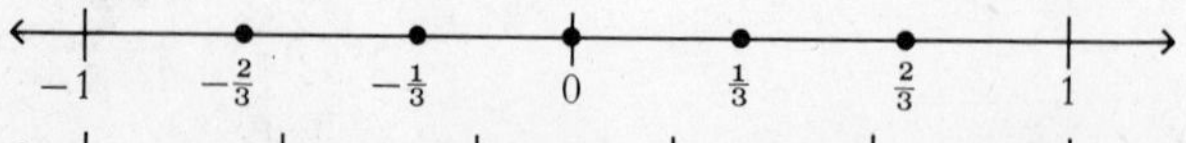

7.
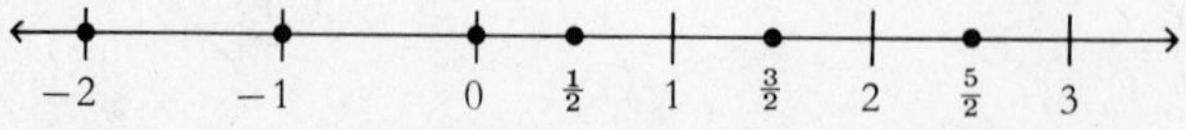

9. 0.3
11. 1.5
13. 0.15

15. 0.091
17. -0.8
19. 0.75
21. -0.625
23. $-0.\overline{6}$
25. 0.35
27. $-0.91\overline{6}$
29. 0.68
31. 0.2
33. -0.24
35. $0.4\overline{6}$
37. $\frac{43}{100}$
39. $\frac{72}{100}$
41. $\frac{27}{100}$
43. $\frac{264}{100}$
45. $-\dfrac{8}{1,000}$
47. $-\dfrac{4,008}{100}$
49. $-\dfrac{582}{1,000}$
51. $-\dfrac{209}{1,000}$
53. $-\dfrac{527}{10,000}$
55. $-\dfrac{52}{100,000}$

PROBLEM SET 1.4, PAGE 26

1. -9
3. 19
5. 13
7. 3
9. -4
11. -17
13. 100
15. -179
17. 9
19. -6
21. 31
23. 41
25. -35
27. -65

29. 0
31. 3
33. 9
35. 2
37. 5
39. 8 or −8
41. 15 or −15
43. 1
45. 3 or −3
47. true
49. true

PROBLEM SET 1.5, PAGE 31

1. 16
3. 38
5. 4
7. −12
9. 31
11. −91
13. −43
15. 0
17. −9
19. −6
21. 9
23. −57
25. 0
27. −14
29. −18
31. −9
33. −1
35. 3
37. −78
39. 5
41. 9
43. 12
45. −6
47. 27
49. 36
51. −29
53. −72
55. 26
57. 20

PROBLEM SET 1.6, PAGE 37

1. 15
3. 133
5. −14
7. −155
9. −112
11. −90
13. 5
15. 20
17. 66
19. −4
21. −1
23. −15
25. 30
27. −42
29. −270
31. 2
33. 2
35. −2
37. −7
39. −7
41. −3
43. −17
45. −36
47. 4
49. 2
51. 5
53. 0
55. 3
57. 81
59. undefined

PROBLEM SET 1.7, PAGE 47

1. no
3. closed under multiplication and addition
5. 9
7. 84
9. 45
11. −84
13. 208
15. −12
17. 385
19. −234

21. 128
23. $-1{,}316$
25. 234
27. 90
29. $-12, \frac{1}{12}$
31. $3, -\frac{1}{3}$
33. $-\frac{1}{5}, 5$
35. $\frac{1}{6}, -6$
37. commutative property for addition
39. commutative property for multiplication
41. closure property for multiplication
43. associative property for multiplication
45. distributive property
47. identity property for multiplication
49. Property 12
51. multiplication property of equality
53. addition property of equality
55. Property 10

PROBLEM SET 1.8, PAGE 51

1. 23
3. -7
5. 23
7. 133
9. -49
11. 260
13. 384
15. 50
17. -10
19. -9
21. 42
23. 131
25. 2

REVIEW PROBLEM SET, PAGE 52

1. 19
3. 43
5. 22
7. $\frac{17}{10}$
9. 23
11. 43
13. 71

15. $2x + 3$

17. $x + 10$

19. true

21. false

23. true

25. $\{x \mid x = 2n \text{ and } n \in \{0,1,2,3,4\}\}$

27. $\{x \mid x = 5n \text{ and } n \in \{0,1,2,3,4,5,6\}\}$

29. $\{1,2,3,4,5\}$

31. $\{2,3\}$

33. $\{2\}$

35. $\{2,3,4,5,7,8\}$

37. $\{1,2,3\}$

39. 2,4,6,8,10

41.

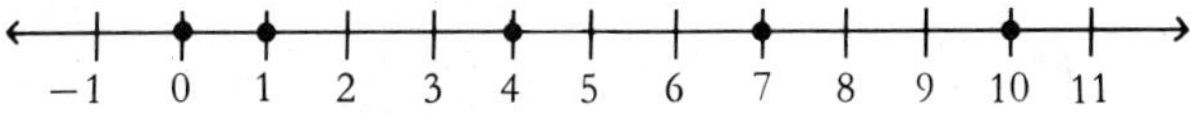

43.

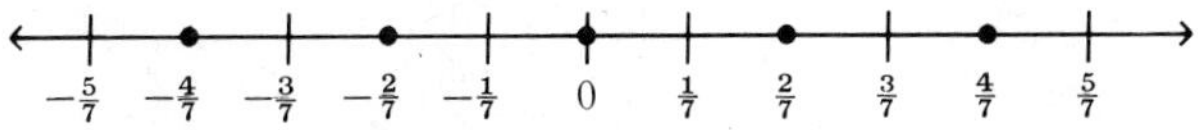

45.

47. 0.275

49. −0.68

51. $0.\overline{7}$

53. $0.3\overline{571428}$

55. $\frac{97}{100}$

57. $-\frac{75}{100}$

59. $\frac{1,111}{10,000}$

61. 4

63. 16

65. −1,282

67. 198

69. 529

71. 81

73. −146

75. 28

77. 9,975

79. 1,774,182

81. −2

83. 16

85. −56

87. 4

89. −8

91. 14

93. 13

95. 12
97. 110
99. 8
101. 20
103. commutative property for addition
105. commutative property for multiplication
107. additive inverse
109. Property 12
111. Property 11c
113. addition property of equality
115. Property 11c
117. cancellation property for addition
119. multiplication property of equality
121. -177
123. 9

CHAPTER 2

PROBLEM SET 2.1, PAGE 66

	Base	*Exponent*
1.	3	16
3.	-4	5
5.	-1	25
7.	$2x$	5

9. 81
11. 125
13. 10
15. 0
17. 19
19. 5^6
21. $(-1)^5$
23. $3^2 + (-6)^3$
25. a^8
27. $-x^6y^4$
29. 2^5
31. 7^6
33. $(-2)^8$
35. $(-4)^5$
37. x^{12}
39. $(-b)^3$

41. 3^6
43. $(-5)^{12}$
45. x^8
47. z^{45}
49. x^{24}
51. $8x^3$
53. u^2v^2
55. $625x^4$
57. $27x^3$
59. $16r^4s^4$
61. x^6y^{12}
63. $625x^{12}$
65. $-16x^7y^{12}$
67. $16a^2x^{18}y^{34}$
69. $9r^{14}s^{11}$

PROBLEM SET 2.2, PAGE 72

1. yes
3. yes
5. no
7. yes
9. no
11. trinomial; degree = 2; 1, −5, 6
13. degree = 3; 2, 6, −5, 4
15. monomial; degree = 1; −5
17. binomial; degree = 5; −8, 15
19. monomial; degree = 0; 11
21. monomial; degree = 1; $-\frac{1}{2}$
23. binomial; degree = 1; $-\frac{7}{9}$, $\frac{1}{3}$
25. 7
27. 10
29. 7
31. 33
33. 61
35. 6
37. 1
39. 9

PROBLEM SET 2.3, PAGE 79

1. $17x$
3. $8y$
5. $2xy$

7. $13ab$
9. $3x^2y^3 + 2x^3y^3$
11. $8x + 11$
13. $3x^2 + 3x + 2$
15. $5x^3 - 3x^2 + 3x + 5$
17. $3y^4 - 5y^3 + 7y + 7$
19. $x^2y^2 + 5x^2y + 2xy^2$
21. $2x$
23. $25y$
25. $-6z^3$
27. $-8uv$
29. $15x^2y$
31. $3y + 2$
33. $x^2 + x + 4$
35. $5x^2 - 2xy - 7y^2$
37. $3x^3 - 6x^2 + 2x - 2$
39. $-x^4 + 4x^3 + 3x^2 + 2$
41. $6x^2 - 5x + 10$
43. $4y^3 - 6y - 3$
45. $6x^3 + 3x^2 - 6x + 10$
47. $z^4 - 5z^2 + 3$
49. $2x^5 + 2x^4 + 7x^3 - 8x^2 + 4x - 5$

PROBLEM SET 2.4, PAGE 86

1. $8x^7$
3. $-21y^7$
5. $15x^3y^5$
7. $-40x^8$
9. $-14x^6y^3$
11. $5x^2 + 2x$
13. $6x^2 - 12x$
15. $3x^3 - 3x^2 + 15x$
17. $4a^3b^3 - 8a^2b^3 + 12ab^4$
19. $-16x^3y^2 + 24x^2y^3 - 8x^3y^4$
21. $x^2 + 6x + 5$
23. $3x^2 + 4x - 15$
25. $6x^2 - 5x - 6$
27. $4x^4 - 3x^2 - 1$
29. $2x^3 - 7x^2 + 11x - 4$
31. $x^3 - x^2y - 7xy^2 + 15y^3$
33. $x^4 - 2x^3 + 5x^2 - 4x + 6$
35. $12y^4 - 3y^3z + 7y^2z^2 + 2yz^3 - 10z^4$
37. $x^4 - 2x^3 - 4x^2 + 2x + 3$

39. $3xy^2 + 5x^2y + 3x^3 - xy^3 - 2x^2y^2 - 3x^3y + y^3$

PROBLEM SET 2.5, PAGE 90

1. $x^2 + 3x + 2$
3. $x^2 + 2x - 15$
5. $y^2 - 6y + 8$
7. $x^2 + 15x + 56$
9. $z^2 - 121$
11. $x^2 - 12x + 36$
13. $2x^2 + 11xy + 15y^2$
15. $3x^2 + 11x - 4$
17. $12y^2 + 13yz - 14z^2$
19. $x^2 + 9xy + 20y^2$
21. $6x^2 - 13xy + 6y^2$
23. $6x^2 - 5xy - 6y^2$
25. $6x^2 + 5xy - 6y^2$
27. $54y^2 - 15y - 56$
29. $50x^2 - 35xy + 6y^2$
31. $x^2 + 8x + 7$
33. $x^2 - 18x + 80$
35. $y^2 - 11y - 26$
37. $x^2 + 9xy + 14y^2$
39. $z^2 - 4yz - 45y^2$
41. $2x^2 + 17x + 21$
43. $6x^2 + x - 40$
45. $12y^2 - 69y + 99$
47. $6x^2 + 11xy + 3y^2$
49. $35y^2 + 24yz - 35z^2$

PROBLEM SET 2.6, PAGE 101

1. 4
3. 9
5. x^5
7. y^5
9. $8x^3$
11. $\frac{1}{25}$
13. $\frac{1}{x^4}$
15. 1
17. $\frac{8}{27}$
19. $\frac{x^5}{y^5}$

21. $\frac{y^4}{z^4}$
23. $-\frac{x^3}{z^3}$
25. $\frac{8x^3}{y^6}$
27. $4x^3$
29. $3x^3y^2$
31. $\frac{-5xy^2}{z^2}$
33. $2x^2 + 5x$
35. $5xy - 7x^3y^3 + 3x^5y^2$
37. $2x^2z^2 - \frac{3x}{z} + \frac{5z}{x}$
39. $\frac{1}{zy} - \frac{2}{z^2} + \frac{3y}{z^3}$
41. $x + 3$
43. $x - 6$
45. $y + 1$
47. $x^2 + 2x + 1$
49. $x^2 + 2x + 3$
51. $x - 3$
53. $3x^2 + 6x - 6 + \frac{-17x + 8}{x^2 - x + 2}$
55. $x^2 - 4xy + y^2$
57. $4x^2 + 6xy + 9y^2$
59. $x^4 - x^3y + x^2y^2 - xy^3 + y^4$
61. $x^2 + 2$

REVIEW PROBLEM SET, PAGE 103

1. 64
3. 3
5. 25
7. $3x^3y$
9. $-4y^2z + 7y^2z^3$
11. $y^3 + 2yx^2 - 3x^3$
13. 3^6
15. x^{13}
17. y^7
19. 2^6
21. x^{12}
23. y^{20}
25. $9y^2$

27. x^4y^4
29. $x^{12}y^8$
31. $8x^7y^8$
33. yes
35. no
37. yes
39. binomial; degree = 2; 2, −7
41. monomial; degree = 2; −7
43. trinomial; degree = 2; 1, 4, −3
45. degree = 3; −1, 6, −2, 4
47. 3
49. 4
51. 6
53. −9
55. $20xy^2$
57. $7xyz$
59. $20xy^3$
61. $11x + 25$
63. $3x^2 + 4x - 6$
65. $5y^3 - 8y^2 + 5y + 20$
67. $9x^2$
69. $9y^3$
71. $2x - 15$
73. $x^3 - 2x^2 + 11x - 5$
75. $4x^3y^2 + 2x^2y - 5xy^3$
77. $5y^2 + 4y - 6$
79. $5x^2 - 8x - 9$
81. $-21x^4y^2z^3$
83. $6y^4 + 15y^2 - 12y$
85. $x^3 + x^2 - x + 2$
87. $6x^3 + 7x^2y - xy^2 + 3y^3$
89. $x^4 - x^3 - 3x^2 + x + 2$
91. $x^2 + 13x + 42$
93. $y^2 - 10y + 9$
95. $2x^2 - 9x - 56$
97. $3y^2 - 7yz - 20z^2$
99. $30x^2 - 43xy + 15y^2$
101. 49
103. $\frac{1}{x^2}$
105. $\frac{x^7}{y^7}$
107. 16
109. $9x^2$

111. $5xy^2$
113. $4xy^4 - 5x^3y^2$
115. $4x^3 + 3 - \frac{1}{x^4}$
117. $x + 9$
119. $x^2 - 2x + 3$
121. $x^2 + 7x + 3$
123. $x^3 + 2x^2y - xy^2 + y^3$

CHAPTER 3

PROBLEM SET 3.1, PAGE 114

1. $x(x - 1)$
3. $y(y - 7)$
5. $3t^2(t + 2)$
7. $5x(2x - 1)$
9. $a(x - y)$
11. $ab(a - b)$
13. $\pi rh(r + 2)$
15. $6rs(s + 5r)$
17. $6pq(p + 4q)$
19. $2(x^2 + 2x + 3)$
21. $x(x^2 + 3x + 1)$
23. $x^3(x^2 + 2x + 3)$
25. $3x^2(x - 2x^3 + 5)$
27. $xy(4x + 12 - 7y)$
29. $5(x - 2x^2y^2 - 5y^2)$
31. $-mn(m + n + 3mn + 6)$
33. $11xy^2(11xy + 5x^2 + 7y^2)$
35. $8xy^3(x^2y^3 - 2y^2 + 3x)$
37. $3a^2x^3(8a^3 - 7x + 6)$
39. $-5rt(7r^2 + 5rt + 3t^2)$

PROBLEM SET 3.2, PAGE 118

1. $x^2 + 10x + 25$
3. $y^2 + 12y + 36$
5. $4x^2 + 28x + 49$
7. $16 + 24yz + 9y^2z^2$

9. $9x^2 + 12xy + 4y^2$
11. $x^2 - 4x + 4$
13. $y^2 - 14y + 49$
15. $9x^2 - 30x + 25$
17. $x^2 - 18xy + 81y^2$
19. $25x^2 - 70xy + 49y^2$
21. $4x^2y^2 - 36xy + 81$
23. $36x^2 - 84xy + 49y^2$
25. $x^2 - 16$
27. $x^2 - 100$
29. $y^2 - 64$
31. $9x^2 - y^2$
33. $25 - x^2$
35. $36x^2 - 49y^2$
37. $64y^2 - 49z^2$
39. $225x^2 - 81y^2$

PROBLEM SET 3.3, PAGE 124

1. $(x + 1)^2$
3. $(x - 2)^2$
5. $(2x + 5)^2$
7. $(x - 3y)^2$
9. $(4y + 5)^2$
11. $(3z + 7)^2$
13. $(6x - 1)^2$
15. $(4y + 3z)^2$
17. $(8x - 3)^2$
19. $(6x + 5y)^2$
21. $(ab - 3)^2$
23. $(2a^2 + 3)^2$
25. $(x - 3)(x + 3)$
27. $(2 - x)(2 + x)$
29. $(2x - 1)(2x + 1)$
31. $(5 - 9y)(5 + 9y)$
33. $(2a - b)(2a + b)$
35. $(5 - 7a)(5 + 7a)$
37. $(2x - 4)(2x + 4)$
39. $(ab - 1)(ab + 1)$
41. $(xz - 4)(xz + 4)$
43. $(xy - z)(xy + z)$
45. $(3m - 12n)(3m + 12n)$
47. $(3x - 4z)(3x + 4z)$
49. $(9ab - 5c)(9ab + 5c)$

PROBLEM SET 3.4, PAGE 128

1. $(x + 1)(x + 2)$
3. $(x + 1)(x + 5)$
5. $(y - 3)(y - 7)$
7. $(z - 2)(z - 5)$
9. $(x + 3)(x - 1)$
11. $(x - 6)(x + 2)$
13. $(y - 5)(y - 6)$
15. $(x + 7)(x + 8)$
17. $(x + 8)(x - 5)$
19. $(z - 2)(z - 10)$
21. $(x - 6)(x + 5)$
23. $(y + 9z)(y - 7z)$
25. $(x - 8y)(x + 4y)$
27. $(x + 6z)(x + 7z)$
29. $(y - 9z)(y + 5z)$
31. $(x - 9)^2$
33. $(x + 11y)(x - 3y)$
35. $(x + 11)(x - 10)$
37. $(x + 14y)(x - 3y)$
39. $(x - 12z)(x + 3z)$

PROBLEM SET 3.5, PAGE 133

1. $(2x + 3)(x + 1)$
3. $(3x + 1)(x + 2)$
5. $(2y - 3)(y - 2)$
7. $(5x - 2)(x + 1)$
9. $(3x + 2)(2x + 1)$
11. $(4z - 1)(z + 2)$
13. $(2x - 5)(x - 3)$
15. $(3y + 4)(2y - 3)$
17. $(4x + 3y)(3x - 2y)$
19. $(2z - 3)(z + 8)$
21. $(6y + z)(3y - 2z)$
23. $(5x - 7y)(3x + 2y)$
25. $(6x - 1)(4x - 5)$
27. $(3x - 7y)(2x + 5y)$
29. $(7y - 2)(6y + 1)$
31. $(2x + 5)(x - 6)$
33. $(7x + 3y)(x + 4y)$
35. $(3y - 4)(3y - 5)$
37. $(8z - 3)(6z - 7)$
39. $(5x - 6y)(4x + 5y)$

PROBLEM SET 3.6, PAGE 138

1. $5x(x - y)(x + y)$
3. $xy(x - 3)(x + 1)$
5. $ax(x - 2)^2$
7. $x(2x + y)(2x - 5y)$
9. $x(3x + y)(x - y)$
11. $(x + 2)(a + b)$
13. $(2x + y)(3 - z)$
15. $m(1 + n)(x - 4)$
17. $(a + 3)(b - 5)$
19. $(a - x)(a + b)$
21. $(a^2 - 5)(x - 2)(x + 2)$
23. $(b + 3)(c - 4)$
25. $(x + y)(x^2 - xy + y^2)$
27. $(2x + y)(4x^2 - 2xy + y^2)$
29. $(3x + y)(9x^2 - 3xy + y^2)$
31. $(x - 1)(x^2 + x + 1)$
33. $(3m - 1)(9m^2 + 3m + 1)$
35. $(5a + b)(25a^2 - 5ab + b^2)$
37. $(1 - 10x)(1 + 10x + 100x^2)$
39. $(ab + c)(a^2b^2 - abc + c^2)$

REVIEW PROBLEM SET, PAGE 139

1. $x(3x - 5)$
3. $4xy(x + 2y^2)$
5. $3xy^2(x^2 + 2xy^2 - 3y)$
7. $7m^2n(m^2n - 2m + 3n^2)$
9. $6a^3b^2c^2(b^2 - 2ac - 3b)$
11. $(8x + 3)(y - z)$
13. $x(7x + 1)(a + b)$
15. $x^2 + 4x + 4$
17. $y^2 - 6y + 9$
19. $4x^2 + 4x + 1$
21. $9y^2 + 24yz + 16z^2$
23. $4x^2y^2 - 20xy + 25$
25. $x^2 - 4$
27. $x^2 - 25y^2$
29. $9a^2 - 64b^2$
31. $16y^2 - 25z^2$
33. $121 - 9y^2$
35. $49 - 16x^2y^2z^2$
37. $(1 - 7xy)^2$
39. $(10 - 3ab)^2$

41. $(x + 2)^2$
43. $(y - 3)^2$
45. $(2x + 7y)^2$
47. $(4xy - 3)^2$
49. $(x - 9)(x + 9)$
51. $(5 - 2y)(5 + 2y)$
53. $(4a - 9b)(4a + 9b)$
55. $(2xy - 3z)(2xy + 3z)$
57. $(1 - 12xy)(1 + 12xy)$
59. $(10 - 9abc)(10 + 9abc)$
61. $(x + 1)(x + 4)$
63. $(x + 4)(x - 1)$
65. $(y - 1)(y - 7)$
67. $(x - 9y)(x + 6y)$
69. $(a + 11b)(a + 7b)$
71. $(2x + 3)(x - 1)$
73. $(2x + 3)(3x - 1)$
75. $(4y - 1)(3y - 2)$
77. $(4x + 3y)(3x + 4y)$
79. $(6y - 5z)(5y + 4z)$
81. $(2x - 3)(a + b)$
83. $(x^2 + y^2)(2x - 1)$
85. $(a + b)(x^2 + 2x + 3)$
87. $xy^2(x - 3y)(x + 3y)$
89. $xy(x + 3y)(x + y)$
91. $(2x - 3)(2x + 3)(4x^2 + 9)$
93. $(a - 2b - 3)(a - 2b + 3)$
95. $(x + 2y - a + 1)(x + 2y + a - 1)$
97. $(x^2 + 5)(x^2 - 3)$
99. $(x + 3)(x^2 - 3x + 9)$
101. $(2y - z)(4y^2 + 2yz + z^2)$
103. $(3 + ab)(9 - 3ab + a^2b^2)$

CHAPTER 4

PROBLEM SET 4.1, PAGE 152

1. yes
3. yes
5. yes

7. no
9. $\frac{1}{3}$
11. $\frac{7}{11}$
13. $-\frac{5}{7}$
15. $\frac{5}{7}$
17. $-\frac{5}{3}$
19. 84
21. 54
23. -77
25. $\frac{2}{3}$
27. $\frac{5}{16}$
29. $-\frac{9}{2}$
31. $-\frac{1}{3}$
33. -1
35. $\frac{2}{33}$
37. $\frac{9}{20}$
39. $\frac{21}{20}$
41. $\frac{20}{27}$
43. $\frac{25}{21}$
45. $-\frac{4}{5}$
47. $-\frac{11}{78}$
49. $\frac{2}{15}$
51. $-\frac{8}{9}$
53. $\frac{1}{2}$
55. 2
57. $\frac{5}{9}$

PROBLEM SET 4.2, PAGE 164

1. $\frac{3}{5}$
3. $\frac{9}{10}$
5. $6\frac{1}{2}$
7. $5\frac{7}{8}$
9. $11\frac{8}{9}$
11. $\frac{2}{11}$
13. $\frac{4}{15}$
15. $\frac{1}{5}$
17. $\frac{3}{4}$
19. $3\frac{1}{2}$
21. $\frac{7}{8}$
23. $\frac{13}{18}$
25. $\frac{8}{15}$
27. $\frac{9}{8}$
29. $\frac{17}{32}$

31. $7\frac{5}{8}$
33. $\frac{1}{6}$
35. $\frac{8}{15}$
37. $\frac{5}{6}$
39. $\frac{53}{63}$
41. $\frac{25}{24}$
43. $\frac{41}{24}$
45. $\frac{5}{18}$
47. $9\frac{7}{45}$
49. $\frac{23}{45}$
51. $\frac{41}{10}$
53. $\frac{29}{24}$
55. $\frac{19}{24}$
57. $\frac{43}{36}$
59. $\frac{91}{40}$

PROBLEM SET 4.3, PAGE 174

1. $\frac{7}{12}$
3. undefined
5. 0
7. yes
9. yes
11. no
13. yes
15. yes
17. no
19. $\frac{x}{y}$
21. $\frac{2x}{3y^3}$
23. $\frac{2b}{3ax}$
25. $\frac{ax^2}{3}$
27. $\frac{5}{9}$
29. $\frac{x+1}{x-2}$
31. $\frac{2x+1}{2x-1}$
33. $\frac{y}{3}$
35. $5x - 1$
37. $2x$

39. $\dfrac{2x}{3 - 2y}$

41. $\dfrac{x - 4}{5}$

43. $\dfrac{x}{x - 2}$

45. $\dfrac{x - 1}{x + 2}$

47. $\dfrac{x + 3}{x + 8}$

49. $\dfrac{m - 2n}{2(m + 3n)}$

51. x

53. x^2

55. $abxy$

57. $3(a^2 - b^2)$

59. $x^3 - 27x$

61. $\dfrac{5}{y - x}$

63. $\dfrac{-xy}{y - x}$

65. $\dfrac{1}{y - x}$

PROBLEM SET 4.4, PAGE 183

1. $\dfrac{2}{3a}$

3. $\dfrac{1}{ab}$

5. $\dfrac{11}{8y^4}$

7. $\dfrac{9(3x - 5)}{16x^2}$

9. $\dfrac{3y}{y + 4}$

11. $\dfrac{3}{x + y}$

13. $\dfrac{10x}{7}$

15. x

17. $\dfrac{x^2}{x - 4}$

19. $\dfrac{x + 5}{x}$

21. $x + 3$

23. $\dfrac{a - b}{2a}$

25. m

27. $\dfrac{a + 4}{a + 3}$

29. $\dfrac{2(x + 3)}{3(x + 5)}$

31. $\dfrac{(a + b)(3a - b)}{(a - b)(3a + b)}$

33. 1

35. $\dfrac{1}{a^2x}$

37. $\dfrac{27}{50a^2}$

39. $\dfrac{2x}{x + y}$

41. $\dfrac{c - b}{x + y}$

43. 3

45. $\dfrac{6(x - 3)}{x^2}$

47. $\dfrac{(x + 3)^2}{x + 1}$

49. $\dfrac{3x + 1}{2}$

51. $\dfrac{(x - 1)(x + 3)}{(2x + 3)(x - 3)}$

53. $\dfrac{6(3x + 1)}{x(x + 1)}$

55. $\dfrac{x + y}{2x(x - y)}$

57. $\dfrac{(x - 6)(x + 4)}{(x + 2)(x + 7)}$

PROBLEM SET 4.5, PAGE 196

1. $\dfrac{10}{x}$

3. $\dfrac{1}{x}$

5. $\dfrac{4}{a}$

7. $\frac{1}{a^2b}$

9. 3

11. $\frac{1}{3x}$

13. a

15. -1

17. $\frac{1}{x}$

19. $2x + 4$

21. $\frac{5x^2}{6}$

23. $\frac{6x^2 - 5y^2}{30}$

25. $-\frac{5a^2}{44}$

27. $\frac{2a^2 + ab + 15a - 20b}{5a}$

29. $\frac{2xy - 3y + zx + 2z}{yz}$

31. $\frac{7y - xy + 2x}{y(y - 2)}$

33. $\frac{9y + 6x - 48}{(x - 5)(y - 2)}$

35. $\frac{8x - 38}{(x - 6)(x - 4)}$

37. $\frac{3b}{ab}, \frac{4a}{ab}$

39. $\frac{5a^2}{a^3b^2}, \frac{3b}{a^3b^2}$

41. $\frac{3x + 6}{12}, \frac{2x - 10}{12}$

43. $\frac{5a - 10}{30}, \frac{10 - 4a}{30}$

45. $\frac{5b^2}{a^2b^2}, \frac{4ab}{a^2b^2}, \frac{3a^2}{a^2b^2}$

47. $\frac{2}{x^2 - 9}, \frac{4x - 12}{x^2 - 9}$

49. $\frac{2x + 2}{x^2 - 1}, \frac{2x}{x^2 - 1}$

51. $\frac{6x - 6}{(x - 1)(x^2 - 36)}, \frac{3x + 18}{(x - 1)(x^2 - 36)}$

53. $\frac{16x}{12(x + 2)}, \frac{15x^2}{12(x + 2)}$

55. $\frac{2(2x-1)(3x+2)}{12(2x-3)^2(3x+2)}, \frac{3(x-2)(2x-3)}{12(2x-3)^2(3x+2)}$

57. $\frac{3x}{4}$

59. $-\frac{2y}{15}$

61. $-\frac{m}{26}$

63. $\frac{5y}{144}$

65. $-\frac{1}{72x}$

67. $\frac{20x-7}{84}$

69. $\frac{3b^2-2ab+2a^2b-5a^2}{a^2b^2}$

71. $\frac{2x-y-5x^2-2xy}{x^2}$

73. $\frac{-x^2}{x^2-x-2}$

75. $\frac{2x^2+8x}{x^2+x-6}$

77. $\frac{x-4y}{x-2y}$

79. $\frac{19x+27}{(2x+1)(x-3)(x+3)}$

81. $\frac{20+14x-x^2}{2x(x^2-4)}$

PROBLEM SET 4.6, PAGE 205

1. $\frac{28}{5}$
3. $\frac{28}{15}$
5. $\frac{10}{7}$
7. $\frac{31}{34}$
9. $-\frac{1}{4}$
11. $\frac{6}{5}$
13. $\frac{9a}{2y}$
15. $\frac{xy}{2a^2}$
17. $\frac{1}{2+x}$
19. $\frac{2x}{x+y}$

21. $\dfrac{y - x}{y^2}$

23. $\dfrac{7x - y}{x - 7y}$

25. xy

27. $\dfrac{1}{1 + x}$

29. $\dfrac{3x - 1}{x}$

31. $\dfrac{y(x - y)}{x + y}$

33. $\dfrac{2}{x - 2}$

35. $\dfrac{y}{y - 2x}$

REVIEW PROBLEM SET, PAGE 208

1. yes
3. yes
5. no
7. $\frac{5}{9}$
9. -5
11. $-\frac{1}{9}$
13. $\frac{2}{5}$
15. $\frac{52}{105}$
17. $\frac{1}{2}$
19. $\frac{7}{6}$
21. $-\frac{11}{4}$
23. $\frac{7}{12}$
25. $\frac{5}{12}$
27. $\frac{3}{2}$
29. 3
31. $\frac{3}{2}$
33. $-\frac{7}{24}$
35. $-\frac{15}{13}$
37. $\frac{4}{11}$
39. $\frac{23}{70}$
41. $\frac{25}{3}$
43. $\frac{541}{90}$
45. $\frac{25}{24}$
47. $\frac{13}{24}$
49. $\frac{7}{66}$
51. $-\frac{2}{105}$
53. $\frac{99}{182}$

55. $\frac{83}{168}$
57. $\frac{2}{77}$
59. $-\frac{127}{66}$
61. $-\frac{25}{4}$
63. yes
65. yes
67. yes
69. $-\frac{1}{3}$
71. $\frac{2ac}{3b}$
73. $\frac{4}{y}$
75. $\frac{2(x - y)}{3}$
77. $\frac{7a}{9(a - b)}$
79. $\frac{2}{x - 1}$
81. $\frac{3x - y}{6(x - 3y)}$
83. $\frac{2y - 3x}{2x - 3y}$
85. $\frac{x + y}{y - x}$
87. $\frac{x - 5}{x + 3}$
89. $\frac{y}{10x^2}$
91. $9a^3b$
93. $\frac{5b^2}{16a^2}$
95. $\frac{4}{15}$
97. $(x + y)(a + b)$
99. $\frac{x(ab + 3)}{ab - 3}$
101. $(x + 12)(x - 4)$
103. $\frac{x - 4}{x}$
105. 1
107. $\frac{(x + 3)(x - 4)}{(2x + 1)(x - 1)}$
109. $\frac{2}{y}$

111. $\frac{x^3}{y^2}$

113. $\frac{2}{3xy}$

115. $\frac{6y^3}{de^2x^2}$

117. $\frac{7x}{x+2}$

119. $-\frac{y^2}{6(cd+3c)}$

121. $(x+3)(x-5)$

123. $\frac{x-2}{x-7}$

125. $\frac{(x-y)(a+b)}{(x+y)(a-b)}$

127. $2x$

129. $\frac{20(x^2-1)}{3(5x+1)^2}$

131. $\frac{5a^2}{8}$

133. $\frac{7y}{8}$

135. $\frac{a+b}{x^3y}$

137. $-\frac{1}{x}$

139. $\frac{5}{12x}$

141. $\frac{18z-16x+15y}{12xyz}$

143. $\frac{2x+1}{(x+3)(x-2)}$

145. $\frac{42y+15x^2y^2+2x}{36x^3y^3}$

147. $\frac{5x+3y+1}{x(x+y)}$

149. $\frac{5x-3y}{7x+2y}$

151. $\frac{3x^2-4x+7}{3(x-1)}$

153. $\frac{2x}{(x+y)^2(x-y)}$

155. $\frac{12x-46}{(x-3)(x+2)(x-5)}$

157. $\dfrac{5x^2 - 14x + 24}{x^2(x^2 - 7x + 12)}$

159. $\dfrac{5}{2(x + 2)}$

161. $\frac{16}{15}$

163. $\frac{13}{14}$

165. $\dfrac{y^2}{2}$

167. $\dfrac{y^4}{x(x + 1)}$

169. $\dfrac{x^2 + 1}{x^2 - 1}$

171. $\dfrac{x}{y - x}$

173. $-\dfrac{1}{x + y}$

CHAPTER 5

PROBLEM SET 5.1, PAGE 230

1. 2
3. 1
5. 2
7. $\{2\}$
9. $\{25\}$
11. $\{4\}$
13. $\{-3\}$
15. $\{3\}$
17. $\{-10\}$
19. $\{2\}$
21. $\{-1\}$
23. $\{4\}$
25. $\{4\}$
27. $\{5\}$
29. $\{6\}$
31. $\{3\}$
33. $\{1\}$
35. $\{0\}$
37. $\{1\}$
39. $\{2\}$

41. $\{\frac{13}{27}\}$
43. $\{2\}$
45. $\{1\}$
47. $\{3\}$
49. $\{-\frac{5}{2}\}$
51. $\{-5\}$

PROBLEM SET 5.2, PAGE 235

1. $\{\frac{3}{2}\}$
3. $\{\frac{5}{2}\}$
5. $\{\frac{7}{12}\}$
7. $\{-6\}$
9. $\{\frac{15}{2}\}$
11. $\{2\}$
13. $\{36\}$
15. $\{65\}$
17. $\{-\frac{4}{3}\}$
19. $\{3\}$
21. $\{\frac{23}{4}\}$
23. $\{7\}$
25. $\{3\}$
27. $\{-\frac{5}{2}\}$
29. $\{-\frac{1}{3}\}$
31. $\{4\}$
33. $\{\frac{10}{3}\}$
35. $\{25\}$
37. $\{15\}$
39. $\{\frac{1}{6}\}$
41. $\{6\}$

PROBLEM SET 5.3, PAGE 240

1. $\{a\}$
3. $\{4c\}$
5. $\{4\}$
7. $\left\{\frac{a+b}{5c}\right\}$
9. $\left\{\frac{a-1}{2}\right\}$
11. $\left\{\frac{b+2}{3a}\right\}$
13. $\left\{\frac{c+7}{5b}\right\}$

15. $l = \dfrac{A}{w}$

17. $r = \dfrac{d}{t}$

19. $r = \dfrac{C}{2\pi}$

21. $t = \dfrac{I}{pr}$

23. $h = \dfrac{V}{lw}$

25. $R = \dfrac{P}{I^2}$

27. $F = \frac{9}{5}C + 32$

29. $V = \dfrac{C}{P}$

31. $n = \dfrac{L - a + d}{d}$

33. $h = \dfrac{S - 2\pi r^2}{2\pi r}$

PROBLEM SET 5.4, PAGE 247

1. 27
3. 22 and 24
5. no
7. 20, 22, and 24
9. 23, 25, and 27
11. 28, 30, 32, and 34
13. 32 and 34
15. 19, 21, and 23
17. 21, 23, 25, and 27
19. 36
21. 2
23. 5
25. 52
27. $150,000
29. 4
31. 10 and 14
33. 14 and 42
35. 12 and 17
37. 42 dimes and 84 nickels
39. 8 quarters and 32 nickels
41. 25 dimes and 87 nickels
43. 27 dimes and 43 nickels

45. 14 quarters, 42 dimes, and 168 nickels
47. length 17 inches, width 10 inches
49. length 60 inches, width 15 inches
51. 10 feet
53. 32 inches
55. \$3,000
57. \$29,000
59. \$16,000 at $3\frac{1}{2}\%$, \$32,000 at 4%, \$64,000 at 5%
61. $32\frac{8}{11}$ minutes

PROBLEM SET 5.5, PAGE 253

1. 11
3. 11
5. 30
7. 36
9. 43
11. $\{-5,5\}$
13. $\{-2,2\}$
15. $\{-7,7\}$
17. $\{-3,3\}$
19. $\{-9,9\}$
21. $\{-4,4\}$
23. $\{-11,1\}$
25. $\{-6,6\}$

REVIEW PROBLEM SET, PAGE 254

1. $\{\frac{3}{4}\}$
3. $\{3\}$
5. $\{-\frac{11}{2}\}$
7. $\{\frac{1}{2}\}$
9. $\{-40\}$
11. $\{4\}$
13. $\{3\}$
15. $\{-\frac{2}{7}\}$
17. $\{16\}$
19. $\{-\frac{150}{103}\}$
21. $\{a - b + 2\}$
23. $\left\{\dfrac{d-b}{a-c}\right\}$
25. $\left\{\dfrac{ab-cd}{a+b-c-d}\right\}$

27. $\left\{\dfrac{16a}{49}\right\}$

29. $\left\{\dfrac{37a - 7b}{7}\right\}$

31. $\{\frac{12}{7}\}$

33. $\{14\}$

35. $\{4\}$

37. $\{8\}$

39. $\{-1\}$

41. $\{14\}$

43. $\{4\}$

45. $\{10\}$

47. $\{-\frac{6}{7},\frac{6}{7}\}$

49. $\{-\frac{1}{2},\frac{3}{2}\}$

51. $\{0,6\}$

53. $\{\frac{4}{3},\frac{20}{3}\}$

55. $\{-1,1\}$

57. $\{-\frac{4}{3},2\}$

59. $\{-2,\frac{10}{3}\}$

61. 20 pounds

63. \$2,400

65. \$800

67. 6, 7, and 8

CHAPTER 6

PROBLEM SET 6.1, PAGE 269

1. (a) $8 - 4 = 4$ (c) $-1 - (-2) = 1$ (e) $4 - (-3) = 7$

3. (a) $4 < 6$ (c) $-2 < 0$ (e) $3x < 3y$

4. (a) addition property (c) transitive property (e) trichotomy property

5. (a) true (c) true (e) true

8. yes, since $-27 < -8$ implies $-3 < -2$ and $8 < 27$ implies $2 < 3$

10. (a)

(c)

(e) 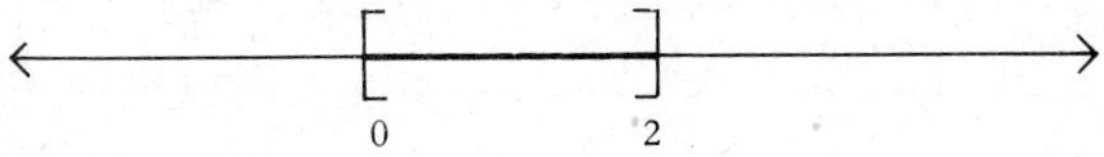

PROBLEM SET 6.2, PAGE 273

1. $\{x|x < 2\}$
3. $\{x|x \geq 6\}$
5. $\{x|x \leq -4\}$
7. $\{x|x < 1\}$
9. $\{x|x \geq 7\}$
11. $\{x|x < 1\}$
13. $\{x|x > -3\}$
15. $\{x|x > \frac{15}{2}\}$
17. $\{x|x > 4\}$
19. $\{x|x > 1\}$
21. $\{x|x \geq -2\}$
23. $\{x|x \leq -1\}$
25. $\{x|x \geq -3\}$
27. $\{t|t \leq 4\}$
29. $\{x|x \leq -1\}$
31. $\{x|x \geq -8\}$
33. $\{x|x \leq 6\}$
35. $\{x|x < -4\}$
37. $\{x|x \leq 11\}$
39. $\{x|x < \frac{22}{3}\}$
41. $\{x|x > -\frac{23}{11}\}$
43. $\{x|x \geq 14\}$

PROBLEM SET 6.3, PAGE 276

1. $\{x|-1 < x < 1\}$
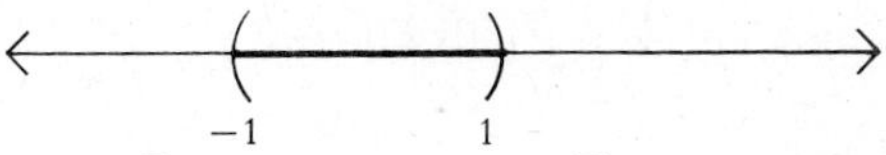

3. $\{x|-4 \leq x \leq 4\}$

5. $\{x|x < -3 \text{ or } x > 3\}$
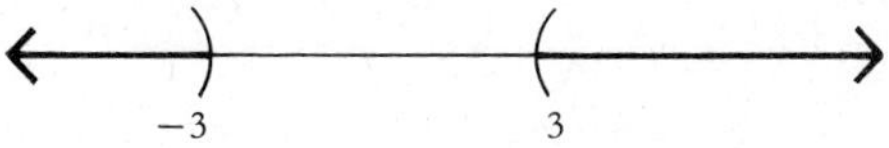

7. $\{x|x \leq -\frac{15}{2} \text{ or } x \geq \frac{15}{2}\}$
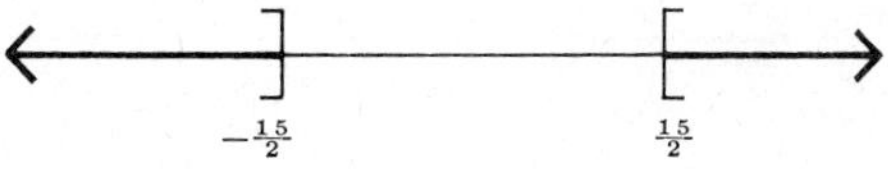

9. $\{x|-4 \leq x \leq 4\}$
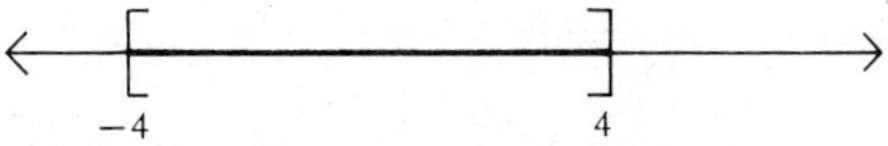

11. $\{x|x \leq -\frac{2}{3} \text{ or } x \geq \frac{2}{3}\}$
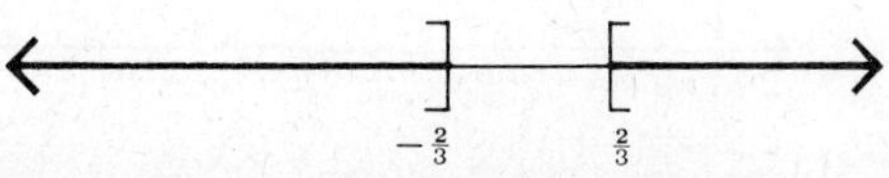

13. $\{x|-3 < x < 3\}$

15. $\{x|x \leq -\frac{3}{2} \text{ or } x \geq \frac{3}{2}\}$

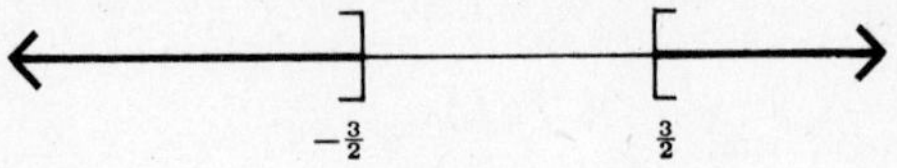

REVIEW PROBLEM SET, PAGE 276

1. false
3. false
5. true
7. true
9. false
11. true
13. true
15. $\{x|x < -4\}$
17. $\{x|x < \frac{7}{4}\}$
19. $\{x|x < 0\}$
21. $\{x|x < 4\}$
23. $\{x|x > 0\}$
25. $\{x|x \leq -3\}$
27. $\{x|x \geq -17\}$
29. $\{x|x > 2\}$
31. $\{x|x \geq 2\}$
33. $\{x|x \leq 7\}$
35. $\{x|x \leq 1\}$
37. $\{x|x \leq \frac{17}{21}\}$
39. $\{x|x \leq \frac{15}{8}\}$

41. $\{x|-7 < x < 7\}$

43. $\{x|x < -7 \text{ or } x > 7\}$

45. $\{x|-2 < x < 2\}$

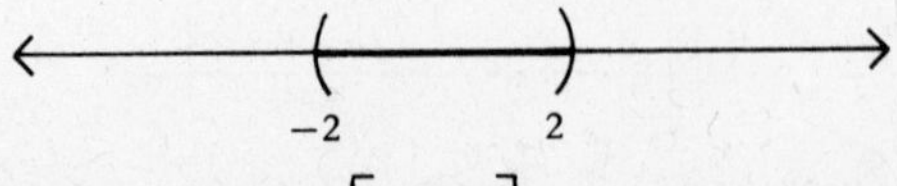

47. $\{x|-\frac{1}{3} \leq x \leq \frac{1}{3}\}$

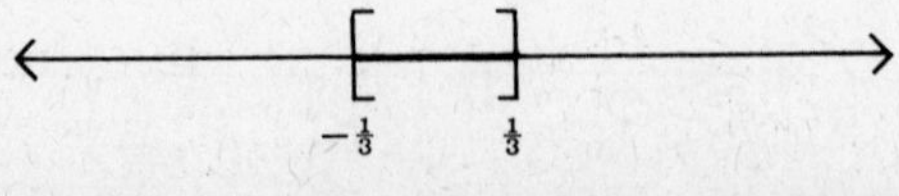

CHAPTER 7

PROBLEM SET 7.1, PAGE 289

1. (a) not equal (c) equal (e) equal
2. (a) 2 (c) 4
3. (a) false (c) true (e) true
5. (a) $(3,3) \in Q_{\text{I}}$ (c) $(5,\sqrt{2}) \in Q_{\text{I}}$ (e) $(-1,-5) \in Q_{\text{III}}$ (g) $(-4,2) \in Q_{\text{II}}$ (i) $(\frac{1}{2},-\frac{3}{2}) \in Q_{\text{IV}}$
6. (a) $x > 0$ and $y > 0$ (c) $x < 0$ and $y < 0$ (e) $x > 0$ and $y = 0$
7.

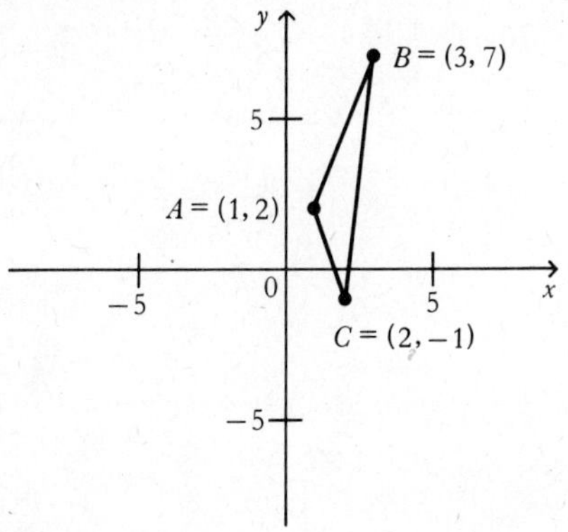

9. yes

PROBLEM SET 7.2, PAGE 293

1. (a) linear (c) linear (e) not linear
2. (a) $(5,-9)$ is a solution (c) $(3,-3)$ is a solution (e) $(-2,12)$ is a solution
3. (a) $\{(-1,-7),(2,8),(3,13),(5,23)\}$ (c) $\{(-1,-11),(0,-7),(1,-3),(2,1),(3,5)\}$ (e) $\{(-3,-1),(-2,-\frac{4}{3}),(-1,-\frac{5}{3}),(0,-2),(1,-\frac{7}{3}),(2,-\frac{8}{3}),(3,-3)\}$
5. (a) $2x + y = -2$ (c) $3x - 2y = 2$

x	y
-1	0
-2	2
0	-2
$-\frac{3}{2}$	1
3	-8
4	-10

x	y
0	-1
1	$\frac{1}{2}$
2	2
$\frac{8}{3}$	3
$-\frac{2}{3}$	-2
4	5

PROBLEM SET 7.3, PAGE 302

1.

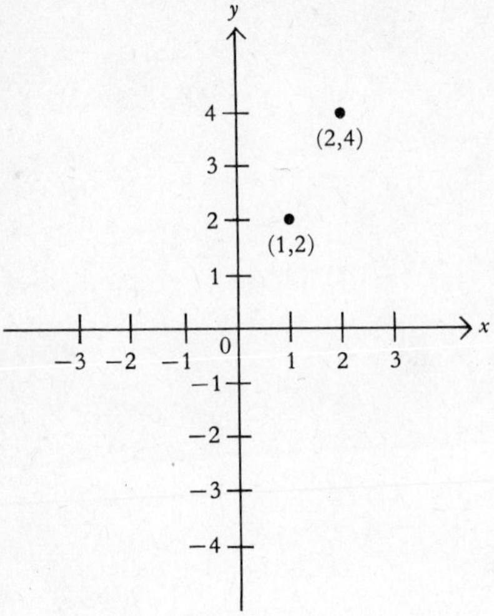

3.

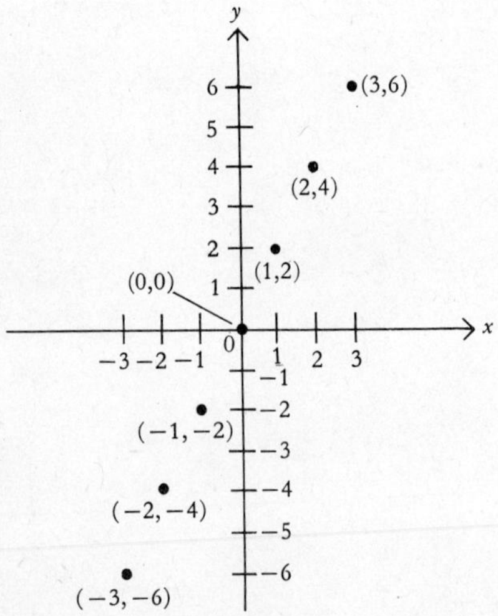

5.

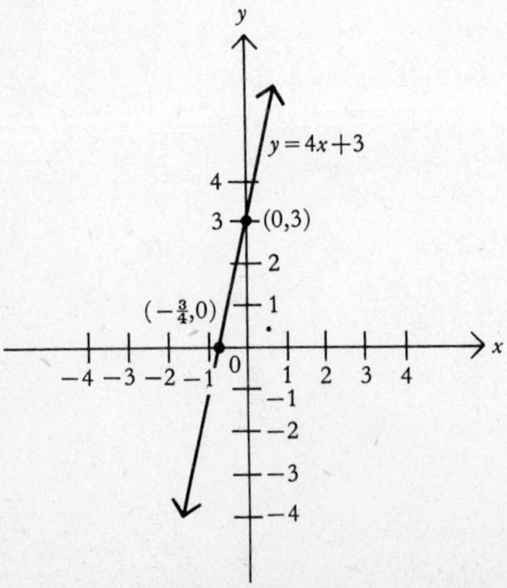

7.

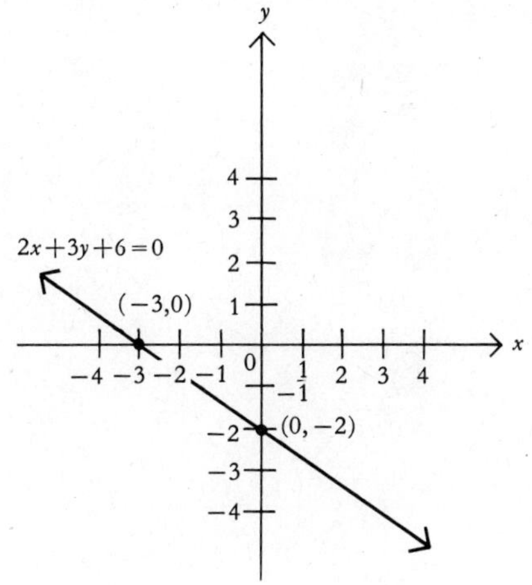

9.

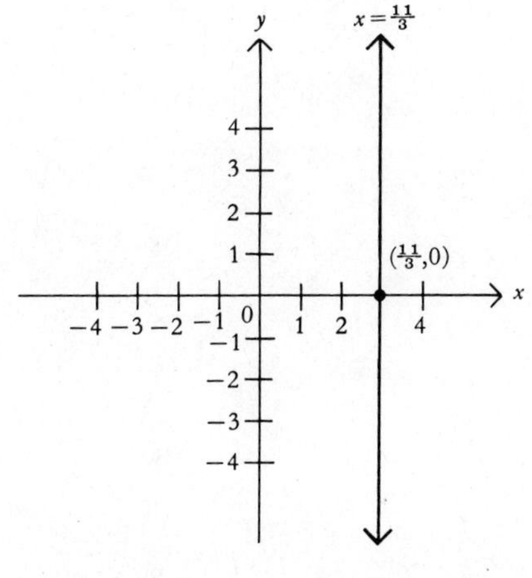

11.

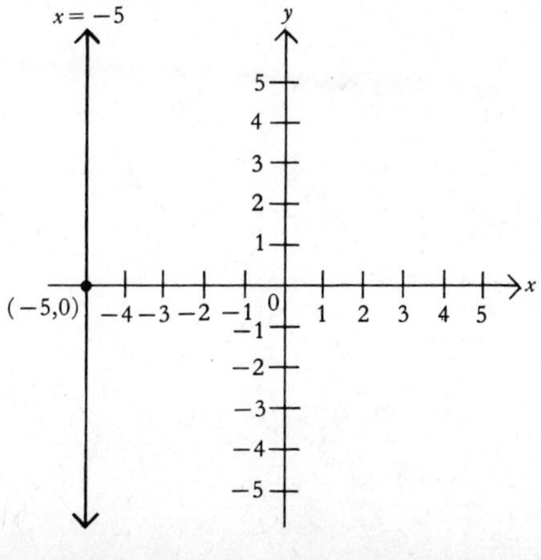

13.

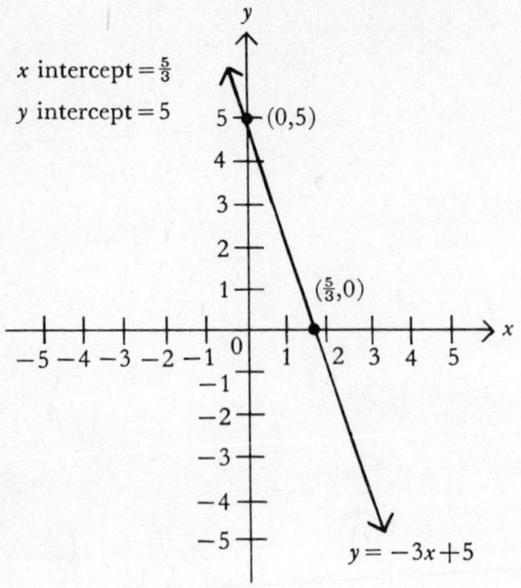
x intercept = 5/3
y intercept = 5
(0,5)
(5/3,0)
y = −3x+5

15.

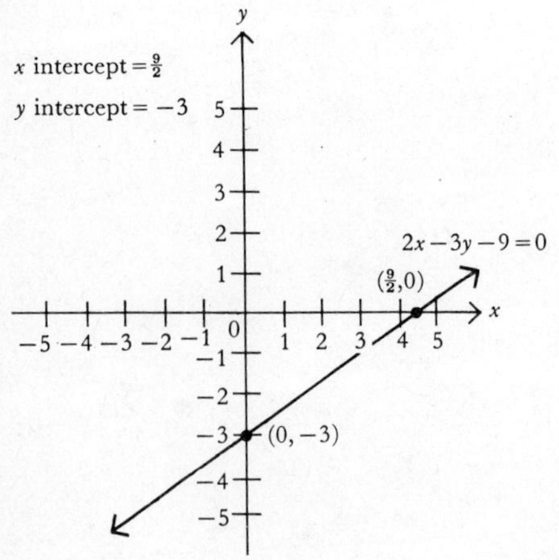
x intercept = 9/2
y intercept = −3
2x − 3y − 9 = 0
(9/2,0)
(0, −3)

17.

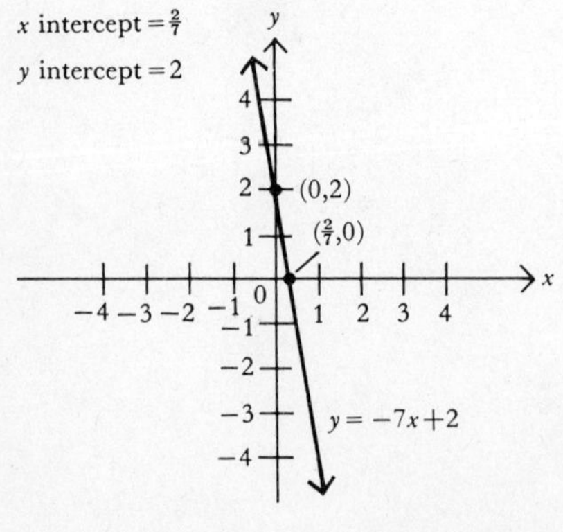
x intercept = 2/7
y intercept = 2
(0,2)
(2/7,0)
y = −7x+2

19.

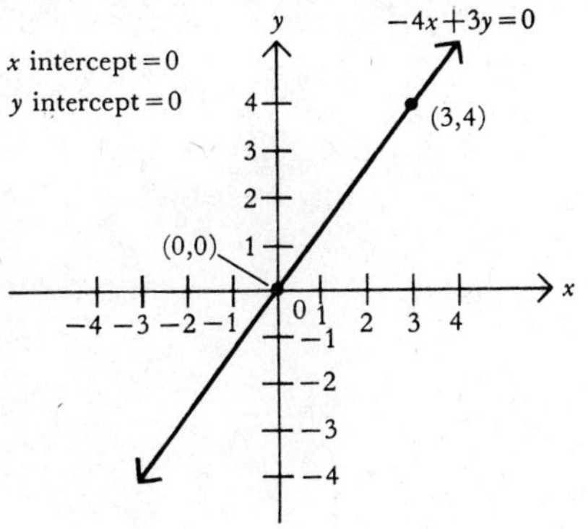
x intercept = 0
y intercept = 0
−4x + 3y = 0
(3,4)
(0,0)

21.

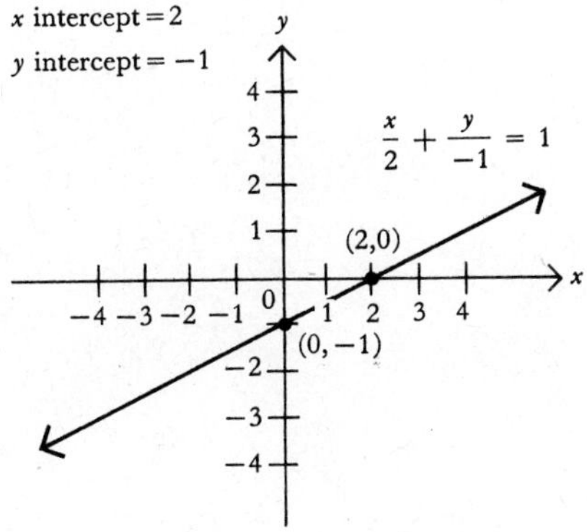
x intercept = 2
y intercept = −1
x/2 + y/−1 = 1
(2,0)
(0, −1)

23.

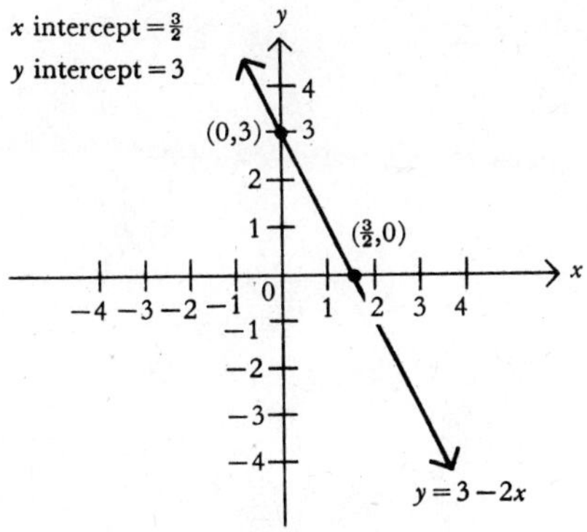
x intercept = 3/2
y intercept = 3
(0,3)
(3/2,0)
y = 3 − 2x

25.

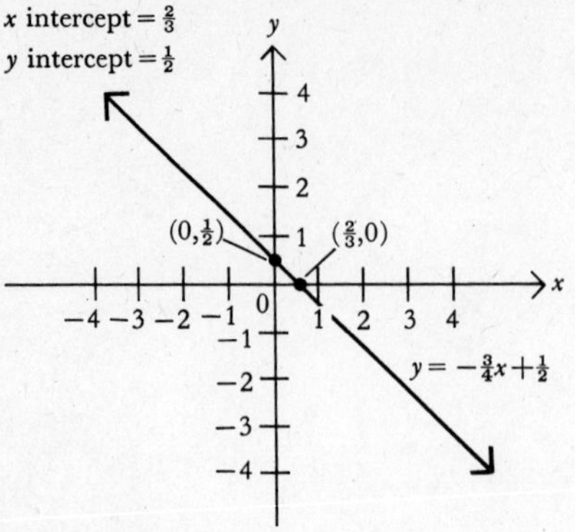
x intercept = 2/3
y intercept = 1/2
(0,1/2)
(2/3,0)
y = −3/4x + 1/2

27.

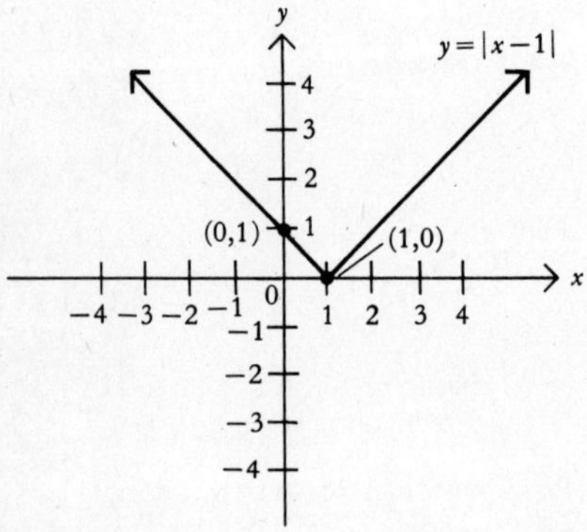
y = |x − 1|
(0,1)
(1,0)

29.

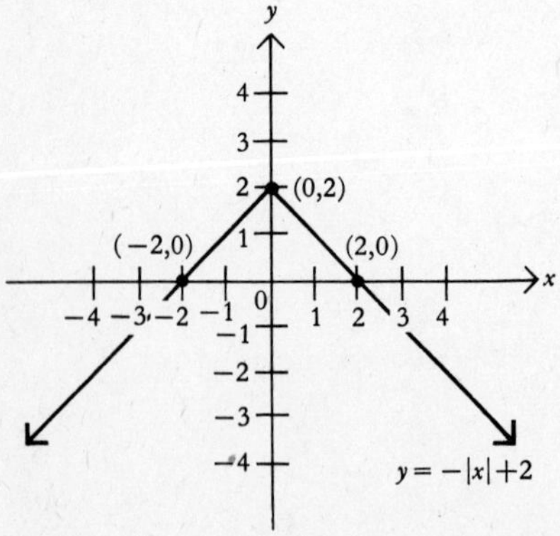
(0,2)
(−2,0)
(2,0)
y = −|x| + 2

PROBLEM SET 7.4, PAGE 307

1.

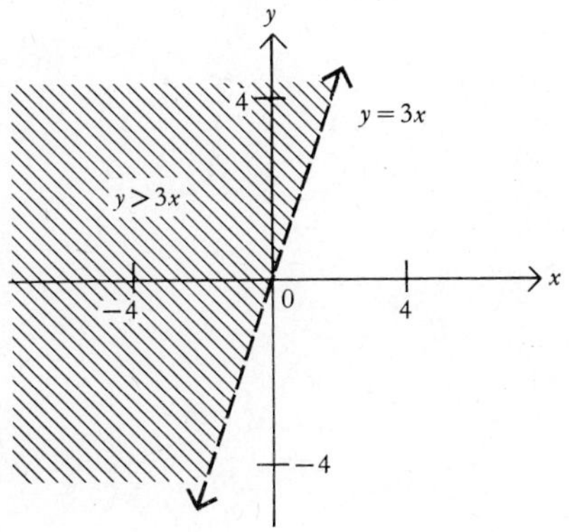

3.

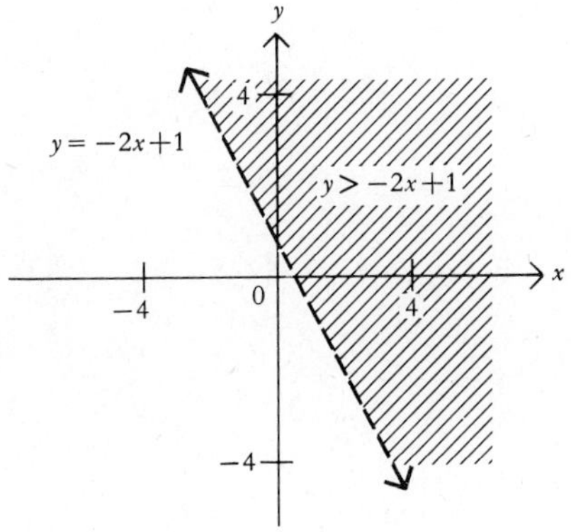

5.

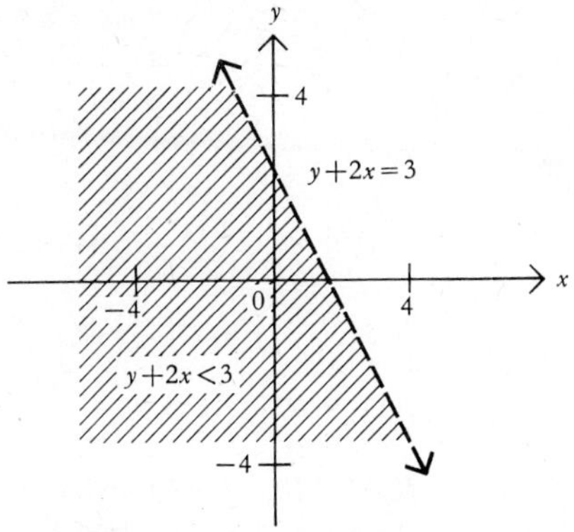

7.

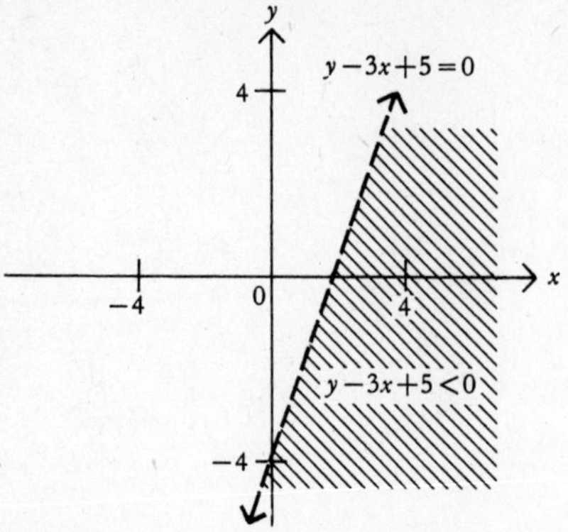

9.

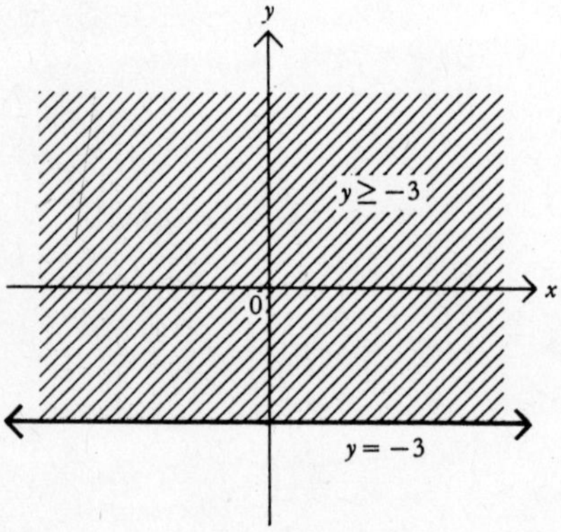

PROBLEM SET 7.5, PAGE 314

1. (a) function (c) not a function
2. (a) $f(x) = 3x + 1$
   $f(1) = 4,\ f(-1) = -2,\ f(2) = 7,\ f(-2) = -5,\ f(5) = 16,\ f(-5) = -14$
   (c) $f(x) = 5x + 1$
   $f(1) = 6,\ f(-1) = -4,\ f(2) = 11,\ f(-2) = -9,\ f(5) = 26,\ f(-5) = -24$
   (e) $f(x) = |x + 1|$
   $f(1) = 2,\ f(-1) = 0,\ f(2) = 3,\ f(-2) = 1,\ f(5) = 6,\ f(-5) = 4$
   (g) $f(x) = |x| + x$
   $f(1) = 2,\ f(-1) = 0,\ f(2) = 4,\ f(-2) = 0,\ f(5) = 10,\ f(-5) = 0$
3. (a) $D = \{x|x \in R\},\ R = \{y|y \in R\}$ (c) $D = \{x|x \in R\},\ R = \{y|y \in R\}$
   (e) $D = \{x|x \in R\},\ R = \{y|y \geq 0\}$ (g) $D = \{x|x \in R\},\ R = \{y|y \geq 0\}$
4. (a) $A = x^2$ (c) $V = x^3$

5. (a) $D = \{x | x \in R\}, \quad R = \{y | y \in R\}$

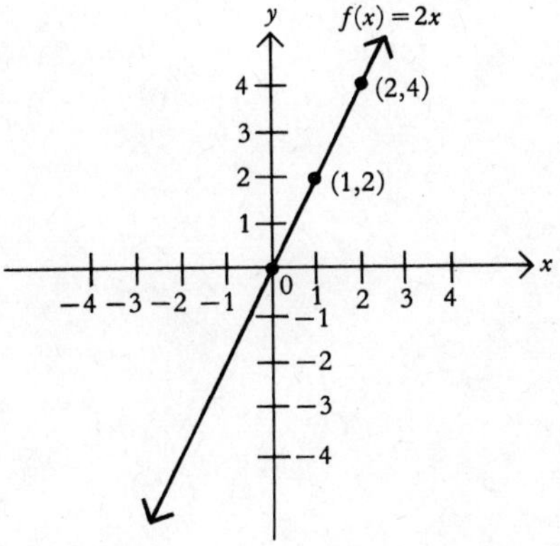

(c) $D = \{x | x \in R\}, \quad R = \{y | y \in R\}$

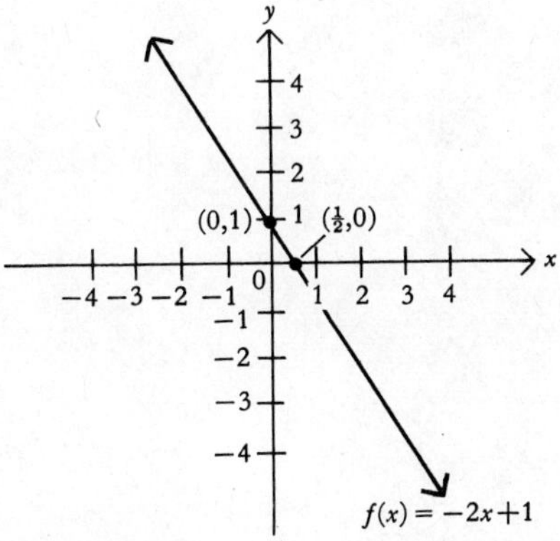

(e) $D = \{x | x \in R\}, \quad R = \{y | y \in R\}$

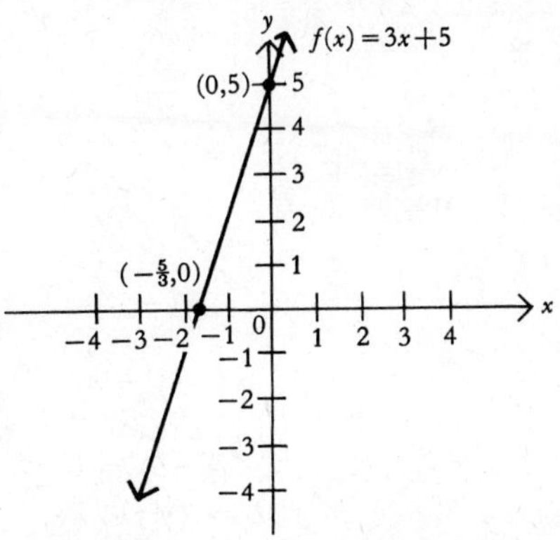

(g) $D = \{x|x \in R\}, \quad R = \{y|y \in R\}$

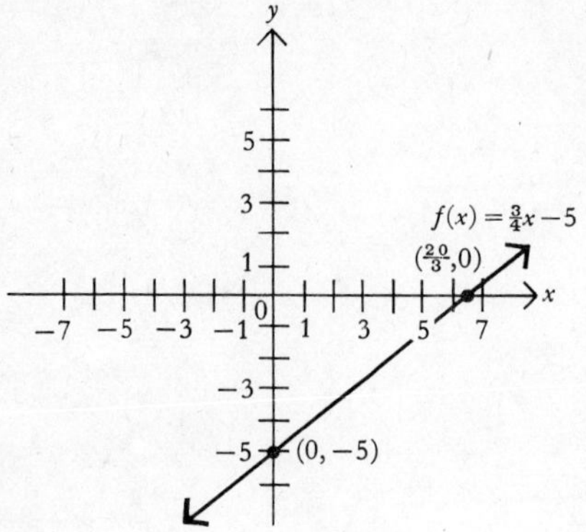

REVIEW PROBLEM SET, PAGE 314

1. $y = 3$
3. $x = 2, \ y = 4$
5. $(-1,3) \in Q_{\text{II}}$
7. $(-3,-2) \in Q_{\text{III}}$
9. $(3,-9) \in Q_{\text{IV}}$
11. (3,0) on x axis
13.

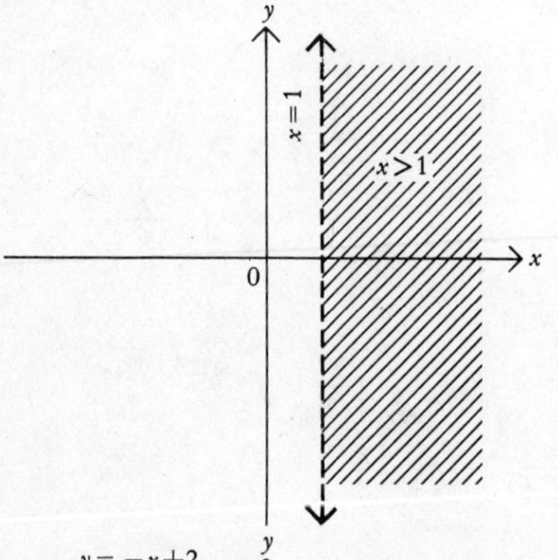

15.

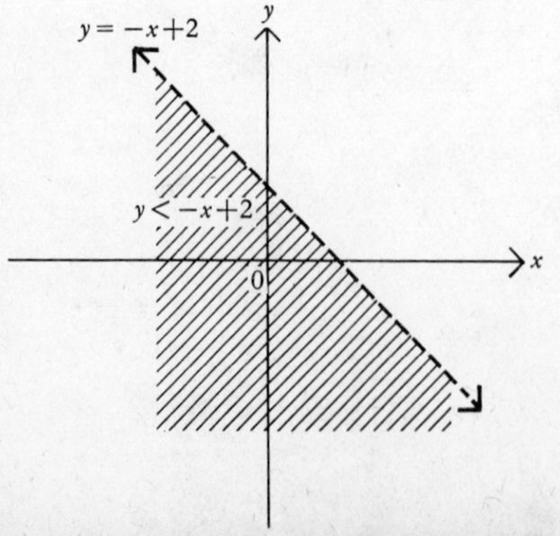

17. linear
19. not linear
21. not a solution
23. a solution
25. a solution
27. $\{(-2,3),(-1,5),(0,7),(1,9),(2,11)\}$
29. $\{(-1,\frac{11}{3}),(0,3),(1,\frac{7}{3})\}$
31.

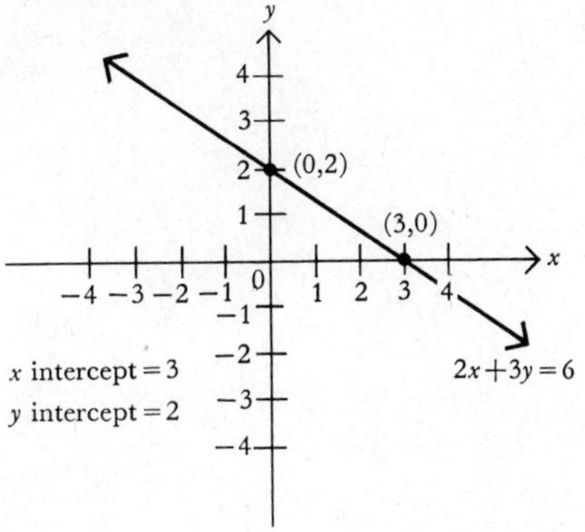

33.

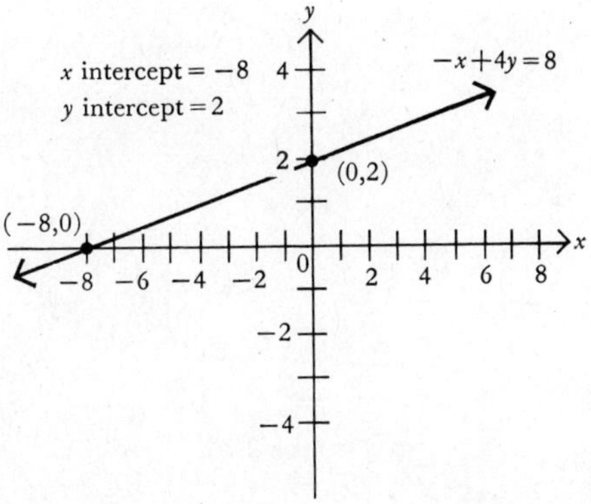

35.

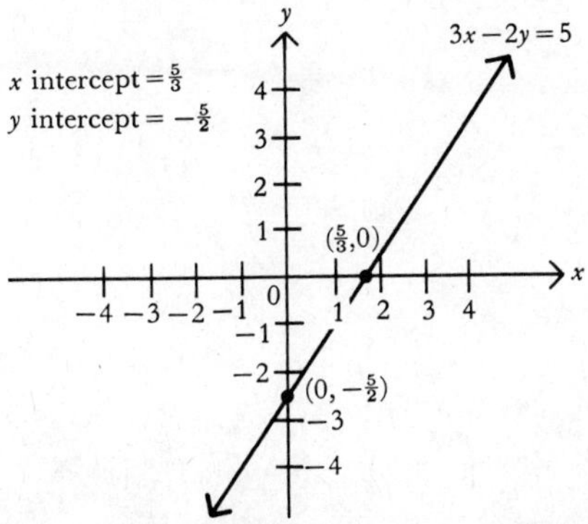

37.

x intercept $= 3$
y intercept $= 4$

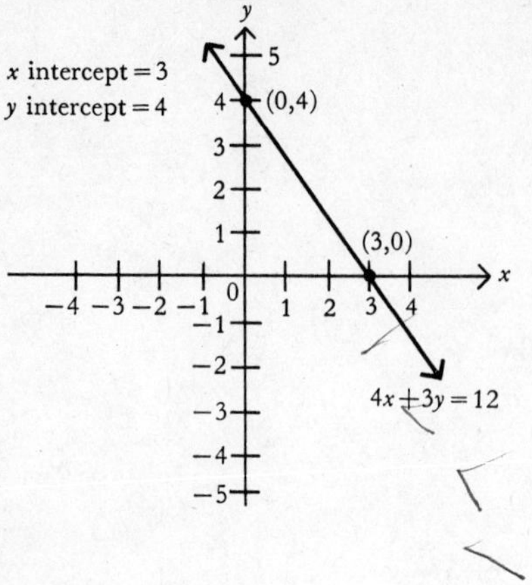

39.

x intercept $= \frac{7}{4}$
y intercept $= -\frac{7}{5}$

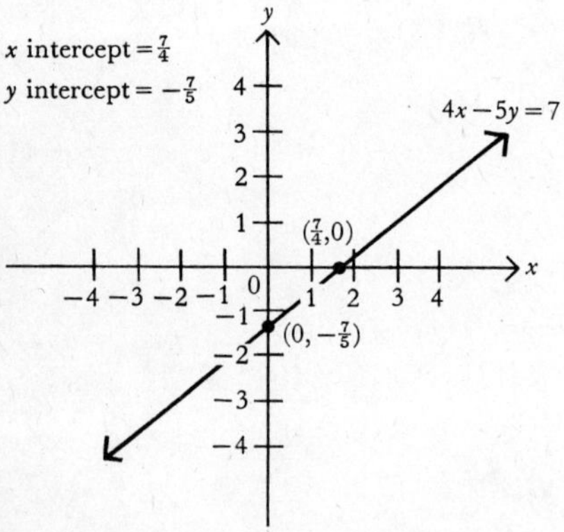

41.

x intercept $= 5$
y intercept $= -\frac{5}{2}$

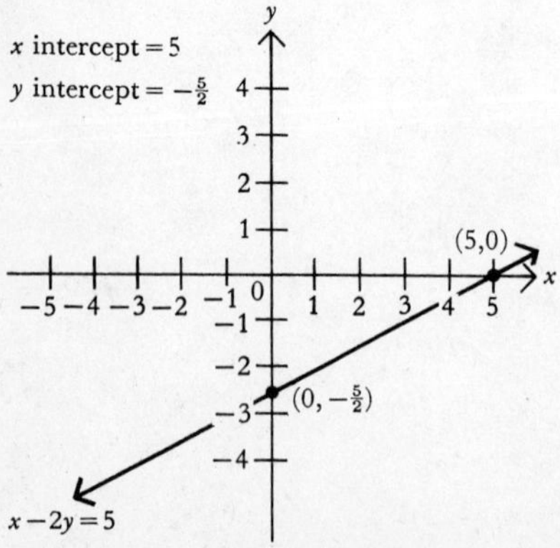

43.

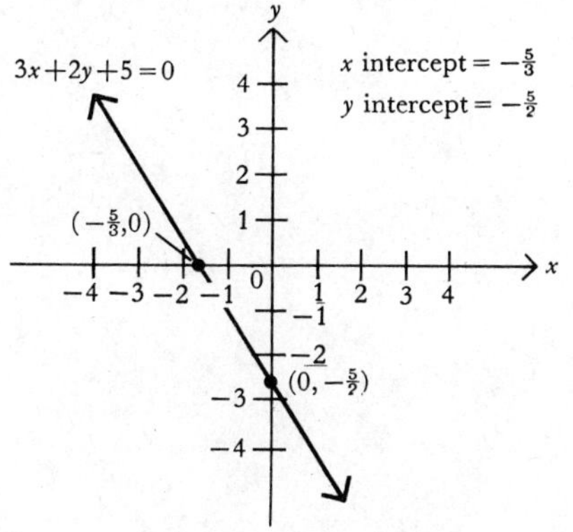

45. y intercept is 3; when $x = 4$, $y = 9$

47. a function, $D = \{x|x \in R\}$, $R = \{y|y \in R\}$

49. not a function

51. 5

53. 5

55. $3a^2 + 2$

57. $3a^2 + 6ab + 3b^2 + 2$

59. 4 or -10

61. $x > \frac{15}{2}$

63. function, $D = \{x|x \in R\}$, $R = \{y|y = 2\}$

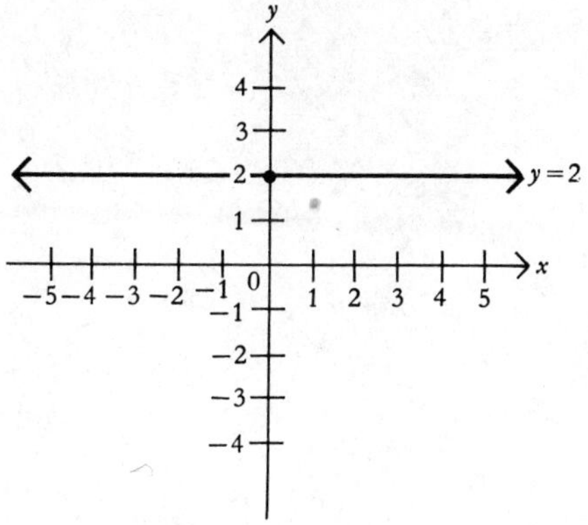

65. not a function

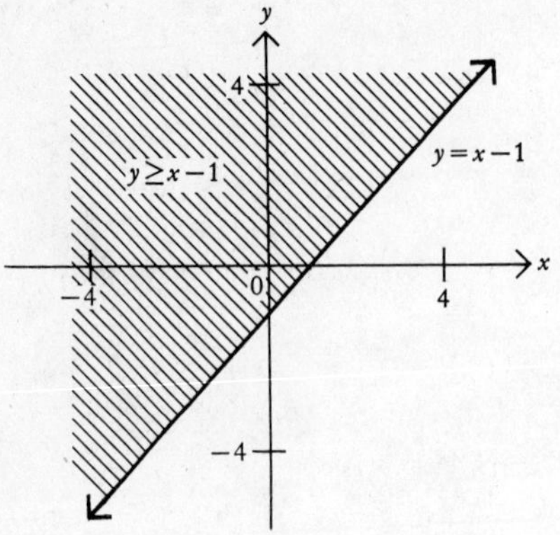

67. not a function

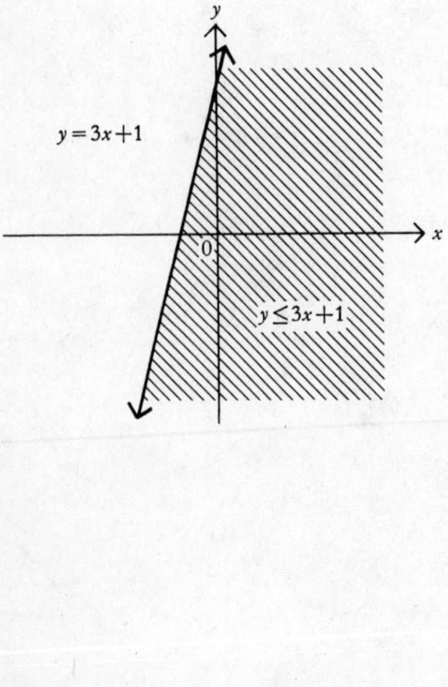

CHAPTER 8

PROBLEM SET 8.1, PAGE 323

1. (4,3)
3. (4,4)
5. (−1,−1)
7. (7,3)
9. $(-\frac{8}{5},-\frac{11}{5})$

Madhao Vaidya Math 007 Name: R. Smith

Tentative Schedule

No.	Day	Date	Sec.	Pages	Topics	H.W. Assignments
1	M	3-31	4.3	168-173	Rational expressions	P.174-178: ODDS
			4.4	178-182	Multiplication, Division	P.183-186: ODDS
2	W	4-2	4.5	186-195	Addition, Subtraction	P.196-200: ODDS
3	M	4-7	5.2	232-234	Fractional Equations	P.235-237: ODDS
			5.4	241-247	Word Problems	P.248-251: 18, 19 23, 25, 27, 29, 61.
4	W (test due!)	4-9	------ REVIEW ----- Practice Test -------			
5	M	4-14	TEST -4: 45 minutes, covers all of the above topics.			
6.	M	4-14	7.1	281-288	Cartesian co-ordinate system	P.289: ALL
7	W	4-16	7.2	290-292	Linear Equations	P.293-294: ALL
			7.3	294-302	Graph of a linear equation	P.302-303: ODDS
8	M	4-21	7.4	304-307	Graph of linear inequalities	P:307-308: ALL
9	W	4-23	8.1	321-323	Systems of linear equations	P.323-324: ALL
			8.2	324-330	Graphical solutions of linear systems	P.330-332: ODDS
10	M	4-28	8.3	332-335	Algebraic solutions by +, -	P.335-337: ODDS
			8.4	337-342	Solution by substution	P.342-343: ODDS

------- TEST-5: Take Home, covers topics 6 through 10.

| | W | 4-30 | REVIEW FOR THE FINAL EXAMINATION. PRACTICE TEST ETC. | | | |

EITHER TAKE FINAL EXAM TODAY OR

work 956-0599
home 264-9709

MADHAO V
Spring '80

I. Properties of

1. $2^4 * 2^3 * 2 = 2^8$ | 2.

5. $((-x)^2)^3 =$ | 6. $(-3(-a)$

9. $-(-1)^6 =$ | 10. $\frac{y^{12}}{y^{12}} =$

13. $\frac{-(-3)^{70}}{(-3)^{70}} =$ | 14. $(\frac{-x}{2y})^3 =$ | 15.

II. Evaluate only those expressions which are polyno

1. $3x^2 + \frac{7}{x}$, $x = 3$. | 2. -10, $x = 0$ | 3.

4. $x^2 - 5x + 6$, $x = 2$ | 5. $3x^2 + xy - zy =$ $x = \frac{1}{3}$, $y = 1$, $z = -\frac{2}{3}$ | 6. $-\frac{1}{2}x$,

neg x neg = pos

III. Identify the following polynomials as monomials, binomials, or trinomials. Then find the degree of each polynomial, and numerical coefficient of each term:

POLYNOMIAL	NAME	DEGREE	COEFFICIENT
1. 0			
2. $\frac{1}{3}y^2$			
3. $\frac{x+7}{10}$	B,	1	$\frac{1}{10}$, $\frac{7}{10}$
4. $3x^2 + 14x - 5$			

IV. Addition and Subtraction:

1. $-13x^3 - (-7x^3) =$
2. $(-11x + 10) - (3x - 4) =$
3. $-11m^2 + (-9m^2) =$
4. $(x^3 - 2x^2 + 7x - 8) - (-2x^3 + 4x^2 + 5x - 6) =$
5. $4ab + 6a - (-ab) =$
6. $(2x^2 - 3x + 11) + (3x - 11) =$

V. Find the following products:

1. $(2x^5)(4x^2) =$
2. $(-9y)(-7y^2) =$
3. $-2x(3x-6) =$
4. $(x+1)(x^2-x+1) =$
5. $(x-10)(x+3) =$
6. $(3x+2)(4x+1) =$
7. $(2x+3)(x+7) =$
8. $(2x-5)(2x+5) =$
9. $(4-x)(4+x) =$

[redacted] $12x^4$ by $-3x$. .2. $(-4x^3 + 10x^2) \div 2x$

3. $\dfrac{6x^4z^4 - 9x^3z^2 + 15xz^3}{3xz^2} =$ .4. $2x-3\overline{)2x^3-9x^2+19x+15}$

5. $(2x^2+9x+9) \div (x+3))$

ANSWERS

I. Properties of exponents:

1. 2^8 or 256 2. $-b^3$ .3. $16r^4s^4$ .4. 1 .5. x^6

6. $27a^3$ .7. $-3r^7s^4$ .8. $8u^3v^3$ .9. -1 .10. 1

11. $\frac{1}{16}$ .12. $8x^3$ .13. -1 14. $-\frac{x^3}{8y^3}$ .15. 1

II. Evaluate polynomials:

1. not a poly. .2. -10 .3. 0 .4. 0 .5. 1 .6. 2

III. Identify polynomials etc...

1. Zero poly., --, --. 2. Mono., Deg.=2, Coeff: $\frac{1}{3}$ .3. Bino., Deg:1, Coeff: $\frac{-1}{10}$, $\frac{7}{10}$

4. Tri., Deg: 2, Coeff: 3, 14, -5

IV. Addition and Subtraction:

1. $-6x^3$, .2. $-14x+14$, .3. $-20m^2$ .4. $3x^3-6x^2+2x-2$, .5. $5ab+6a$

6. $2x^2$.

V. Products:

1. $8x^7$, .2. $63y^3$, .3. $-6x^2+12x$, .4. x^3-1, .5. $x^2-7x-30$

6. $12x^2+11x+2$, .7. $2x^2+17x+21$, .8. $4x^2-25$, .9. $16-x^2$

VI. Division:

1. $-4x^3$, .2. $-2x^2+5x$, .3. $2x^3z^2-3x^2+5z$, .4. Q= x^2-3x+5, R= 30 .5. Q= $2x+3$, R= 0

MV/S'80 Eat Shit!
500,000,000,000,000
flies Cant be wrong Math 007/page2

Test-2:Part-1: Sample Test.

VI. Division:

1. Divide

Name:

PROBLEM SET 8.2, PAGE 330

1. $\{(6,1)\}$
3. $\{(30,-9)\}$
5. no solution (inconsistent)
7. dependent
9. $\{(2,3)\}$
11. $\{(-1,1)\}$
13. no solution (inconsistent)
15. $\{(3,2)\}$
17. $\{(7,9)\}$
19. $\{(-3,-4)\}$
21. $\{(5,5)\}$

PROBLEM SET 8.3, PAGE 335

1. $\{(2,0)\}$
3. $\{(7,2)\}$
5. $\{(4,1)\}$
7. $\{(2,3)\}$
9. $\{(3,1)\}$
11. $\{(\frac{85}{2},32)\}$
13. $\{(\frac{2}{41},\frac{50}{41})\}$
15. $\{(2,2)\}$
17. $\{(1,1)\}$
19. $\{(1,3)\}$
21. $\{(4,10)\}$

PROBLEM SET 8.4, PAGE 342

1. $\{(2,1)\}$
3. $\{(1,2)\}$
5. $\{(0,0)\}$
7. no solution (inconsistent)
9. dependent
11. $\{(13,17)\}$
13. $\{(-2,-3)\}$
15. $\{(7,9)\}$
17. $\{(88,-93)\}$
19. $\left\{\left(\frac{2b-18a}{3}, b-12a\right)\right\}$
21. $\left\{\left(\frac{44}{19a}, -\frac{9}{19b}\right)\right\}$

PROBLEM SET 8.5, PAGE 348

1. 5 and 7
3. 72
5. 47
7. 84
9. 93
11. 84
13. Rate he rows = 3 mph, current rate = 1.5 mph
15. 24 and 4
17. length = 36 feet, width = 6 feet
19. 3 cents and 7 cents
21. \$2,000 at 5% and \$2,000 at 4%
23. Joan's age is 10 years, Doreene's age is 7 years
25. Slower car traveled 5 hours, faster car traveled 3 hours
27. 34
29. 35 mph and 45 mph
31. 0 and 2
33. length = 12 feet, width = 10 feet
35. 16 and 48
37. \$85,000 at 5%, \$40,000 at 7%
39. $35\frac{5}{9}$ gallons and $4\frac{4}{9}$ gallons of milk

PROBLEM SET 8.6, PAGE 354

1.

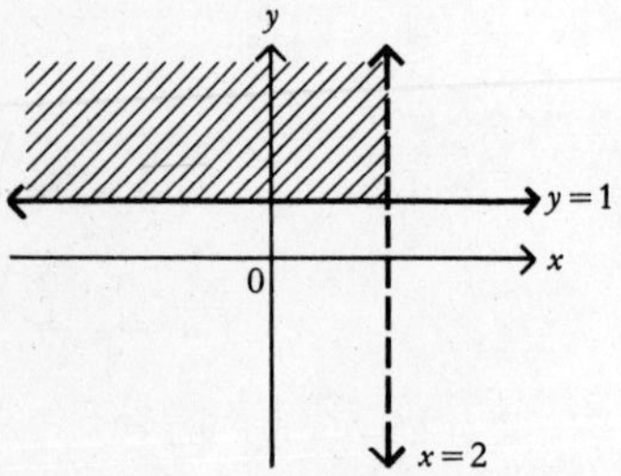

3.

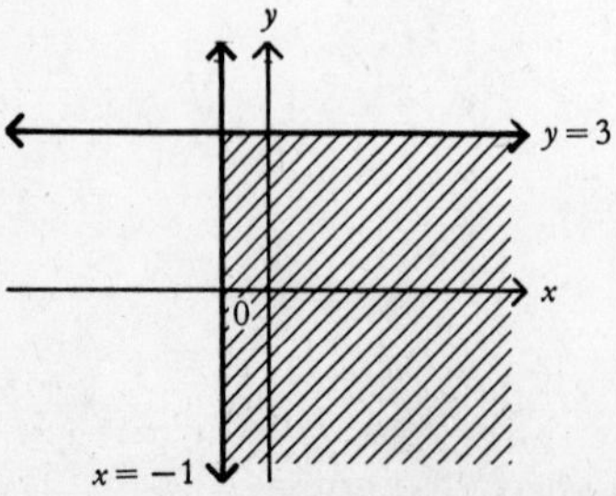

5.

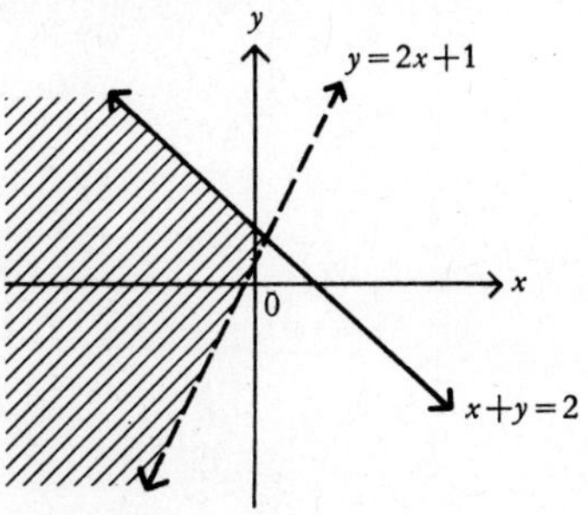

7.

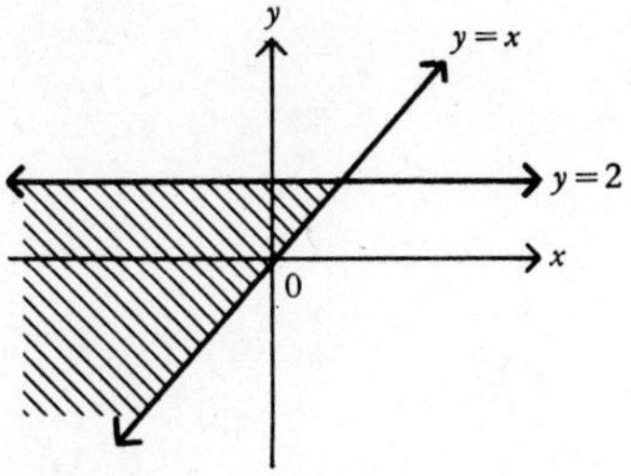

9.

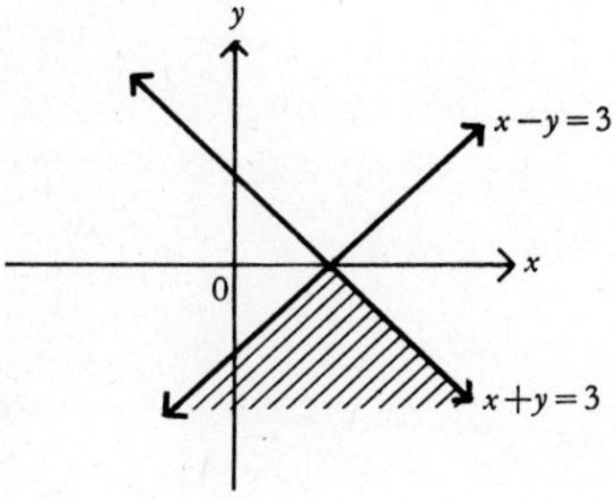

11.

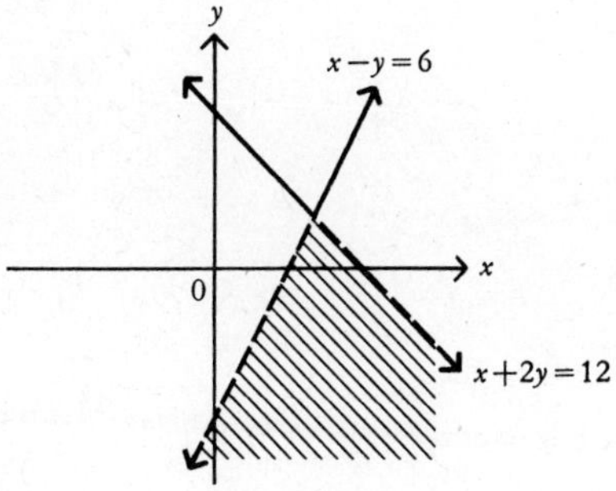

13.

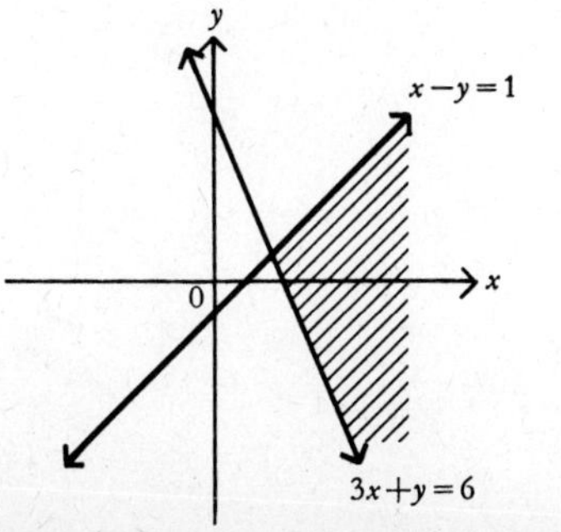

15.

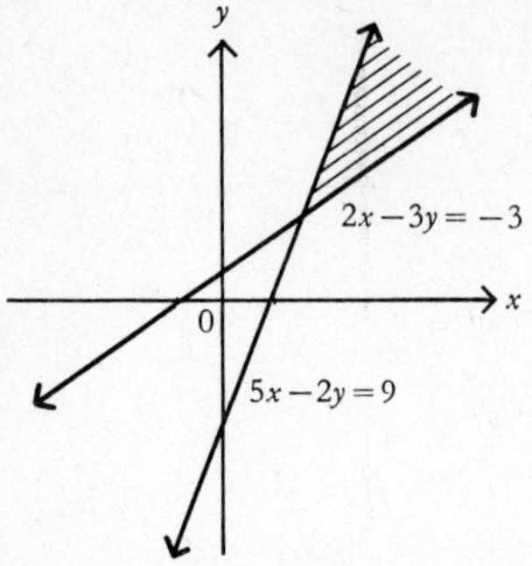

REVIEW PROBLEM SET, PAGE 355

7. $\{(1,2)\}$
9. $\{(-1,1)\}$
11. $\{(18,-13)\}$
13. $\{(4,5)\}$
15. $\{(-2,3)\}$
17. $\{(6,-5)\}$
19. $\{(1,1)\}$
21. $\{(3,2)\}$
23. $\{(-1,4)\}$
25. $\{(1,2)\}$
27. $\{(\frac{11}{3},-\frac{2}{3})\}$
29. $\{(-\frac{2}{5},\frac{29}{5})\}$
31. $\{(\frac{32}{19},\frac{4}{19})\}$
33. $\{(3,8)\}$
35. $\{(2,\frac{17}{7})\}$
37. $\{9,5\}$
39. $\{(-2,3)\}$
41. $\{(\frac{48}{17},\frac{4}{17})\}$
43. $\{(\frac{8}{5},-\frac{6}{5})\}$
45.

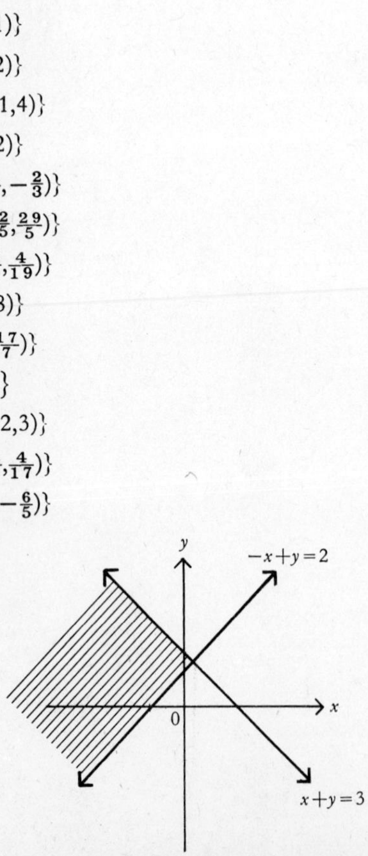

47.

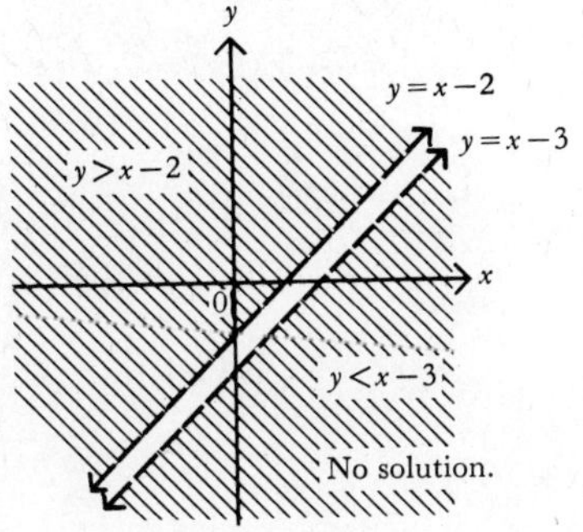

49.

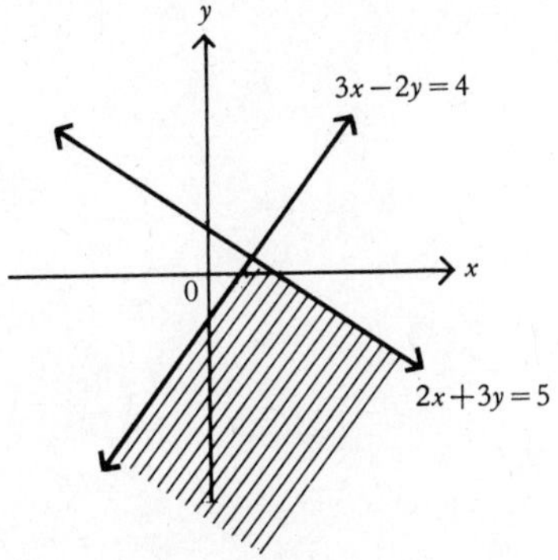

51. 13 and 20

53. \$4,000 at 4%, \$6,000 at 4.5%

55. 80 at 10 cents, 40 at 8 cents

CHAPTER 9

PROBLEM SET 9.1, PAGE 363

1. $\sqrt[4]{16} = 2$

3. $\sqrt[4]{81} = 3$

5. $\sqrt[4]{7} = x$

7. $\sqrt[3]{64} = x$

9. $\sqrt[5]{-243} = -3$

11. 3

13. -7

15. -12

17. -9

19. -21

21. x^5

23. $11x^2$
25. 0.19
27. 0.17
29. -2
31. -3
33. -11
35. 2
37. $5x^2$

PROBLEM SET 9.2, PAGE 366

1. $4\sqrt{2}$
3. $3\sqrt{5}$
5. $4\sqrt{10}$
7. $5\sqrt{6}$
9. $5\sqrt{37}$
11. $2\sqrt[3]{9}$
13. $4\sqrt[3]{5}$
15. $3x^2$
17. $2\sqrt[4]{2}$
19. $2\sqrt[5]{7}$
21. $2x^2\sqrt[5]{7}$
23. $xy^2\sqrt[6]{x^3}$
25. $2y^2\sqrt[4]{2}$
27. $5x\sqrt[3]{x^2}$
29. $-2xy^2\sqrt[5]{xy^2}$
31. $2xy^2\sqrt[4]{3x^3y^3}$
33. $-3z^3\sqrt[3]{9z}$

PROBLEM SET 9.3, PAGE 370

1. $\frac{\sqrt{3}}{5}$
3. $\frac{\sqrt{5}}{3}$
5. $\frac{2\sqrt{2}}{7}$
7. $-\frac{1}{2}$
9. $-\frac{3}{4}$
11. $\frac{\sqrt{x}}{y}$

13. $\frac{\sqrt[6]{3}}{2}$

15. $\frac{\sqrt[3]{3}}{10}$

17. $\frac{5}{2x^2}$

19. 2

21. $\frac{2}{x}$

23. $\frac{3}{2x}$

PROBLEM SET 9.4, PAGE 372

1. $5\sqrt{2}$
3. $5\sqrt{2}$
5. $43\sqrt{5}$
7. $2\sqrt{2}$
9. $3\sqrt{3}$
11. $5\sqrt{2}$
13. $14\sqrt{6}$
15. $24\sqrt[3]{3}$
17. $\sqrt[3]{2}$
19. $4\sqrt{3}$
21. $12\sqrt{3x}$
23. $(6x + 3)\sqrt{x}$

PROBLEM SET 9.5, PAGE 375

1. $3\sqrt{2}$
3. $11\sqrt{2}$
5. $40\sqrt{2}$
7. 18
9. $3 - 2\sqrt{2}$
11. $140 - 48\sqrt{6}$
13. $2\sqrt{6}$
15. $3\sqrt[3]{2}$
17. $2\sqrt[4]{5}$
19. $\sqrt{2}$
21. $3 - 8\sqrt{2}$
23. $6 + \sqrt{6}$
25. $x - \sqrt{xy}$

27. $2x + \sqrt{x} - 3$
29. $3x + 5\sqrt{xy} - 2y$
31. $4x - 4y\sqrt{x} + y^2$

PROBLEM SET 9.6, PAGE 380

1. $\frac{2\sqrt{3}}{3}$
3. $\frac{3\sqrt{21}}{7}$
5. $\sqrt{3y}$
7. $\frac{5\sqrt{2x}}{2x}$
9. $\frac{3}{2}$
11. $\frac{5\sqrt{5} + 5}{2}$
13. $2 - \sqrt{2}$
15. $\frac{3\sqrt{x} - 3\sqrt{y}}{x - y}$
17. $\frac{x - 2\sqrt{xy} + y}{x - y}$
19. $\frac{1 + 2\sqrt{y} + y}{1 - y}$
21. $\frac{y(\sqrt{y} - 1)}{y - 1}$
23. $-5 - 2\sqrt{6}$
25. $\frac{x(x - \sqrt{y})}{x^2 - y}$
27. $19 - 5\sqrt{15}$
29. $2 - 3\sqrt{2} + 2\sqrt{6} - \sqrt{3}$

REVIEW PROBLEM SET, PAGE 382

1. 25
3. 6
5. 3
7. $5\sqrt{6}$
9. $3\sqrt{7}$
11. $ab\sqrt[3]{ab^2}$
13. $y\sqrt[4]{y}$
15. $\frac{5}{8}$
17. $\frac{5}{2}$

19. $\frac{5}{2}$
21. $\dfrac{\sqrt[5]{3x}}{2}$
23. $4\sqrt{2}$
25. $6\sqrt[3]{5}$
27. $60\sqrt{2}$
29. $5\sqrt{2}$
31. $14\sqrt{3}$
33. 10
35. 27
37. $x - y$
39. $\sqrt{7} - 1$
41. $\dfrac{4\sqrt{3}}{3}$
43. $\dfrac{2\sqrt{3x}}{5x}$
45. $\sqrt{3} - \sqrt{2}$
47. $\dfrac{x(\sqrt{x} + \sqrt{y})}{x - y}$

CHAPTER 10

PROBLEM SET 10.1, PAGE 390

1. $7x^2 - 3x + 5 = 0, a = 7, b = -3, c = 5$
3. $5x^2 - 18x + 13 = 0, a = 5, b = -18, c = 13$
5. $3x^2 - 32x + 17 = 0, a = 3, b = -32, c = 17$
7. $\{-3,3\}$
9. $\left\{-\dfrac{\sqrt{2}}{2}, \dfrac{\sqrt{2}}{2}\right\}$
11. $\{-\frac{2}{3},\frac{2}{3}\}$
13. $\{-1,3\}$
15. $\{6 - \sqrt{5}, 6 + \sqrt{5}\}$
17. $\{-5,7\}$
19. $\{-\frac{5}{2},-\frac{1}{2}\}$
21. $\{\frac{1}{2},\frac{7}{6}\}$
23. $\{-\frac{1}{3},\frac{17}{3}\}$
25. $\{-\frac{1}{4},\frac{7}{4}\}$
27. $\left\{-\dfrac{\sqrt{30}}{2}, \dfrac{\sqrt{30}}{2}\right\}$

29. $\{-5,5\}$
31. $\{-\frac{11}{2},7\}$

PROBLEM SET 10.2, PAGE 394

1. $\{-2,-1\}$
3. $\{-\frac{1}{2},\frac{1}{2}\}$
5. $\{-\frac{2}{5},\frac{2}{5}\}$
7. $\{-\sqrt{3}, \sqrt{3}\}$
9. $\{-\frac{7}{3},-\frac{2}{3}\}$
11. $\{-2,\frac{7}{3}\}$
13. $\{-\frac{2}{3},\frac{1}{7}\}$
15. $\{-\frac{3}{5},\frac{5}{8}\}$
17. $\{0,\frac{7}{3}\}$
19. $\{2,4\}$
21. $\{-1,\frac{5}{3}\}$
23. $\{-\frac{7}{2},3\}$
25. $\{\frac{3}{5},\frac{2}{3}\}$
27. $\{-3,1\}$
29. $\{-\frac{1}{3},1\}$
31. $\{-1,\frac{3}{2}\}$
33. $\{-3,3\}$
35. $\{-\frac{7}{2},\frac{1}{9}\}$
37. $x^2 - 11x + 30 = 0$
39. $x^2 + 3x + 2 = 0$
41. $9x^2 + 3x - 2 = 0$

PROBLEM SET 10.3, PAGE 400

1. $4;\ (x + 2)^2$
3. $4;\ (x - 2)^2$
5. $6;\ (x + 3)^2$
7. $\frac{225}{4};\ (x - \frac{15}{2})^2$
9. $12x;\ (x + 6)^2$
11. $\left\{\frac{4 - \sqrt{10}}{3}, \frac{4 + \sqrt{10}}{3}\right\}$
13. $\left\{\frac{7 - \sqrt{205}}{6}, \frac{7 + \sqrt{205}}{6}\right\}$
15. $\{3,5\}$
17. $\{\frac{3}{4}\}$
19. $\{4 - \sqrt{21}, 4 + \sqrt{21}\}$
21. $\{-\frac{7}{5},1\}$

23. $\left\{\frac{3-\sqrt{89}}{10}, \frac{3+\sqrt{89}}{10}\right\}$

25. $\{4-\sqrt{3}, 4+\sqrt{3}\}$

27. $\{-3,1\}$

29. $\left\{\frac{-4-2\sqrt{7}}{3}, \frac{-4+2\sqrt{7}}{3}\right\}$

31. $\{-\frac{2}{5}, \frac{7}{5}\}$

33. $\left\{\frac{-3-\sqrt{2}}{7}, \frac{-3+\sqrt{2}}{7}\right\}$

35. $\left\{\frac{5-\sqrt{13}}{6}, \frac{5+\sqrt{13}}{6}\right\}$

37. $\left\{\frac{-11-\sqrt{21}}{10}, \frac{-11+\sqrt{21}}{10}\right\}$

39. $\{-\frac{1}{3}, -\frac{1}{4}\}$

41. $\left\{\frac{39-\sqrt{1937}}{26}, \frac{39+\sqrt{1937}}{26}\right\}$

43. $\left\{\frac{9-\sqrt{97}}{6}, \frac{9+\sqrt{97}}{6}\right\}$

45. $\left\{\frac{-13-\sqrt{193}}{4}, \frac{-13+\sqrt{193}}{4}\right\}$

47. $\left\{\frac{4-\sqrt{15}}{2}, \frac{4+\sqrt{15}}{2}\right\}$

PROBLEM SET 10.4, PAGE 404

1. $\left\{\frac{5-\sqrt{17}}{4}, \frac{5+\sqrt{17}}{4}\right\}$

3. $\{3-2\sqrt{2}, 3+2\sqrt{2}\}$

5. $\{-1,-2\}$

7. $\{-\frac{1}{2},\frac{5}{3}\}$

9. $\{-6,-\frac{1}{6}\}$

11. $\{-\frac{7}{2},\frac{2}{3}\}$

13. $\{-\frac{4}{5},\frac{2}{3}\}$

15. $\{\frac{5}{7},\frac{4}{3}\}$

17. $\{-1,\frac{5}{2}\}$

19. $\{-4,-12\}$

21. $\{-7,9\}$

23. $\{-\frac{11}{4},\frac{1}{3}\}$

25. $\{-\frac{3}{4},\frac{7}{8}\}$

27. $\left\{\frac{8-\sqrt{19}}{9}, \frac{8+\sqrt{19}}{9}\right\}$

29. $\{-\frac{4}{3},2\}$

31. $\left\{\frac{-11-\sqrt{377}}{2}, \frac{-11+\sqrt{377}}{2}\right\}$

33. $\left\{\frac{-3-\sqrt{85}}{2}, \frac{-3+\sqrt{85}}{2}\right\}$

35. $\left\{\frac{-5-\sqrt{73}}{12}, \frac{-5+\sqrt{73}}{12}\right\}$

37. $\left\{\frac{-1-\sqrt{33}}{4}, \frac{-1+\sqrt{33}}{4}\right\}$

39. $\left\{\frac{6-\sqrt{26}}{5}, \frac{6+\sqrt{26}}{5}\right\}$

REVIEW PROBLEM SET, PAGE 405

1. $\{-2,2\}$

3. $\left\{\frac{-2-\sqrt{5}}{3}, \frac{-2+\sqrt{5}}{3}\right\}$

5. $\{1,4\}$

7. $\{-\frac{3}{7},1\}$

9. $\{-\frac{7}{3},3\}$

11. $\{-\frac{1}{3},\frac{1}{2}\}$

13. $\{\frac{4}{3},-5\}$

15. $\{\frac{1}{2},\frac{3}{2}\}$

17. $\{-\frac{3}{2},\frac{1}{3}\}$

19. $\{-7,12\}$

21. $\{\frac{3}{4},2\}$

23. $\left\{\frac{-10}{3}, 4\right\}$

25. $\{-19,2\}$

27. $\{\frac{1}{5},7\}$

29. $\{-\frac{2}{5},\frac{2}{5}\}$

31. $\{-1,\frac{5}{3}\}$

33. $\{-\frac{3}{2},\frac{2}{3}\}$

35. $\{-2,12\}$

37. $\{-2,1\}$

39. $\{-5-2\sqrt{3}, -5+2\sqrt{3}\}$

41. $\left\{\frac{1-\sqrt{17}}{8}, \frac{1+\sqrt{17}}{8}\right\}$

43. $\left\{\frac{-5-\sqrt{161}}{4}, \frac{-5+\sqrt{161}}{4}\right\}$

45. $\left\{\frac{-3-\sqrt{65}}{8}, \frac{-3+\sqrt{65}}{8}\right\}$

47. $\left\{\frac{7-\sqrt{13}}{3}, \frac{7+\sqrt{13}}{3}\right\}$

49. $\left\{\frac{-19 - \sqrt{41}}{8}, \frac{-19 + \sqrt{41}}{8}\right\}$

51. $\{-2,-\frac{1}{3}\}$

53. $\{-5,2\}$

55. $\left\{\frac{5 - \sqrt{17}}{4}, \frac{5 + \sqrt{17}}{4}\right\}$

57. $\{-4,6\}$

59. $\left\{\frac{-3 - \sqrt{65}}{4}, \frac{-3 + \sqrt{65}}{4}\right\}$

61. $\{-13,1\}$

63. $\left\{\frac{1 - \sqrt{41}}{10}, \frac{1 + \sqrt{41}}{10}\right\}$

65. $\left\{\frac{1 - \sqrt{13}}{2}, \frac{1 + \sqrt{13}}{2}\right\}$

67. $\{-\frac{2}{3},\frac{1}{2}\}$

69. $\{-2,-1\}$

71. $\{4\}$

73. $\{2,3\}$

75. $\left\{\frac{1 - \sqrt{21}}{2}, \frac{1 + \sqrt{21}}{2}\right\}$

APPENDIX B

PROBLEM SET, PAGE 428

1. (a) 15 square inches
2. (a) 24.624 square feet
3. (a) 30 square inches
4. (a) 12.25 square inches
5. (a) 45 square feet
6. (a) 25π square inches
7. (a) 20 inches
8. (a) 40 inches
9. (a) 32 inches
10. (a) 18 inches
11. (a) 24 inches
12. (a) 8π inches
13. (a) $V = 60$ cubic meters, $S = 94$ square meters
14. (a) $L = 16\pi$ square inches, $S = 24\pi$ square inches, $V = 16\pi$ cubic inches
15. (a) $S = 16\sqrt{10} + 16$ square inches, $V = 32$ cubic inches
16. (a) L.S. $= 15\pi$ square inches, $S = 24\pi$ square inches, $V = 12\pi$ cubic inches
17. (a) $S = 100\pi$ square centimeters, $V = \frac{500\pi}{3}$ cubic centimeters

INDEX